Advances in Sustainability Science and Technology

Series Editors

Robert J. Howlett, Bournemouth University and KES International, Shoreham-by-Sea, UK

John Littlewood, School of Art & Design, Cardiff Metropolitan University, Cardiff, UK

Lakhmi C. Jain, KES International, Shoreham-by-Sea, UK

The book series aims at bringing together valuable and novel scientific contributions that address the critical issues of renewable energy, sustainable building, sustainable manufacturing, and other sustainability science and technology topics that have an impact in this diverse and fast-changing research community in academia and industry.

The areas to be covered are

- Climate change and mitigation, atmospheric carbon reduction, global warming
- Sustainability science, sustainability technologies
- Sustainable building technologies
- Intelligent buildings
- Sustainable energy generation
- Combined heat and power and district heating systems
- Control and optimization of renewable energy systems
- Smart grids and micro grids, local energy markets
- Smart cities, smart buildings, smart districts, smart countryside
- Energy and environmental assessment in buildings and cities
- Sustainable design, innovation and services
- Sustainable manufacturing processes and technology
- Sustainable manufacturing systems and enterprises
- Decision support for sustainability
- Micro/nanomachining, microelectromechanical machines (MEMS)
- Sustainable transport, smart vehicles and smart roads
- Information technology and artificial intelligence applied to sustainability
- Big data and data analytics applied to sustainability
- Sustainable food production, sustainable horticulture and agriculture
- Sustainability of air, water and other natural resources
- Sustainability policy, shaping the future, the triple bottom line, the circular economy

High quality content is an essential feature for all book proposals accepted for the series. It is expected that editors of all accepted volumes will ensure that contributions are subjected to an appropriate level of reviewing process and adhere to KES quality principles.

The series will include monographs, edited volumes, and selected proceedings.

More information about this series at https://link.springer.com/bookseries/16477

Mohammad Aslam · Shrikant Shivaji Maktedar ·
Anil Kumar Sarma

Editors

Green Diesel: An Alternative to Biodiesel and Petrodiesel

 Springer

Editors
Mohammad Aslam
Department of Chemistry
National Institute of Technology Srinagar
Srinagar, J&K, India

Shrikant Shivaji Maktedar
Department of Chemistry
National Institute of Technology Srinagar
Srinagar, J&K, India

Anil Kumar Sarma
Sardar Swaran Singh National Institute
of Bio-Energy
Kapurthala, Punjab, India

ISSN 2662-6829   ISSN 2662-6837 (electronic)
Advances in Sustainability Science and Technology
ISBN 978-981-19-2237-4   ISBN 978-981-19-2235-0 (eBook)
https://doi.org/10.1007/978-981-19-2235-0

This Springer imprint is published by the registered company Springer Nature Singapore Pte Ltd.
The registered company address is: 152 Beach Road, #21-01/04 Gateway East, Singapore 189721,
Singapore

Preface

The global energy scenario of the twenty-first century clearly unveils the fact of transition of fuel market and economy from fossil fuels to carbon neutral fuels or to a decarbonized fuel system. Recently, Green diesel (also known as renewable diesel) is considered as one of the most efficient alternative fuel substitutes for petroleum diesel for CI engines due to its more than two times cetane number (Cetane No.~ 90) as compared to biodiesel and petrodiesel. Many scientific communities have a misconception to consider fatty acid methyl ester or fatty acid ethyl esters (Biodiesel) as green diesel attributed to their origin from fats or lipids. However, Green diesel significantly differs from biodiesel with reference to structural chemistry, composition, and production routes. The fundamental difference between biodiesel and green diesel is that the former is produced by the transesterification process and the latter is through hydroprocessing technology and subsequent distillation using similar fractional distillation as used for petroleum refining. Renewable diesel is a non-oxygenated fuel wherein carbon and hydrogen are the major constituents. It has a comparable composition to petroleum diesel with a very high cetane number attributed to a very high percentage of straight-chain hydrocarbons equivalent to cetane ($C_{16}H_{34}$). Unlike petroleum diesel, green diesel has no aromatic contents which is a positive attribute but the acid value is on the higher side which requires special attention. The pour point of green diesel is far better when compared to petrodiesel, making green diesel a versatile fuel for cold climate applications in IC engine. The fuel characterization of green diesel is very similar to petroleum diesel standards and follows appropriate methods of ASTM, IP methods or EN standards. Currently, green diesel has entered in most of the European countries and USA markets by Neste Oil Corporation (Finland), Honeywell, UOP (USA), Diamond Green Diesel (USA), etc. Unfortunately, many of the young generations are also not aware of the term "Green Diesel"; a next generation green and sustainable transportation fuel.

Globally, biodiesel and petrodiesel have been covered in the wide scientific literature including academic books, and hence require no further introduction. Nevertheless, there is not a single specific book available on "green diesel" in International markets. Keeping in view of significant contribution in the field of Bioenergy, biofuels, green diesel, and catalysis, the editors are highly motivated and enthusiastic to step forward with an initiative to edit the first specific book on "Green Diesel". The editors feel proud to edit and publish such a dedicated book on green diesel that can make a forceful foundation for our graduate engineers, technocrats, and researchers in the field of Mechanical, Chemical, Automobile, Energy Science and Engineering, etc. In addition, this book aims to help to attract professionals, entrepreneurs, and corporate leaders working in the domain of green energy to set up new green diesel refineries in the world and hence contribute toward a sustainable and greener future for our next generations.

This book envisaged the most important and fundamental aspects of green diesel from its beginning to commercial applications. Chapters 1 and 2 discuss the origin and history of green diesel, fundamental properties, and feedstocks including first, second, and third generation required for the production of green diesel. Chapter 3 emphasizes on the catalytic materials employed for the production of green diesel and their effects on the fuel properties. Hydroprocessing technology; a well-established petroleum refining technology, plays a key role in green diesel production and is considered as the heart of the green diesel refineries. Recent Advances in the catalytic hydroprocessing techniques of various feedstocks like Jatropha curcas oil, Palm oil, used cooking oil and animal fats, etc., to produce renewable diesel have been deliberated in Chaps. 4, 5 and 6. Chapter 7 discusses about the technological advancements for the production of green diesel from biomass.

Moreover, Chap. 8 focuses on the existing standards for the characterization of green diesel fuel. An overview of the current status of the green diesel industry includes Neste Oil Corporation (Finland), Honeywell/UOP (USA), Diamond Green Diesel (USA), USA, etc., emphasized in Chap. 9. Further, how a green diesel is a better fuel than biodiesel and petrodiesel, their comparative analyses have been discussed in Chap. 10. In addition, the most important aspect such as techno-economic analyses of biodiesel and green diesel is discussed in Chap. 11. Further, the application of green diesel in CI engines for performance, combustion, and emission characteristics studies has been elaborated in Chap. 12. A techno-economic and environmental perspective on the role of green diesel in a prospective fuel production for surface transport has been discussed in Chap. 13. Finally, the last chapter of the book, Chap. 14 dedicatedly addresses the policies, techno-economic analysis, and future perspective of green diesel.

Srinagar, Jammu and Kashmir, India   Mohammad Aslam
Srinagar, Jammu and Kashmir, India   Shrikant Shivaji Maktedar
Kapurthala, Punjab, India   Anil Kumar Sarma

Acknowledgements

We express our deep sense of gratitude to our parents whose blessings have always prompted us to pursue academic activities deeply. We are indebted to Springer Nature for giving this opportunity to editors and the Department of Chemistry, NIT Srinagar (J&K), India for all technical support. We would like to thank the staff at Springer Nature, in particular, Aninda Bose and Naresh Mani, for their immense help and support.

I would like to acknowledge the contribution of each of the authors in this book. The editors are highly thankful to all the academicians/scientists whose contributions have enriched this book. Their research efforts have shed considerable light on our understanding of the challenges and opportunities in the area of green diesel. We look forward to their continued contributions toward our progress to ultimate green energy independence.

Srinagar, Jammu and Kashmir, India   Mohammad Aslam
Srinagar, Jammu and Kashmir, India   Shrikant Shivaji Maktedar
Kapurthala, Punjab, India   Anil Kumar Sarma

Contents

Editors and Contributors

About the Editors

Dr. Mohammad Aslam is an Assistant Professor at Department of Chemistry, National Institute of Technology (Ministry of Education, Government of India), Srinagar (India). He obtained his first Master's degree in science "Industrial Chemistry" from Aligarh Muslim University, Aligarh, Uttar Pradesh, followed by Master's degree in interdisciplinary programme "Fuels and Combustion" from Birla Institute of Technology, Mesra (Ranchi), Jharkhand (India). Later, He has awarded Doctorate from National Institute of Technology, Jalandhar (India) for his dissertation entitled, "Hydroprocessing of Seed Oils into Transportation Fuels". He also worked as Post-Doctoral Fellow/Research Associate in Sardar Swaran Singh National Institute of Bioenergy (An Autonomous Institution of Ministry of New and Renewable Energy, Government of India). He has more than 15 years of academic and research experience. His current research is focusing on green fuels derived from second/third generation feedstocks and their application to transportation sectors. He has established Biofuels Research Laboratory, at Department of Chemistry, NIT Srinagar, India.

Dr. Shrikant Shivaji Maktedar is an Assistant Professor at Department of Chemistry, National Institute of Technology (Ministry of education, government of India), Srinagar, J&K, India. He received B.Sc. Degree in Chemistry from Dr. Babasaheb Ambedkar Marathwada University, Aurangabad (Maharashtra) and M.Sc. Degree in Physical Chemistry from Dr. Babasaheb Ambedkar Marathwada University, Aurangabad (Maharashtra). He has completed his Ph.D. from Central University of Gujarat, Gandhinagar, India. In last 10 years he is working in the field of carbonaceous materials with emphasis on their multifunctional applications. Dr. Shrikant has published more than 10 research publications in peer-reviewed international journals of repute, one book chapter and two full length conference papers. He has served as reviewer for few international journals of repute. After his joining to NIT Srinagar, he is serving as Ph.D. supervisor and has established Materials Chemistry Research Laboratory at Department of Chemistry.

Dr. Anil Kumar Sarma is a Scientist-E/Director at Sardar Swaran Singh National Institute of Bio-Energy (Ministry of New and Renewable Energy, Govt of India), Punjab, India. He is an MTech in Energy Technology (2002) and Ph.D. in Energy from Tezpur University (2006). He remains guest faculty at Tezpur University, Research Associate at IIT Guwahati (2007–2008), visiting researcher at Seikei University, Japan (2008–2009), joined and left from faculty at RGIPT before joining his present position at Sardar Swaran Singh National Institute of BioEnergy. He has been working as a scientist at SSS-NIBE since July 2009. He has been rendering his duty as Head of Office at SSS-NIBE and coordinator for Indo US collaboration under SAGE. He delivered invited lectures in several platforms such as training programs and conferences of national and international stature. He was awarded with Bharat Jyoti Award (14th Feb, 2012) by India International Friendship Society, for Meritorious Services, Outstanding performance and Remarkable role in his field. He was selected as the member of the sustainability and development by the Japanese Society. Recently he has been appointed as the honorary Research Advisor, by Nan Yang Acade my of Sciences (NAAS), Singapore and as a College member for GCRF (UK). He is also a reviewer of renewable and bioenergy projects for several bodies of Govt of India and remains a key catalyst of the Institute since the inception. His frontier areas of interest include biodiesel production and catalysis: Biocrude oil production, fractional distillation and characterization including application in IC engines leading to biorefinery development from oilseeds, catalyst development from agrowaste for fuel additive synthesis. He has published over 50 publications in reputed journal and proceedings which are readily available in google scholar citations. He has guided 4 Ph.D., 4 post-doctoral fellows, 15+ Master's students for their dissertation works. Currently, 5 students are working in his group for their Ph.D. thesis. He remains research advisory committee member for a few students in Dr. B. R. Ambedkar NIT Jalandhar.

Contributors

J. Aburto Gerencia de Transformación de Biomasa, Instituto Mexicano del Petróleo, Mexico City, Mexico

M. A. Amezcua-Allieri Gerencia de Transformación de Biomasa, Instituto Mexicano del Petróleo, Mexico City, Mexico

Khursheed B. Ansari Department of Chemical Engineering, Zakir Husain College of Engineering and Technology, Aligarh Muslim University, Aligarh, Uttar Pradesh, India

Mohammad Aslam Department of Chemistry, National Institute of Technology, Srinagar (J&K), India

Nur Izyan Wan Azelee School of Chemical and Energy Engineering, Faculty of Engineering, Universiti Teknologi Malaysia, UTM Skudai, Skudai, Johor, Malaysia; Institute of Bioproduct Development (IBD), Universiti Teknologi Malaysia, UTM Skudai, Skudai, Johor, Malaysia

Venu Babu Borugadda Catalysis and Chemical Reaction Engineering Laboratories (CCREL), Department of Chemical and Biological Engineering, University of Saskatchewan, Saskatoon, SK, Canada

Satinder Kaur Brar Department of Civil Engineering, Lassonde School of Engineering, York University, North York (Toronto), Canada

Pedro L. Cruz Systems Analysis Unit, IMDEA Energy, Madrid, Spain

Ajay K. Dalai Catalysis and Chemical Reaction Engineering Laboratories (CCREL), Department of Chemical and Biological Engineering, University of Saskatchewan, Saskatoon, SK, Canada

Piyali Das The Energy and Resources Institute (TERI), New Delhi, India

Fahimeh Esmi Catalysis and Chemical Reaction Engineering Laboratories (CCREL), Department of Chemical and Biological Engineering, University of Saskatchewan, Saskatoon, SK, Canada

Danilo Henrique da Silva Santos Laboratório de Processos, Centro de Tecnologia, Universidade Federal de Alagoas, Av. Lourival de Melo Mota, Tabuleiro Dos Martins, Maceió, Alagoas, Brazil

S. A. Farooqui CSIR-Indian Institute of Petroleum, Dehradun, India

Saleem Akhtar Farooqui Hydroprocessed Renewable Fuel Area, Biofuels Division, Indian Institute of Petroleum, Dehradun, India

Diego García-Gusano TECNALIA, Basque Research and Technology Alliance (BRTA), Derio, Spain

Raunaq Hasib Department of Chemical Engineering, Zakir Husain College of Engineering and Technology, Aligarh Muslim University, Aligarh, Uttar Pradesh, India

Saeikh Zaffar Hassan Department of Petroleum Studies, Zakir Husain College of Engineering and Technology, Aligarh Muslim University, Aligarh, Uttar Pradesh, India

Diego Iribarren Systems Analysis Unit, IMDEA Energy, Madrid, Spain

M. K. Jha Department of Chemical Engineering, DR B R Ambedkar National Institute of Technology, Jalandhar, India

Jaspreet Kaur Chemical Conversion Division, Sardar Swaran Singh National Institute of Bio-Energy, Kapurthala, Punjab, India;
Department of Chemical Engineering, DR B R Ambedkar National Institute of Technology, Jalandhar, India

Wan Nor Adira Wan Khalit Department of Chemistry, Faculty of Science, Universiti Putra Malaysia, Selangor, UPM Serdang, Malaysia

Mohd Shariq Khan Department of Chemical Engineering, Dhofar University, Salalah, Oman

Parvez Khan Department of Chemical Engineering, Zakir Husain College of Engineering and Technology, Aligarh Muslim University, Aligarh, Uttar Pradesh, India

Himansh Kumar Mechanical Engineering Department, Teerthanker Mahaveer University, Moradabad, UP, India;

Mechanical Engineering Department, Dr. B R Ambedkar National Institute of Technology, Jalandhar, Punjab, India;
Chemical Conversion Division, Sardar Swaran Singh-National Institute of Bio-Energy, Kapurthala, Punjab, India

Pramod Kumar Department of Mechanical Engineering, Dr. B R Ambedkar National Institute of Technology, Jalandhar, Punjab, India

R. Kumar CSIR-Indian Institute of Petroleum, Dehradun, India

Azizul Hakim Lahuri Department of Science and Technology, Universiti Putra Malaysia Bintulu Campus, Sarawak, Bintulu, Malaysia

Hilman Ibnu Mahdi Chemical and Materials Engineering, College of Engineering, National Yunlin University of Science and Technology, Douliou, Yunlin, Taiwan

Surahim Mahmud Department of Chemistry, Faculty of Science, Universiti Putra Malaysia, Selangor, UPM Serdang, Malaysia

Sunil K. Maity Department of Chemical Engineering, Indian Institute of Technology Hyderabad, Kandi, Sangareddy, Telangana, India

Tengku Sharifah Marliza Department of Science and Technology, Universiti Putra Malaysia Bintulu Campus, Sarawak, Bintulu, Malaysia

Mario Martín-Gamboa Chemical and Environmental Engineering Group, Rey Juan Carlos University, Madrid, Spain

Shima Masoumi Catalysis and Chemical Reaction Engineering Laboratories (CCREL), Department of Chemical and Biological Engineering, University of Saskatchewan, Saskatoon, SK, Canada

Lucas Meili Laboratório de Processos, Centro de Tecnologia, Universidade Federal de Alagoas, Av. Lourival de Melo Mota, Tabuleiro Dos Martins, Maceió, Alagoas, Brazil

Zaira Navas-Anguita Systems Analysis Unit, IMDEA Energy, Madrid, Spain

A. R. Shakeelur Rahman Department of Applied Science, Shri Vile Parle Kelavani Mandal's Institute of Technology, Dhule, India

A. Ray CSIR-Indian Institute of Petroleum, Dehradun, India

M. Safa-Gamal Department of Chemistry, Faculty of Science, Universiti Putra Malaysia, Selangor, UPM Serdang, Malaysia

Baishakhi Sarkhel Thermo-Catalytic Processes Area, Material Resource Efficiency Division, CSIR-Indian Institute of Petroleum, Dehradun, UK

Anil K. Sarma Chemical Conversion Division, Sardar Swaran Singh-National Institute of Bio-Energy, Kapurthala, Punjab, India

Saurabh Jyoti Sarma Department of Biotechnology, Bennett University, Greater Noida (UP), India

Mohd Razali Shamsuddin Preparatory Centre for Science and Technology, Universiti Malaysia Sabah, Kota Kinabalu, Sabah, Malaysia

Sumit Sharma Department of Biotechnology, Bennett University, Greater Noida (UP), India

Shikha Singh Department of Biotechnology, Bennett University, Greater Noida (UP), India

A. K. Sinha CSIR-Indian Institute of Petroleum, Dehradun, India

Priyanka Tirumareddy Catalysis and Chemical Reaction Engineering Laboratories (CCREL), Department of Chemical and Biological Engineering, University of Saskatchewan, Saskatoon, SK, Canada

Quang Thang Trinh Cambridge Centre for Advanced Research and Education in Singapore (CARES), Campus for Research Excellence and Technological Enterprise (CREATE), 1 Create Way, 138602, Singapore

Uplabdhi Tyagi Chemical Conversion Division, Sardar Swaran Singh National Institute of Bio-Energy, Kapurthala, Punjab, India

Tresylia Ipah Anak Ujai Department of Chemistry, Faculty of Resource Science and Technology, Universiti Malaysia Sarawak, Kota Samarahan, Sarawak, Malaysia

Praveenkumar Ramprakash Upadhyay The Energy and Resources Institute (TERI), New Delhi, India

Sudhakara Reddy Yenumala Thermo-Catalytic Processes Area, Material Resource Efficiency Division, CSIR-Indian Institute of Petroleum, Dehradun, UK

Chapter 1
Introduction to Green Diesel

Priyanka Tirumareddy, Fahimeh Esmi, Shima Masoumi, Venu Babu Borugadda, and Ajay K. Dalai

Abstract First, second, and third-generation green diesel are being considered as promising alternatives to petrodiesel in terms of renewable energy demand, process economic, and environmental concerns. Green diesel is an advanced biofuel that can be produced from different cellulosic biomass such as crop residue, forestry waste or woody biomass. As green diesel has identical chemical properties to petrodiesel, it could be used in its pure form or blended with petrodiesel. This chapter covers the literature related to the different feedstocks (First, second, and third-generation) used for green diesel production, hydroprocessing technology, catalytic materials for such processes, characterization of green diesel, comparison between petro-diesel and green diesel in terms of physicochemical properties, techno-economic analysis, and life cycle assessment. Besides, this chapter covered the current status of the green diesel industry from various commercial plants such as Neste, Honeywell, and ExxonMobil. This chapter discusses several commercial plants that are proposed, under construction or expansion stage, for the production of green diesel.

Keywords Green diesel · Hydroprocessing technologies · Biomass feedstocks · Catalysts

Abbreviations

ASTM	American Society for Testing and Materials
CFPP	Cold Filter Plugging Point
CI	Compression Ignition
CNSL	Cashew Nut Shell Liquid
FCC	Fluid Catalytic Cracking
FT	Fischer-Tropsch

P. Tirumareddy · F. Esmi · S. Masoumi · V. B. Borugadda · A. K. Dalai (✉)
Catalysis and Chemical Reaction Engineering Laboratories (CCREL), Department of Chemical and Biological Engineering, University of Saskatchewan, Saskatoon, SK S7N 59, Canada
e-mail: ajay.dalai@usask.ca

© The Author(s), under exclusive license to Springer Nature Singapore Pte Ltd. 2022   1
M. Aslam et al. (eds.), *Green Diesel: An Alternative to Biodiesel and Petrodiesel*,
Advances in Sustainability Science and Technology,
https://doi.org/10.1007/978-981-19-2235-0_1

GGE   Gasoline Gallon Equivalent
GHG   Greenhouse Gas Emissions
HDO   Hydrodeoxygenation
HDRD   Hydrogen Derived Renewable Diesel
HTL   Hydrothermal Liquefaction
IEA   International Energy Agency
LCA   Life Cycle Analysis
LHV   Lower Heating Value
MFSP   Minimum Fuel Selling Price
MTPD   Metric Tons Per Day
NEB   Net Energy Balance
NER   Net Energy Ratio
TEA   Techno-Economic Analysis
VGO   Vacuum Gas Oil
VO   Vegetable Oil

1 Introduction

Encountering environmental challenges triggered by fossil fuel consumption, increase in crude oil prices, reduction in fossil fuel reserves, and associated environmental problems have been large contributing factors toward the usage of alternative energy sources, which are cleaner and renewable. Biofuels are promising substitutes, which are anticipated by International Energy Agency (IEA) to be the dominant energy sector by 2030 [1, 2]. The trend of energy consumption from 2008 until now, along with future trend predictions are presented in Table 1.1.

Biofuels derived from biomass such as biogas, bioethanol, biodiesel, and green diesel could have a large influence on reducing the emission of CO_2 and the rate of global warming. A wide range of biomass raw materials could be employed for biofuel generation, which can be categorized into three groups, such as triglycerides, sugars and starches, and lignocelluloses [4].

Green diesel, also known as bio-hydrogenated diesel, contains a saturated hydrocarbon chain comprising of approximately 15–18 carbon atoms, obtained from

Table 1.1 Past and future global energy consumption and demand [3]

Year	Consumption (Quadrillion Btu)
2008	510
2015	550
2020	605
2025	670
2030	725
2035	780

various biomass feedstocks such as vegetable oils and animal fats. This fuel can then be utilized in compression ignition engines without the requirement for engine modification in both pure and blended forms of diesel [4, 5].

Green diesel is a non-corrosive biofuel and does not have aromatics in its composition, resulting in cleaner combustion than petroleum diesel which contains aromatics. To compare the properties of green diesel with biodiesel as an attractive biofuel, green diesel has a higher cetane number and better stability due to the absence of oxygen in its composition. Furthermore, it does not emit NO_X into the atmosphere [6, 7].

2   Feedstocks for Green Diesel (Including 1st, 2nd, and 3rd Generation Feedstocks)

Plant oils are suitable raw materials for biofuel production since they contain long fatty acid carbon chains. There are some factors that play a significant role in the selection of appropriate feedstocks such as local accessibility, feedstock price, and food vs. fuel competition issues.

In the US, soybean oil is the major raw material used due to its availability in large quantities [8]. In Europe, palm oil, sunflower oil, and rapeseed oil could be employed as potential raw materials for biofuel production as they are produced largely in this region [9]. Scaldaferri and Pasa [10] investigated the production of bio-jet fuel, green diesel, and gasoline through a single catalyzed-reaction step using soybean oil as a feedstock. This single-step reaction led to the production of approximately 62% of bio-jet fuel, 40% of green diesel, and 18% of gasoline under optimum reaction conditions [10].

Second-generation green diesel is derived from residual (lignocellulosic) biomass such as wood, husk, stem, waste cooking oils, and organic waste in order to raise the capacity of biofuel production besides the first-generation biofuels, which are produced from food crops [11].

Various researchers have explored the production of green diesel using first-generation edible vegetable oils (Table 1.2) such as sunflower oil [12], rapeseed oil [13], soybean oil [14], and palm oil [15] and nonedible vegetable oils such as karanja oil [16], rubber seed oil [17], jatropha oil [18], and castor oil [5].

In a study conducted by Scaldaferri and Pasa (2019), cashew nut shell liquid (CNSL) was employed as raw material for the production of second-generation green diesel [10]. CNSL is a waste material in the cashew industry that can be beneficial for biofuel production in terms of an economic and environmental point of view. Outcomes of their experimental investigations revealed that 98% yield was obtained under the mild reaction condition of 4 MPa H_2 at 300 °C reaction temperature. The utilization of waste cooking oil as a low-cost feedstock has also been investigated by Vázquez-Garrido et al. [19] for green diesel production, which can significantly decrease the cost of production and is not a threat to food supply [19]. In addition to green diesel production in the cashew biorefinery, other valuable products can

Table 1.2 Overview of biofuel generations in terms of feedstock, method, merits and demerits [22, 24]

Generation	Feedstock	Fuel production techniques	Merits	Demerits
First	Edible oils, starch, sugar	Esterification, Transesterification, Fermentation	Decreased CO_2 emission, reduced fossil fuel consumption	Negative influence on food supply, agricultural land, and raise in food price
Second	Non-edible oils, residual biomass, waste cooking oil	Enzymatic hydrolysis, Fischer–Tropsch, Hydroprocessing, Gasification	Low-cost feedstock, no threat to hunger	More processing steps, land requirement
Third	Algae	Algae cultivation, Harvesting, Oil extraction	Less space and simple requirements for growth, ecofriendly, high oil content	High cost, less stable biofuel
Fourth	Microalgae, yeast, fungi, cyanobacteria	Genetic modification of algae	High growth rate, high oil content, low structural complexity	Still in early improvement stage

be generated for use in polymer, chemicals, and nanotechnology markets [10, 18, 20–22]. The flow chart of this process is shown in Fig. 1.1.

Third-generation green diesel is obtained from microalgae resources, which contain triglycerides and are available widely in aquatic areas. They grow at a significant rate and are produced in non-agricultural areas which do not consider a threat to food security [9, 10]. Catalytic hydrodeoxygenation of *Nannochloropsis salina* microalgae oil to green diesel using 1% Pt, 0.5% Rh, and presulfided NiMo over Al_2O_3 as catalysts were investigated by Zhou and Lawal [23]. Further, the highest hydrocarbon yield of about 76% was achieved by using 1%Pt/ Al_2O_3. It was reported that the growth and cultivation of microalgae and catalyst deactivation due to coke formation and feedstock impurities are the main challenges of using microalgae oils as feedstocks which necessitates more development.

Fourth-generation biofuels could be achieved by modification of algae genetically to obtain better oil yield [24]. Di Visconte et al. [25] explored the microalgae biorefinery with respect to the current and future strategies for microalgae modification for biofuel production with lower processing cost [25]. Table 1.2 summarizes different biofuel generations with regard to the feedstock used, production technique, and their advantages and disadvantages.

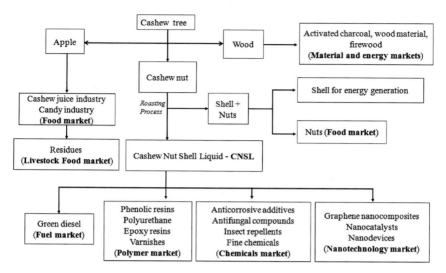

Fig. 1.1 Flowchart of converting biomass to value products in a potential cashew biorefinery. Reproduced with permission from [10]

3 Catalytic Materials for Green Diesel Production

The general process of producing green diesel is the deoxygenation of plant oils, animal fats, and other sources of triglycerides and fatty acids. Three primary routes of the catalytic reactions include decarboxylation, decarbonylation, and hydrodeoxygenation, which lead to the production of alkane hydrocarbons in the range of green diesel [24, 26].

Several types of metal catalysts can be employed for green diesel production. Noble metal catalysts, such as palladium (Pd) and platinum (Pt), are beneficial due to their high activity [27]; however, high cost of these catalysts limits their application [15, 28]. Bimetallic catalysts such as Co–Mo [29–31] and NiMo [32, 33] possess high activity and stability. Furthermore, metal oxide catalysts and conventional metal catalysts such as Ni, W, Mo, and Co were employed for green diesel production [24]. Some metal sulfides, which are in the metal oxide catalysts group comprising cobalt [32] nickel [33], molybdenum [34], platinum [13], and palladium [35], are reported as suitable catalyst materials for hydrotreating processes. Furthermore, metal phosphides [36, 37], metal carbides [38, 39], metal nitrides [40], and metal borides [41] are other types of metal oxide catalysts tested for the production of green diesel. Different supports and active metals for HDO are reported in Fig. 1.2.

The most commonly used catalysts among conventional and metal catalysts are Ni-based and Pd-based due to their higher catalytic activity in reaction, which are discussed in reported studies [36, 37].

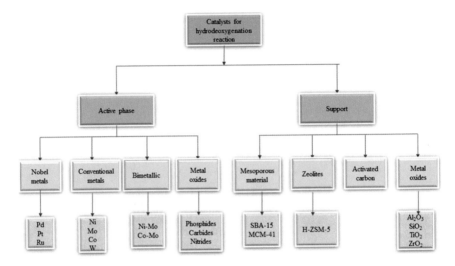

Fig. 1.2 Potential catalysts for hydrodeoxygenation reaction, including active metals and different supports

3.1 Supports

The types of supports used for hydrodeoxygenation processes play a significant role in determining the activity of catalysts. Some favorable features, which need to be considered for the selection of a support material, include appropriate inter-action between the support and active phase, high surface area, suitable pore size, pore structure, active metal distribution, and high stability [38, 39]. Among several supports, mesoporous materials, activated carbon, zeolites, and various metal oxides such as γ-Al$_2$O$_3$, SiO$_2$, Nb$_2$O$_5$, TiO$_2$, ZrO$_2$, and CeO$_2$ are some support materials that have been used for HDO process (Fig. 1.2) [34, 35].

γ-Al$_2$O$_3$ is broadly used as a support material in HDO due to its high stability and acidic feature. However, one of the main problems associated with alumina support material is the formation of coke, which results in catalyst deactivation [36]. In another study conducted by Toba et al. (2011), sulfided catalysts including Ni–Mo/Al$_2$O$_3$, Ni–W/Al$_2$O$_3$, and Co–Mo/Al$_2$O$_3$ were employed for the production of C$_{15}$–C$_{18}$ hydrocarbon chain using waste cooking oil as feedstock in hydrotreatment process [29]. All catalysts showed high catalytic activity with a slight difference and their activity followed the sequence of Ni–Mo > Ni–W > Co–Mo. The catalytic performance of different active metals involving Co, Ni, Pd, and Pt on alumina support material for HDO has been examined and the order of Co > Pd > Pt > Ni for catalytic activity was reported [15].

The performance of Pt–MoOx/ZrO$_2$ catalyst was investigated for green diesel production by Alvarez-Galvan et al. [37]. Owing to the strong metal-support electronic interaction, this catalyst is more efficient and selective for green diesel production with 100% oleic acid conversion (10 wt.% catalyst, 240 °C, 20 bar H$_2$, and 5 h) as compared with Pt on unmodified ZrO$_2$.

Wang et al. [38] compared the performance of various catalysts in terms of activity for production of green diesel from vegetable oil [38]. The order of catalytic activity was found to be Mo$_2$C/AC > MoO/Al$_2$O$_3$ > MoS$_2$/Al$_2$O$_3$ > Mo/Al$_2$O$_3$ > MoO/AC > NiP/Al$_2$O$_3$ > Ni/Al$_2$O$_3$ and with conversion of 100%, 85.6%, 83.5%, 68.0%, 56.1%, 48.7%, and 18.1%, respectively.

Ni-based catalysts using three different carriers including ZrO$_2$, H-ZSM-5, and activated carbon were explored for green diesel production by Wang et al. [39]. High dispersion was reported for all supported catalysts, however, Ni on ZrO$_2$ indicated the highest palmitic acid conversion with 98.3% as compared with Ni/H-ZSM-5 and Ni/AC with 86.2% and 44.7%, respectively [40].

The catalytic performance of three metallic activities of Ni, Co, and Ni–Co on SBA-15 in the deoxygenation reaction of palm fatty acid distillate is discussed under reaction conditions of 10 wt.% of catalyst at 350 °C for 3 h. Utilization of Ni/SBA-15 and Ni–Co/SBA-15 resulted in the selectivity of more than 85% and yield of 85.5% and 88.1%, respectively, for diesel-range hydrocarbons. With regard to Co/SBA-15, coke was formed due to the larger particle size that led to the deactivation of catalyst active sites [41].

Kon et al. [42] investigated the catalytic activity of Pt–MoO$_x$/TiO$_2$ through hydrodeoxygenation process of different fatty acids and triglycerides with 89–99% yield [42]. Several studies on green diesel production with different catalysts are reported in Table 1.3.

4 Characterizations of Green Diesel

The composition of green diesel is primarily paraffin, lacks aromatics, and additionally contains low amount of sulfur (<10 ppm) [60]. One advantage of catalytic hydroprocessing of green diesel production is that no by-product is generated [61]. Table 1.4 shows the features of various renewable diesels. The green diesel cetane number is more than 80, which is higher than other renewable diesel types. This measures the diesel fuel ignition performance. The higher cetane number of green diesel means that the ignition delay is less which leads to a faster combustion process. Consequently, less fuel is consumed and the engine works with better thermal efficiency. The cetane index of different renewable diesels is illustrated in Fig. 1.3. The more the cetane index of the fuel, the quicker it combusts in the engine. Although the density of green diesel is lower than biodiesel and fossil diesel, which result in an increase in fuel consumption, the more lower heating value (LHV) of green diesel as compared to fossil diesel can some extent compensate for the disadvantage of the

Table 1.3 Green diesel production from different catalyst systems

Metal	Supported catalyst	Feedstock	Condition	Result	References
Pd	Pd/C	Palmist fat	T = 300 °C, P_{H2} = 10 bar, t = 5 h	Conversion = 96%	[42]
	5 wt.% Pd/C	Cashew nutshell liquid	T = 300 °C, P_{H2} = 40 bar, t = 10 h	Conversion = 98%	[43]
	0.5% Pd/Al$_2$O$_3$	Crude palm kernel oil	T = 477 °C, P = 5.6 MPa, LHSV = 1.5 h^{-1}	Diesel selectivity = 19.19%	[44]
	2 wt.% Pd/Ni–Ce0.6Zr0.4O$_2$	Oleic acid	T = 300 °C, P = 1 bar, t = 3 h	Conversion = 90.8%	[45]
	5 wt.% Pd/C	Waste cooking oil	T = 375 °C, P = 4.8 MPa, LHSV = 0.5 h^{-1}	Conversion = 100%	[46]
Pt	Pt/SAPO-11-Al$_2$O$_3$	Soybean oil	T = 375–380 °C, P = 30 atm, LHSV = 1 h^{-1}	Conversion = 100%	[14]
	1% Pt/Al$_2$O$_3$	Microalgae	T = 310 °C, P = 500 psig, 1000 SmL mL^{-1} gas/oil ratio, 1.5 s residence time	Hydrocarbon yield = 76.5%	[23]
	2 wt.% Pt/Ni–Ce0.6Zr0.4O$_2$	Oleic acid	T = 300 °C, P = 1 bar, t = 3 h	Conversion = 98.7%	[45]
	4Pt/SiO$_2$–Al$_2$O$_3$	Oleic acid	T = 260 °C P_{H2} = 20 bar, t = 5 h	Conversion = 69%	[10]
	4Pt/ZrO$_2$	Oleic acid	T = 220 °C, P_{H2} = 20 bar H2, t = 5 h	Conversion = 23%	[46]
	Pt/SAPO	Castor oil	T = 400 °C, P = 4 MPa, LHSV = 1 h^{-1}, t = 2 h	Deoxygenation rate = 84.82%, C8–C16 n-alkane = 49.96%	[47]
Ni	15 wt.% Ni/γ-Al$_2$O$_3$	Rubber seed oil	T = 350 °C, P = 3.5 MPa, t = 3 h	Conversion = 99%	[48]

(continued)

Table 1.3 (continued)

Metal	Supported catalyst	Feedstock	Condition	Result	References
	30 wt.% Ni/palygorskite	Waste cooking oil	$T = 310\ °C$, $P_{H2} = 40$ bar, t = 9 h	Conversion = 100%	[49]
Mo	15 wt.% Mo/γ-Al$_2$O$_3$	Rubber seed oil	$T = 350\ °C$, P = 3.5 MPa, t = 3 h	Conversion = 98%	[48]
	Mo$_2$C/AC	Fatty acids	$T = 360\ °C$, $P_{H2} = 3.0$ MPa, t = 3 h	Conversion = 96.01%	[38]
	10 wt.% Mo/AC	PFAD*	$T = 350\ °C$, t = 1 h, under N$_2$	Hydrocarbons yield = 41%,	[50]
	15 wt.% Mo/γ-Al$_2$O$_3$	Rubber seed oil	$T = 350\ °C$, P = 3.5 MPa, t = 3 h	Conversion = 98%	[48]
	8 wt.% MoOx/ZrO2	Oleic acid	$T = 220\ °C$, t = 5 h, $P_{H2} = 20$ bar	Conversion = 12%	[37]
	10 wt.% Mo/γ-Al$_2$O$_3$	Oleic acid	$T = 300\ °C$, $P_{H2} = 4$ MPa, t = 2 h	Conversion = 91.39%	[51]
Co	10 wt.% Co/endocarp AC	Macauba acid oil	$T = 250\ °C$, $P_{H2} = 30$ bar, t = 2 h	Conversion = 96.25%	[52]
	5 wt.% Co/SBA-15	PFAD*	$T = 350\ °C$, t = 3 h, under N$_2$	Hydrocarbons yields = 16.3%,	[40]
	10 wt.% Co/AC	PFAD*	$T = 350\ °C$, t = 1 h, under N$_2$	Hydrocarbons yield = 38%,	[53]
Ni	5 wt.% Ni/SBA-15	PFAD*	$T = 350\ °C$, t = 3 h, under N$_2$	Hydrocarbons yields = 88.1%,	[40]
	10 wt.% Ni/ZrO$_2$	Palmitic acid	$T = 260\ °C$, $P_{H2} = 10$ bar, t = 6 h	Conversion = 98.33%	[39]
	10 wt.% Ni/H-ZSM-5	Palmitic acid	$T = 260\ °C$, $P_{H2} = 10$ bar, t = 6 h	Conversion = 86.17%	[39]
	10 wt.% NiOx/zeolite Y	Triolein	$T = 380\ °C$, t = 2 h, under H$_2$ free	Conversion = 76.21%,	[54]

(continued)

Table 1.3 (continued)

Metal	Supported catalyst	Feedstock	Condition	Result	References
	7.5 wt.% Ni/DsDA-β-zeolite	Methyl palmitate	$T = 240\ °C$, $P_{H2} = 2.5\ MPa$, $t = 4\ h$	Conversion = 77%	[55]
	15 wt.% Ni/γ-Al$_2$O$_3$	Rubber seed oil	$T = 350\ °C$, $P = 3.5\ MPa$, $t = 3\ h$	Conversion = 99%	[49]
	30 wt.% Ni/palygorskite	Waste cooking oil	$T = 310\ °C$, $P_{H2} = 40\ bar$, $t = 9\ h$	Conversion = 100%	[50]
	5 wt.% NiO/ZnO	PFAD*	$T = 350\ °C$, $t = 2\ h$, under N$_2$	Hydrocarbons yield = 77.8%	[56]
	10 wt.% Ni/SAPO-11	Stearic acid	$T = 290\ °C$, $t = 3\ h$, $P_{H2} = 4\ MPa$	Conversion > 97%,	[57]
NiMo	NiMo-alumina	Karanja oil	$T = 340\ °C$, $t = 4\ h$, 0.9 mmol Ni and 3.4 mmol Mo	Conversion = 100%	[58]
	NiMo/γ-Al$_2$O$_3$	Rubber seed oil	$T = 350\ °C$, $P = 35\ bar$, $WHSV = 1\ h^{-1}$	Diesel range hydrocarbons = 80.87 wt.%	[17]
	NiMo/Mn–Al$_2$O$_3$	Waste soybean oil	$T = 380\ °C$, $P_{H2} = 40\ bar$, 1 mol% Mn	Diesel yield = 60–65%	[19]
	NiMo/Mn–Al$_2$O$_3$	Soybean oil	$T = 380\ °C$, $P_{H2} = 40\ bar$, 1 mol% Mn	Diesel yield = 80–82%	[19]
	NiMo/Al$_2$O$_3$	Microalgae	$T = 360\ °C$, $P = 500\ psig$, 1000 SmL mL^{-1} gas/oil ratio, 1 s residence time	Hydrocarbon yield = 62.7%	[23]
	NiMo/γ-Al$_2$O$_3$	Waste cooking oil	$T = 400\ °C$, $P = 60\ bar$, $t = 4\ h$	Green diesel yield = 77.97%	[59]

[a] Palm Fatty Acid Distillate

Table 1.4 Comparison of the properties of different types of renewable diesel. Reproduced with permission from [60]

Analysis	Units	White diesel	FT diesel	FAME biodiesel	Green diesel (HDO VO)	Hybrid diesel (VCO + VO)	Fossil diesel	Diesel standard Min/Max	
Density	**g/ml**	0.79	0.72–0.82	0.855–0.9	0.77–0.83	0.781–0.85	0.85	Min 0.8	Max 0.845
Sulphur	mg/kg (ppmwt)	**1.54**	**<10**	0–0.012	<10	3–13	12		Max 10
Cetane Index		77.23	70	58.3	50–105	51–64	54.57	Min 46	
Cetane number			55–99	45–72.7	80–99	50–101	50	Min 51	
Flash point	mg/kg	**116**	55–78	96–188	S8–120	74–105	52–136	Min 60	Max 170
Water	mg/kg	**13**	**19**	28.5–500	42–95	10–50	0.5	200	
MCRT carbon residue	(Wt%) 5£m/m	**0.0066**	0.02–4.5	0.02–0.3		85.8			Max 0.3
VISC0 40 C	cSt	3.5	2.1–3.5	3.89–7.9	2.5–4.15	2.7–5.5	2.71	Min 2	Max 4.5
Cooper strip corrosion	(3 h in 50 °C)	1b		1			<3	Class 1	–
Colour	{ASTM}	**0**			-2				–
HPLC	/twr (%m/m)		0		<0.1	0.1–1.2		<11	–
Induction time (oxidation time) (110 C)	h	**>22**	**>22**	0.9–10.9	>22	>22		Min 6	–

(continued)

Table 1.4 (continued)

Analysis	Units	White diesel	FT diesel	FAME biodiesel	Green diesel (HDO VO)	Hybrid diesel (VCO + VO)	Fossil diesel	Diesel standard Min/Max	
Distillation 90 vol% C		302.6	295–335		298–342	300–332	341	85–360	–
Net heating value	MJ/kg	49	43–45	37.1–40.4	42–44	43.3–47	34.97	Min 35	–
CFPP		20	(−22)–0	{-13)–15	>20	(−24 J-22	−6	−5	+5
Cloud point			(−255)–0	(−3)–17	(−25)–30	(−23)–20	−5	Min −5	Max 12
Pour point	°C	23		(−15)–16	(−3)–20	(−26)–2D	−21	Min −13	Max 10

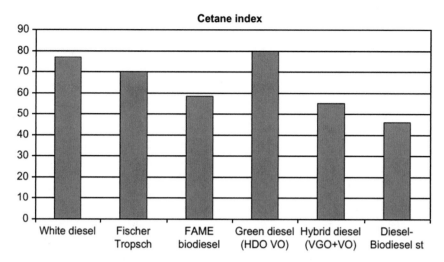

Fig. 1.3 Cetane index of different types of renewable diesel. Reproduced with permission from [60]

lower density. On the other hand, lower density of green diesel is economically beneficial as it can be blended with petroleum products, which are heavy and economical [3, 59].

With regard to cloud point, a fuel with less cloud point is more favorable as the waxes form in the temperature below which cloudy appearance of the fuel occurs with an adverse effect on the engine's performance.

Various studies have been done to specify some properties of green diesel (Table 1.4) including emissions, green diesel produces lower amounts of smoke, SO_X, NO_X, and CO as compared with biodiesel [4].

Orozco et al. [5] investigated an efficient process for green diesel production from castor oil and analyzed the alkanes produced between C_9 and C_{23} via chromatography technique [5]. The mixture of alkanes with carbon chains between C_9 and C_{24} was defined as green diesel. Furthermore, the green diesel features were obtained by ASTM International standards which addressed most of the European standard EN 590 specifications except for cold filter plugging point (CFPP). However, it is reported that 20% vol. of green diesel blended with fossil diesel met the diesel grade B specifications of CFPP.

5 Green Diesel Production by Hydroprocessing Technology

Green diesel or renewable diesel fuel, which is obtained from plant oil, animal fats, and cellulosic biomasses through different conversion technologies, is a biofuel that is chemically the same as petroleum diesel fuel and meets the American Society for Testing and Materials (ASTM) specification for petroleum diesel (D975) [61]. The suitable mechanism and reaction condition for hydroprocessing production will support the commercialization of green diesel. As previously discussed about the different feedstocks and catalysts used for hydroprocessing to produce green diesel, in this section, the effects of hydrogen and process parameters (temperature and pressure) are discussed in detail.

Several studies have reported the need for the production of sustainable and low-cost transport fuel due to the shortcomings related to the usage of biodiesel such as high kinematic viscosity, and corrosiveness [62]. Also, the production of green diesel compared to biodiesel, which is obtained from the transesterification of free fatty acid methyl ester, has offered more benefits due to its lower cost, high emission reduction, higher cetane number, and feedstock flexibility [63].

The global renewable diesel capacity and demand are expected to reach 14.63 and 12.88 million tons in 2024. For the production of green diesel, hydroprocessing is used to saturate the unsaturated double bonds and remove oxygen. The production of green diesel is performed through hydrodeoxygenation, decarboxylation, decarbonylation, cracking, and hydrogenation reactions. Figure 1.4 shows the hydroprocessing reaction path of vegetable oil [64].

Hydrotreating is widely used in refinery unit to remove sulfur, nitrogen, and metals from petroleum-based feedstock. The facilities used for the production of petroleum-based fuel such as catalysts, types of reactors, and distillation facilities can be used for the production of green diesel as well. Figure 1.5 shows the schematic representation of hydroprocessing plant [65].

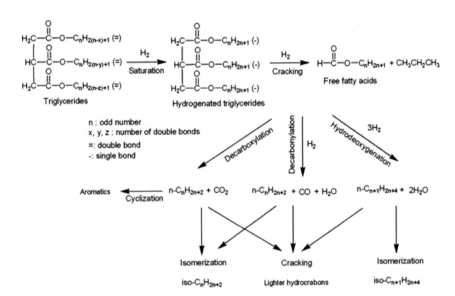

Fig. 1.4 The hydroprocessing reaction path of vegetable oil. Reproduced with permission from [30]

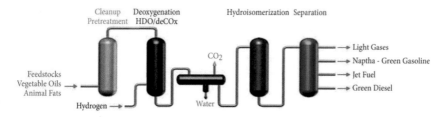

Fig. 1.5 Schematic representation of hydroprocessing plant [65]

Some industries have started to commercialize green diesel production in different countries. For instance, Neste oil has two plants, one in Finland with a total capacity of 170 000 tons/year, and the other with a plant capacity of 800 000 tons/year is in Singapore. Also, Eco fining technology has started producing green diesel by catalytic hydroprocessing vegetable oils [16, 66, 67].

Further, Haldor Topsøe has commercialized green diesel production using a co-processing of vegetable oils (Fig. 1.6) mixed with petroleum feedstocks, which has resulted in a product with lower sulfur content and carbon footprint [68, 69].

Most of the producers have developed proprietary technologies including the pretreatment section for raw materials, the hydrotreatment section, a hydroisomerization reactor, and finally separation columns to obtain green diesel [65]. The choice of feedstock, catalyst selection, reaction time, reaction temperature, and pressure are considered as critical steps for the production of green diesel [70].

Hydrogen plays a significant role in hydroprocessing technology, which can be obtained from different sources such as fossil fuels, biomass, and waste. Currently, hydrogen is mostly generated from non-renewable sources, almost 90% from steam reforming of natural gas and naphtha, and coal gasification [71]. Researches have been focused on the production of hydrogen from renewable resources, biomass, which is plenty uniformly available in all countries and considered sustainable. Technologies such as biomass pyrolysis and gasification have possessed several advantages due to their abundance, low cost, and CO_2 neutrality [4, 30]. Therefore, finding

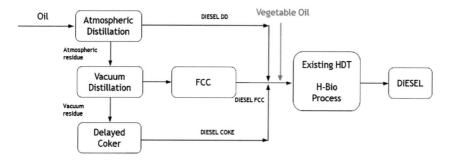

Fig. 1.6 Schematic representation of the co-processing at the refineries. Reproduced with permission from [70]

an alternative method for the production of hydrogen will enhance the hydropro-cessing production efficiencies toward commercialization and lead to a complete green technology for producing green diesel.

The hydrogen pressure and reaction temperature play a pivotal role in the yield and characteristics of producing green diesel [72]. Bezergianni and Dimitriadis (2013) studied the effects of temperature in the range of 330–398 °C on yield, conver-sion, and selectivity of the products as well as the removal of heteroatom and saturation of double bonds through hydrotreating of waste cooking oil [67]. Their results showed that lower reaction temperatures favored diesel production. As the temperature increased, the saturation of double bonds and oxygen removal increased.

Masoumi and Dalai [73] studied the effect of different reaction temperatures (350–450 °C) on oxygen removal through hydroprocessing of algal biofuels mainly in the range of diesel. Their results showed that increasing temperature up to 400 °C resulted in higher oxygen removal from produced biofuel due to the increase of the cracking reaction of larger molecules [73].

Sankaranarayanan et al. [74] studied the co-processing of sunflower oil and gas oil in the diesel fuel range and the effects of different reaction conditions on the conversion. Their results showed that increasing the hydrogen pressure increased the conversion as the hydrogen is required for the hydrocracking reactions [74].

6   Biodiesel, Green Diesel, and Petrodiesel: A Comparison

6.1   Biodiesel

Biodiesel is a renewable fuel produced from animal fats or vegetable oils (edible and nonedible) through the transesterification process. The triglycerides of fatty acids present in the aforementioned feedstock are responsible for their high viscosity and make them undesirable for their use as fuel. Pure oil/fats have less density and heating value than the fossil-derived fuels. Therefore, the transesterification of the oils/fats is carried out to reduce their viscosity and to enhance the density and calorific value [75]. During transesterification, the triglycerides react with alcohols such as methanol or ethanol in the presence of an alkali catalyst such as sodium hydroxide or potassium hydroxide at atmospheric pressure around 50–70 °C temperature [76]. Transesterification produces fatty acid alkyl esters as the main product, which is called biodiesel, while glycerol is formed as the by-product. The length of the carbon chain in the biodiesel varies over a wide range from C_8 to C_{25} [77, 78].

Although biodiesel is chemically different from petrodiesel, it has similar prop-erties, so it could be used as a drop-in fuel (pure form) or could be blended with its fossil-derived counterparts [78]. One of the greatest advantages of using biodiesel is that it can be used in the existing diesel engines (compression ignition engines) with very less or no engine modification. Further, a similar infrastructure used for the storage of petroleum diesel can be used for the storage of biodiesel [75]. Pure biodiesel is designated as B100 which should meet the requirement of US specification ASTM D7651. Raw vegetable oils and animal fats cannot meet these specifications and hence they are not considered biodiesel. The blend of biodiesel with petroleum diesel should meet the specifications for ASTM D7467 diesel motor fuel [79].

6.2 Green Diesel

Green diesel or renewable diesel is produced from the same renewable resources such as biodiesel but uses a different production process. Biodiesel is manufactured by the transesterification process and comprises long-chain fatty acid alkyl esters. On the other hand, green diesel production involves the hydrotreatment process and it consists of components such as alkanes and aromatics [80]. As discussed in the introduction, green diesel is produced by the hydrotreatment of the 1st, 2nd, and 3rd generation feedstock. It is primarily produced by waste and residues and hence called second-generation fuel. During this process, the feedstock reacts with hydrogen in the presence of catalysts at high temperatures (300–400 °C) and pressure of almost 10 MPa. Heterogeneous catalysts like $Ni–Mo/Al_2O_3$ and $Co–Mo/Al_2O_3$ are used in this process to produce hydrocarbons in the range of $C_{15}–C_{18}$. Hydrotreatment also results in the removal of heteroatoms like oxygen, nitrogen, and sulfur [81]. Green Diesel has a similar chemical composition as that of petrodiesel and can be used in existing refinery infrastructure such as pipelines and storage tanks. It has similar advantages to that of biodiesel in terms of its use in diesel engines without any engine modification. It can be either used in its pure form or can be blended with petrodiesel [78].

Although several other methods are being developed to produce green diesel, hydrotreatment is mostly accepted and is also commercially available. Currently, green diesel meets the ASTM D975 specification in the U.S. and EN 590 in Europe.

6.3 Petrodiesel (Conventional Diesel)

Diesel fuel is produced in petroleum refineries through fractional distillation of crude oil. It contains a mixture of hydrocarbon molecules that varies in length from 8 to 21 carbon atoms [82]. It is used in a diesel engine, also known as a compression ignition engine, where ignition takes place due to the elevated temperature of the air caused by mechanical compression. Large trucks, trains, ambulances, buses, cars, boats, farm and construction equipment have diesel engines. Diesel fuel is also used in diesel generators for backup power supply. Conventional diesel should meet the ASTM D975 specification to be used as fuel in diesel engines [76].

7 Comparison: Biodiesel, Green Diesel, and Petrodiesel

Owning to the environmental challenges, several federal and provincial governments around the globe encouraged the use of renewable fuels by introducing renewable fuel mandates that require the addition of biofuel along with fossil-derived fuels. Therefore, it is necessary to evaluate the characteristics of biodiesel and green diesel, and compare them with the standards of conventional diesel (Table 1.5).

The performance of an engine is strongly affected by the physicochemical properties of the fuel being used. The properties of green diesel obtained from most of the vegetable oils are found to be identical to conventional diesel. Density is related to the energy content for a given volume of fuel. The calorific value is directly proportional to the density of the fuel being used [67]. Viscosity in fuels is the property that generally measures their resistance to flow. It is related to the chemical structure of the compounds present in the fuel and the intermolecular forces between them. High viscosity hinders pumping and causes poor atomization, which leads to problems in combustion. The density and viscosity of green diesel are less than the biodiesel due to its less oxygen content [83].

Flashpoint is the measure of flammability of the fuel. It is the lowest possible temperature at which the vapor will be formed near the liquid fuel surface and will ignite on exposure to flame [84]. Diesel fuel consists of hydrocarbons of various chain lengths C_8–C_{21}. As long-chain hydrocarbons have a higher freezing point than shorter ones, they start forming crystals upon decreasing temperature. Cloud point is the maximum temperature at which the oil becomes cloudy (hazy) in appearance [67]. The oil still flows at this temperature as some of the molecules are in a liquid state, while other molecules are in a crystal state. Upon further decreasing the temperature, all the hydrocarbons are converted into crystals and the oil loses its flowing characteristics, this point is called the pour point. Both cloud point and pour point determine the low-temperature fluidity of the fuel. Biodiesel has mostly ester components that do not pose hydrogen bonding and hence they have very few intermolecular forces between them. Therefore, the flashpoint of biodiesel is less than green diesel, which consists of various other compounds like aldehydes, ketones, etc. The cloud point and pour point of green diesel are higher than the biodiesel; therefore, it may not perform well in cold climates [85]. These cold flow properties can be altered through isomerization or hydrocracking processes so that they can be blended with conventional diesel [86]. The calorific value of green diesel is almost similar to conventional diesel and higher than biodiesel due to its less oxygen content.

Cetane number denotes how fast the fuel can be ignited in CI (compression ignition) engines. Higher the cetane number, the lesser the ignition delay which eventually increases the engine performance. The cetane number of green diesel is higher than the biodiesel, though the cetane number of both fuels is more than the standard value required for the use in diesel engines. Both green diesel and biodiesel have minute amounts of sulfur content. The oxygen content of biodiesel is higher than green diesel as the green diesel undergoes hydrotreatment for the removal of oxygen. A distillation curve (temperature vs. percentage volume recovered) is used to characterize the volatility of the fuels and this depends upon the composition of the fuel [67]. The property of lubricity deals with the capacity of reducing friction. It is important as it reduces the wear and tear of engine components caused due to friction. Biodiesel has better lubricity due to the presence of ester components. Furthermore, GHG emissions by green diesel are around 40% lower than biodiesel and about 80% less than fossil diesel [83].

8 Life Cycle Analysis (LCA) of Green Diesel Fuel

One of the major advantages of green diesel fuel production is related to the decrease in carbon dioxide emission and global warming. Life cycle analysis is a systematic means to study the energy consumption and environmental impacts of any process, which is also used to quantify these impacts. LCA of green diesel assists the government and petroleum industries in making decisions on the future implementation of renewable fuels related to sustainability [87]. Generally, biofuels like green diesel are considered carbon–neutral because the plants absorb CO_2 from the atmosphere for performing photosynthesis and the same CO_2 is emitted when they are burnt. However, the steps involved in the production of green diesel led to GHG emissions and have other environmental impacts. The attributes of green diesel and its environmental impacts mainly depend upon the conversion pathways and type of biomass feedstock being used.

9 Production of Green Diesel from Vegetable Oil

Arguelles et al. studied the life cycle assessment of green diesel which is produced through hydrotreatment of palm oil. The key stages considered for green diesel production are shown in Fig. 1.7. In this study, the impact categories like global warming, ozone layer depletion, acidification, eutrophication, photochemical smog, and human toxicity are simulated [88].

For higher growth and yield, palm culture has high water requirement causing a higher water footprint than the fuel produced from other vegetable oils. The study demonstrated that green diesel reduces GHG emissions by around 110% when compared to its petroleum counterparts. Fuel combustion and palm oil processing contribute to the highest emissions. The study showed that green diesel represents a significant reduction in ozone layer depletion, photochemical smog of about 75%, and 36.4% respectively when compared to conventional diesel. When it comes to acidification potential, green diesel has about 27.6% less emissions than its fossil counterparts. Green diesel has a substantial ecological impact of nearly 8.6 and 60 times more than petrodiesel for the human toxicity potential and eutrophication categories, respectively. Photochemical smog and eutrophication mainly happen due to the use of triple phosphate fertilizer in palm culture. Moreover, the use of pesticides

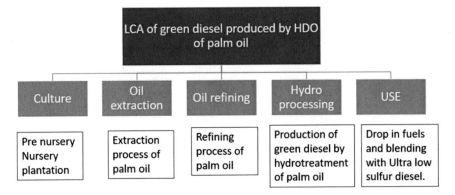

Fig. 1.7 Key stages involved in the green diesel production from hydrotreatment of palm oil

and herbicides generates toxicity problems like contamination of water, soil, and causes respiratory diseases in people. It was concluded that modification of fertilization practices and substitution of agrochemicals with organic products can reduce the ecological impact [88].

10 Production of Green Diesel from Lignocellulosic Biomass

Wong et al. [86] studied the life cycle assessment of green diesel which is produced from agricultural and forest residue [86]. Three different lignocellulosic feedstocks were compared for the production of renewable diesel in this study, i.e., agricultural residue (wheat and barley straw), forest residue (chips from logging operations), and whole tree (chips from cutting the whole tree). Greenhouse gas emissions and Net Energy Ratio (NER) are the two main factors that are considered for this LCA study. Net energy ratio is defined as the energy value of the final product (hydrogen derived renewable diesel—HDRD) per unit amount of fossil fuel energy used in the production process [89]. The key stages for the production of HDRD production from lignocellulosic biomass are given in Fig. 1.8.

The GHG emissions for the production of HDRD from whole tree, forest residue, and agricultural residue were reported as 39.7 $gCO_{2,eq}$/MJ, 42.3 $gCO_{2,eq}$/MJ, and 35.4 g $CO_{2,eq}$/MJ, respectively. The net energy ratio of the above 3 feedstock was reported as 1.71 MJ/MJ, 1.55 MJ/MJ, and 1.9 MJ/MJ, respectively. Different feedstocks demonstrated variation in results due to different energy requirements for harvesting and pre-treatment of biomass [90]. Also, the chemical composition of the biomass affects the pyrolysis yield. The agricultural residue was found to be the most efficient feedstock as they were reported the least GHG emissions and the highest net energy ratio (NER). The energy-intensive hydroprocessing stage was found to be accountable for the vast majority of the GHG emissions produced during the entire

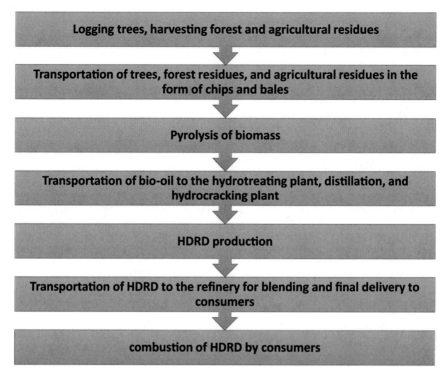

Fig. 1.8 Key stages considered for emissions and energy usage in HDRD production from lignocellulosic biomass [89]

process. This process can be further improved to reduce emissions and improve the Net Energy Ratio. The results demonstrated that GHG emissions from renewable diesel can be 39–47% of those of fossil counterparts [86]. This indicates that renewable diesel can lessen GHG emissions by a particular limit. From the perspective of energy usage, the NER for all feedstocks used is more than 1.5 MJ/MJ, showing the sustainability of hydrogen-derived renewable diesel (HDRD).

11 Production of Green Diesel from Algae

Pragya et al. studied the LCA for producing green diesel from *Chlorella Vulgaris* microalgae. In this study, greenhouse gas emissions from the production process and the net energy balance were calculated [84].

Chlorella algae grow both in fresh water and marine water. Because of its adaptability to different environments, it is called ubiquitous [91]. The stages of the life cycle assessment include algae culture and harvesting, extraction of oil, processing of oil and biomass residue, and combustion of green diesel. The target of the study is to increase the Net Energy Balance (NEB) through a mixture of various agronomical practices and the methods that are used in producing biofuel from microalgae. NEB is defined as a difference between the energy value of the final product (green diesel) and the total energy consumption in manufacturing the fuel [89].

In this study, two different habitats are considered for the algae culture, i.e., open raceway pond and photobioreactor. Different sequences of harvesting and drying techniques were used which are shown in Fig. 1.9. Both dry route (pyrolysis) and wet route (hydrothermal liquefaction) are studied for the conversion of algae culture (after lipid extraction) to bio oil, which is further upgraded through hydrotreatment.

The results showed that both the open raceway pond and photobioreactors (wet and dry routes) yielded a negative net energy balance. The fertilizers used for providing nutrient media to culture, electricity used for harvesting and drying operations were the main energy-intensive activities. The production of green diesel from photobiore-actors has higher negative net energy balance and less net energy ratio compared to the raceway pond (both dry and wet routes). This is because the tube lights and heat exchangers used in photobioreactors require high energy. For the raceway pond, the dry route has a marginally higher NER than the wet route. Even though the wet route saves energy on the drying process, the huge energy requirements for the HTL

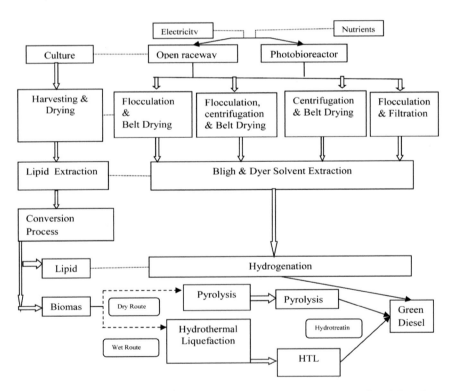

Fig. 1.9 Various possible combinations and routes for green diesel production from Microalgae. Reproduced with permission from [84]

process led to less NER. The open raceway pond dry route is proved to be the best possible route, but it has a negative energy balance of around −4.07 MJ and very less NER of 0.2. Hence, further research in developing efficient culture, harvesting, and drying technologies is needed to lower the huge energy requirements for producing green diesel from microalgae [84].

12 Techno-Economic Feasibility of Green Diesel

The feasibility study of the process used for green diesel production is based on the results obtained from techno-economic analyses (TEA). In general, the cost of the feedstock, product yield, and the capacity of the processing unit mainly affects the economics of the process [92, 93]. About 40–90% of the cost of production of biofuel is attributed to the feedstock cost. Figure 1.10 shows the variation in the cost of the process based on the feedstock cost [93].

Different thermochemical conversion technologies such as hydrothermal lique-faction, pyrolysis, and gasification are used to produce biofuel. The produced bio-oil from hydrothermal liquefaction and pyrolysis has revealed some characteristics such as high viscosity and acidity which require further upgradation in order to be used as transportation fuel [94].

The hydrothermal liquefaction process is a thermochemical process to convert wet biomass such as microalgae to bio-oil at high pressure (10–25 MPa) and

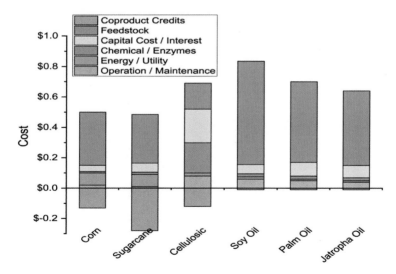

Fig. 1.10 Production cost of biofuel based on the usage of various feedstocks. Reproduced with permission from [93]

moderate temperature (200–400 °C) [95]. Masoumi and Dalai [96] studied the techno-economic analysis of biofuels production (mainly in the range of diesel) obtained from microalgae through hydrothermal liquefaction and hydroprocessing [96]. Their results showed that feedstock cost is one of the most significant factors to determine the selling price of produced biofuel, and a 50% decrease in feedstock price resulted in an approximately 27% decrease in the minimum fuel selling price. Their results also showed that the utilization of by-product could improve the overall economic of the two-stage process. Therefore, utilization of hydrochar as a catalyst support resulted in lowering the minimum fuel selling price which was about 10% compared with the other method which was utilization of hydrochar through combustion process to provide heat for HTL process. Figure 1.11 shows the flow diagram of the processes as well as their reaction conditions.

Regarding the production of green diesel from HTL followed by hydroprocessing, there are some challenges, such as handling the biomass slurry to be processed at high temperature and pressure, high cost required for separating the produced bio-oil and aqueous phase, and also wastewater treatment and recycling [93, 97]. Some researchers have focused on combining wastewater treatment and microalgae cultivation as the feasible way for biofuel production, as the estimated fuel selling price ($2.2/gallon) meets the acceptable level [98].

Magdeldin et al. [99] studied the techno-economic assessment of non-catalytic hydrothermal liquefaction of forest residues by taking advantage of the by-products' contribution. The non-catalytic hydrothermal liquefaction process was carried out at 330 °C and 210 bar [99]. A flowsheet and Aspen Plus® model were developed which included the pre-treatment stage, hydrothermal liquefaction, fuel upgrading, and finally the recovery of residue. The results showed that if the capital costs for the HTL reactor system are reduced by 30% or co-products price could reach 4 USD

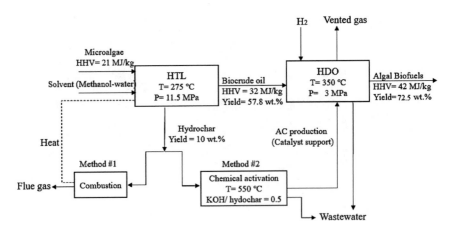

Fig. 1.11 Process flow diagram of algal biofuels production using utilization of hydrochar. Reproduced with permission from [97]

per GGE of liquid fuel, the capital and operational costs are offset by the economical value of selling products [99].

Biomass pyrolysis is another thermo-chemical method to decompose biomass to value-added products such as biochar, bio-oil, and gases that are formed in the absence of oxygen. Wu et al. [100] studied the techno-economic analysis of co-processing of bio-oil (obtained from fast pyrolysis and catalytic pyrolysis) and vacuum gas oil to produce transportation fuel [100]. The minimum fuel selling price which was in the range of $2.63 and $2.60 per gallon, could compete with petroleum-based fuels. According to the sensitivity analysis results, the price of produced fuels extremely depends on the yield. Besides the production of green fuels through pyrolysis process, there are some challenges including the presence of heteroatoms and alkali metals in the bio-oil as well as its high acidity and viscosity. Also, the blending of bio-oil and vacuum gas oil is limited to 20% [93, 97].

Zhang et al. [97] studied the techno-economic feasibility of pyrolysis bio-oil through two upgrading pathways; the first method included two-stage hydrotreating followed by fluid catalytic cracking (FCC) and the second method involved the single-stage hydrotreating followed by hydrocracking [97]. The two pathways were modeled using Aspen Plus® for a capacity of 2000 metric tons/day (MTPD). Figure 1.12 shows the process flow diagram generalized for these two pathways including six areas; biomass pre-treatment, pyrolysis reaction at a temperature of 500 °C in a fluidized bed reactor, solid removal, bio-oil recovery, combustion, and upgradation. The total biofuels yield for the hydrotreating/FCC pathway under option 1 (hydrogen was purchased from the market) and 2 (hydrogen produced through steam reforming of natural gas) were 223.3 MTPD and 88.9 MTPD, respectively, with the estimated installed equipment costs of $130 million, $155 million, respectively. For the hydrotreating/hydrocracking pathway, the yield of diesel fuel was 165 MTPD and estimated installed equipment costs of $190 million. Sensitivity analysis results showed that the price of hydrogen, feedstock cost, and the yield of the product are the main factors in the economic feasibility of the two pathways employed for pyrolysis bio-oil upgradation. Also, their results showed that an investment in the hydrotreating/FCC pathway could present a relatively low risk while hydrogen was produced through steam reforming.

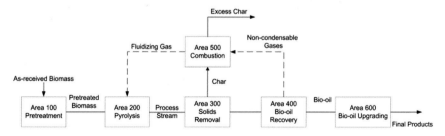

Fig. 1.12 Process diagram for production of biofuels using bio-oil obtained from pyrolysis process. Reproduced with permission from [101]

Meyer et al. [102] studied the process economics of biofuels production through fast pyrolysis of different types of biomasses including lignocellulosic feedstocks such as switchgrass and corn stover as well as upgradation of bio-oil through hydrodeoxygenation [102]. Their results showed that for all of the biomasses, the largest cost contribution (30–40%) of the minimum fuel selling price (MFSP) is related to the capital related costs, and also feedstock costs (approximately 30%). Compared to the usage of the pure feedstocks, employing the blend feedstocks, due to their lower cost, could improve the overall economy. The range of minimum fuel selling price of produced fuels depends on the biomass feedstocks, which were in the range of $3.7/GGE to $5.GGE [102].

The use of gasification-FTS (Fischer–Tropsch synthesis) is another method that could be used for the production of green diesel. Gasification is a thermal process to obtain gaseous products mainly CO and H_2 which are called syngas in the presence of an oxidizing agent. The produced syngas should be cleaned for FTS to produce liquid biofuels [103]. There are some challenges related to this method as well, one such challenge is the formation of tar through gasification process, which causes some issues such as catalysts deactivation and corrosion [93].

Rafati et al. [101] studied the process economics of production of Fischer–Tropsch liquids via biomass gasification using different catalysts and natural gas co-feeding (Fig. 1.13) [101]. The overall thermal efficiency of the process was in the range of 41.3–45.5%. Their results showed that using a reformer to recycle off-gas and co-feeding of natural gas could improve the economics of the process and maximize FT fuel production. The production cost of FT liquids was around $28.8 per GJ using a reformer and in the range of $19–20 per GJ using co-feeding of natural gas, which could not be economically feasible considering an oil price of $60/barrel [101].

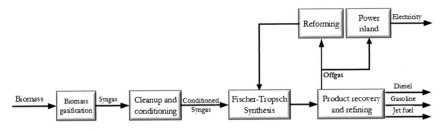

Fig. 1.13 Schematic diagram of the production of FT fuels through biomass gasification [104]

13   Current Status of the Green Diesel Industry

Neste oil: Currently, Neste is considered as the highest producer of green diesel (hydrotreated/hydrogenated vegetable oil) and renewable turbine fuel in the world. Neste manufactures high-quality green diesel at Porvoo (Finland), Rotterdam (the Netherlands), and Singapore refineries; in addition, Neste is also a co-owner of a base-oil plant in Bahrain. Neste low-carbon biofuel is also known as Neste MY Renewable Diesel™ which was produced from 100% renewable feedstocks that result in 90% reduction of greenhouse gas emissions relative to the fossil-derived diesel fuel [105]. Hydrotreatment of vegetable oils, used cooking oils, and fats were transformed into renewable diesel via NEXBTL™ commercial manufacturing technology. During hydrotreatment of triglyceride molecules that are present in fats and oils, hydrogen was consumed to remove the oxygen in the form of water using acidic catalyst to create hydrocarbons that are identical to conventional diesel fuel. Further, produced green diesel is compatible with the existing compression ignition engine infrastructure by blending with conventional diesel at different ratios or as a pure diesel. Figure 1.14 shows the typical routes to produce diesel-range hydrocarbons via hydrotreatment and decarboxylation routes using edible, nonedible vegetable oil feedstock, used cooking oil, and waste fats [105]. In addition, Fig. 1.15 depicts the utilization of triglycerides to produce fatty acid methyl esters and renewable diesel through hydrotreatment routes at higher temperature and pressure. Green diesel is a mixture of different straight chain and branched paraffin types of hydrocarbons in the range of C_{15}–C_{18}. Further, Neste renewable diesel meets the ASTM D975, EN 15,940 requirements for paraffinic diesel fuels and allows blending with EN 590

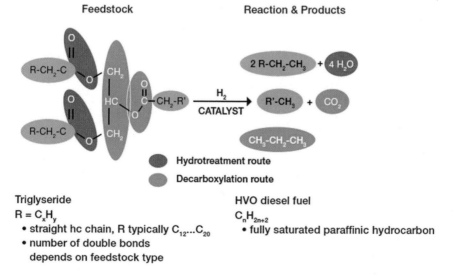

Fig. 1.14 Green diesel production pathways via hydrotreatment and decarboxylation using triglycerides [106]

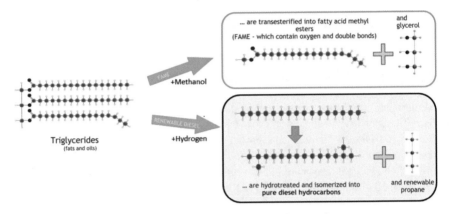

Fig. 1.15 Schematic representation for the production of fatty acid methyl esters and green diesel using triglycerides [105]

B7 diesel fuel without any fixed blend percentage. However, Neste renewable diesel does not meet EN 14,214 standard for fatty acid methyl esters owing to the difference in their chemical composition. Unlike conventional diesel fuel, green diesel doesn't contain aromatics which are not favorable for clean combustion of the fuel. Therefore, green diesel fuel properties are identical to the fuels produced via Gas-to-Liquid and Biomass-to-Liquid diesel fuels produced via Fischer–Tropsch synthesis. Along with Neste, several industries are focused to produce renewable diesel via hydrotreatment of vegetable oils and are reported in Table 1.6 along with their production technology and product names [105].

Table 1.5 Comparison of physicochemical properties of biodiesel, green diesel, and conventional diesel [67, 81]

Property	Biodiesel	Green diesel	Conventional diesel
Density (kg/m^3)	855–900	770–830	800–850
Kinematic viscosity at 40 °C (mm^2/s)	3.89–7.9	2.5–4.15	2–4.5
Flash point (°C)	55–78	68–140	60–170
Pour point (°C)	(−15)–16	(−3)–29	3–15
Cloud point (°C)	(−3)–17	(−25)–30	(−13)–15
Calorific value (MJ/kg)	37.1–40.4	42–44	Min 35
Cetane number	45–72.7	80–99	Min 51
Acid number (mg KOH/g)		33.33	
Sulfur content (mg/kg)	0–0.012	<10	Max 10
Oxygen (wt.%)	10–12 (Manchada)		
Distillation 90 vol% (°C)		265–320	85–360
Lubricity (μm)		360	Max 460

Table 1.6 List of the industries, their green diesel production technology, and final products derived from renewable feedstocks

Name of the company	Production technology	Products
Neste	NEXBTL™	Neste MY Renewable Diesel™
Axnes IFP	Vegan®	Renewable diesel and jet fuels
Honeywell UOP/ENI	Ecofining™	Green diesel
Haldor Topsoe	HydroFlex™	UPM BioVerno
Valero	Hydrotreatment	Renewable diesel
Diamond green diesel	Hydrotreatment	Renewable diesel

Along with the existing refineries to produce green diesel, Table 1.7 reported the upcoming renewable fuel plants that are under construction, proposed, and expansion stages.

13.1 Neste Green Diesel Fuel Properties and Engine Performance

Most of the Neste green diesel is composed of n-paraffins and i-paraffins that are readily blended as drop-in-fuels in compression ignition engines. The properties of the green diesel are superior to sulfur free conventional diesel and fatty acid methyl esters; moreover, Neste renewable diesel exhibited identical properties to the diesel fuel generated from the Gas-to-Liquid conversion technology. A detailed comparison of the Neste green diesel properties with ASTM D975, EN 15,940, and EN 590 standards is reported in the Neste Renewable Diesel Handbook [105]. Upon relative comparison of the Neste renewable diesel properties, it showed lower density compared to European diesel fuels, very high cetane numbers (70–95), higher calorific value, and lower final boiling point relative to fatty acid methyl esters, and identical distillation range of European diesel fuels. Sulfur and ash contents are reported as <1 mg/kg (5 mg/kg is the allowable limit), <0.001 wt.%, respectively; besides, it requires lubricity additive like petrodiesel. When it comes to the stability of the Neste renewable diesel, it is the same as the fossil-derived diesel fuel and there is no "best before date" to consume the Neste renewable diesel.

Engine performance of the Neste green diesel was tested on 300 vehicles (300,000 km/vehicle) and found that no engine modification is needed and on par with the conventional diesel for torque and maximum power was noticed. Besides, excellent cold flow properties (up to −34 °C) were exhibited and tendency to form deposits in fuel injection system, fuel injectors were found to be very low. In addition, Neste renewable diesel offers reduced NO_X, carbon monoxide, hydrocarbons and particulate matter, polyaromatic hydrocarbons, mutagenic emissions, and exhibits a low tendency toward injector fouling for neat and conventional diesel fuel blends.

Table 1.7 Upcoming green diesel production facilities [107]

Name of the facility	Feedstock	Capacity (MMgy)	Status	Location	Products
Bakersfield renewable fuels	Waste fats, used cooking oil, soybean oil and distillers corn oil, camelina oil	230	Under construction	Bakersfield, California	Renewable diesel, liquid propane, and naphtha
CVR Energy Inc.—Wynnewood	Corn oil, animal fats, and used cooking oils	100	Under construction	Wynnewood, Oklahoma	Renewable diesel and naphtha
Diamond Green Diesel—Norco	Recycled animal fats, used cooking oils, and inedible corn oil	675	Under expansion	Norco, Louisiana	Renewable diesel fuel and naphtha
Diamond Green Diesel—Port Arthur	Recycled animal fats, used cooking oils, and inedible corn oil	400	Proposed	Port Arthur, Texas	Renewable diesel fuel and naphtha
HollyFrontier Corp.—Artesia	Lower priced unrefined soybean oil, animal fats, and distillers corn oil	110	Under construction	Artesia, New Mexico	Renewable diesel

(continued)

Table 1.7 (continued)

Name of the facility	Feedstock	Capacity (MMgy)	Status	Location	Products
HollyFrontier Corp.—Cheyenne	Lower priced unrefined soybean oil, animal fats, and distillers corn oil	90	Under construction	Cheyenne, Wyoming	Renewable diesel
Grön fuels LLC	Soybean oil, corn oil, and animal fats	900	Proposed	Baton Rouge, Louisiana	Renewable diesel and jet fuel
Marathon Petroleum—Dickinson	Distillers corn oil, soybean oil, and rendered fats	184	Operational	Dickinson, North Dakota	Renewable diesel
Marathon Petroleum—Martinez	Animal fats, soybean oil, and distillers corn oil	736	Proposed	Martinez, California	Renewable diesel
Next Renewable fuels	Brown grease, animal tallow, soy oil, and a variety of vegetable oils	575	Proposed	Port Westward, Oregon	Renewable diesel and propane
REG Geismar LLC	High and low free fatty acid feedstocks	340	Under expansion	Geismar, Louisiana	Renewable diesel, naphtha, autogas

(continued)

Table 1.7 (continued)

Name of the facility	Feedstock	Capacity (MMgy)	Status	Location	Products
Ryze Renewables—Las Vegas	Inedible agricultural oils and animal fats	100	Under construction	Las Vegas, Nevada	Renewable diesel
World Energy—Paramount	Inedible agricultural wastes	330	Under construction	Paramount, California	Sustainable aviation fuel, renewable diesel, gasoline and propane
Imperial Oil	Canadian grown crops	220	Under construction	Edmonton, Alberta	Renewable diesel

13.2 Honeywell

Honeywell Green Diesel™ is produced via a single-stage innovative UOP Ecofining™ technology using waste cooking oils and animal fats. It can reduce 80% of the greenhouse gas emissions relative to petrodiesel and its cetane value is 80, which is 40–60% higher than the conventional diesel. Therefore, Honeywell Green Diesel™ fuel shows better engine performance with less emissions and it can also be easily blended with low-cetane diesel to meet the liquid transportation fuel standards. Currently, Honeywell is operating renewable diesel and jet fuels production facilities at 20 licensed Ecofining units in nine countries with 12 different types of renewable feedstocks [108]. As discussed in the previous section on Neste green diesel, Honeywell Green Diesel™ also exhibits identical physicochemical properties of the fuel and engine performance and emission properties.

13.3 ExxonMobil

ExxonMobil in collaboration with Imperial Oil Ltd. is planning to produce green diesel at the Strathcona refinery in Edmonton, Canada. This refinery complex is expected to produce 20,000 barrels/day of renewable diesel to reduce the 3 million metric tons/year of greenhouse gas emissions. Strathcona refinery complex will utilize locally grown plant-based feedstocks and blue hydrogen (hydrogen produced from natural gas with carbon capture and storage) to produce the green diesel. A

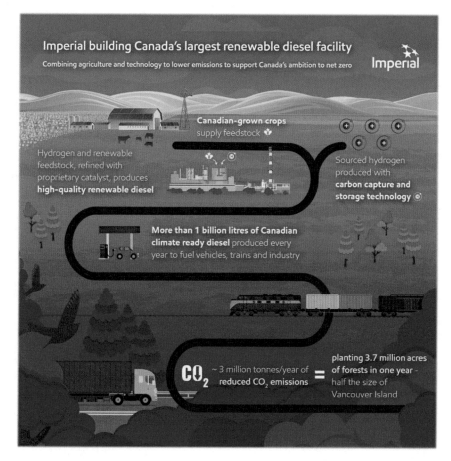

Fig. 1.16 Proposed Canada's largest renewable diesel facility by ExxonMobil and Imperial Oil Ltd [109]

detailed schematic representation of the proposed Canada's largest renewable diesel facility is shown in Fig. 1.16. Besides, these Neste, ExxonMobil, Honeywell and other green diesel production facilities are reported in Table 1.7, World energy is also producing renewable diesel at Massachusetts, USA.

References

1. Esmi F, Nematian T, Salehi Z, Khodadadi AA, Dalai AK (2021) Amine and aldehyde functionalized mesoporous silica on magnetic nanoparticles for enhanced lipase immobilization, biodiesel production, and facile separation. Fuel 291:120126. https://doi.org/10.1016/j.fuel.2021.120126
2. Azadi P, Farnood R (2011) Review of heterogeneous catalysts for sub- and supercritical water

gasification of biomass and wastes. Int J Hydrogen Energy 36:9529–9954. https://doi.org/10.1016/j.ijhydene.2011.05.081

3. International Energy Outlook 2011, Report Number: DOE/EIA-0484 (2011). http://www.eia.gov/forecasts/ieo/pdf/0484(2011).pdf. Accessed 26 Dec 2021

4. Douvartzides SL, Charisiou ND, Papageridis KN, Goula MA (2019) Green diesel: Biomass feedstocks, production technologies, catalytic research, fuel properties and performance in compression ignition internal combustion engines. Energies 12:809. https://doi.org/10.3390/en12050809

5. Orozco LM, Echeverri DA, Sánchez L, Rios LA (2017) Second-generation green diesel from castor oil: development of a new and efficient continuous-production process. Chem Eng J 322:149–156. https://doi.org/10.1016/j.cej.2017.04.027

6. Kordouli E, Pawelec B, Bourikas K, Kordulis K, Fierro JLG, Lycourghiotis A (2018) Mo promoted Ni–Al$_2$O$_3$ co-precipitated catalysts for green diesel production. Appl Catal B Environ 229:139–154. https://doi.org/10.1016/j.apcatb.2018.02.015

7. Kordouli E, Kordulis C, Lycourghiotis A, Cole R, Vasudevan PT, Pawelec B, Fierro JLG (2017) HDO activity of carbon-supported Rh, Ni and Mo–Ni catalysts. Mol Catal 441:209–220. https://doi.org/10.1016/j.mcat.2017.08.013

8. Dupont C, Lemeur R, Daudin A, Raybaud P (2011) Hydrodeoxygenation pathways catalyzed by MoS$_2$ and NiMoS active phases: a DFT study. J Catal 279:276–286. https://doi.org/10.1016/j.jcat.2011.01.025

9. Arun N, Sharma RV, Dalai AK (2015) Green diesel synthesis by hydrodeoxygenation of bio-based feedstocks: strategies for catalyst design and development. Renew Sustain Energy Rev 48:240–255. https://doi.org/10.1016/j.rser.2015.03.074

10. Scaldaferri CA, Pasa VMD (2019) Production of jet fuel and green diesel range biohydrocarbons by hydroprocessing of soybean oil over niobium phosphate catalyst. Fuel 245:458–466. https://doi.org/10.1016/j.fuel.2019.01.179

11. Ewing M, Msangi S (2009) Biofuels production in developing countries: assessing tradeoffs in welfare and food security. Environ Sci Policy 12:520–528. https://doi.org/10.1016/j.envsci.2008.10.002

12. Onyestyák G, Harnos S, Szegedi A, Kalló D (2012) Sunflower oil to green diesel over Raney-type Ni-catalyst. Fuel 102:282–288. https://doi.org/10.1016/j.fuel.2012.05.001

13. Sotelo-Boyás R, Liu Y, Minowa T (2011) Renewable diesel production from the hydrotreating of rapeseed oil with Pt/zeolite and NiMo/Al$_2$O$_3$ catalysts. Ind Eng Chem Res 50:2791–2799. https://doi.org/10.1021/ie100824d

14. Herskowitz M, Landau MV, Reizner Y, Berger D (2013) A commercially-viable, one-step process for production of green diesel from soybean oil on Pt/SAPO-11. Fuel 111:157–164. https://doi.org/10.1016/j.fuel.2013.04.044

15. Srifa A, Faungnawakij K, Itthibenchapong V, Assabumrungrat S (2015) Roles of monometallic catalysts in hydrodeoxygenation of palm oil to green diesel. Chem Eng J 278:249–258. https://doi.org/10.1016/j.cej.2014.09.106

16. Yenumala SR, Kumar P, Maity SK, Shee D (2019) Production of green diesel from karanja oil (Pongamia pinnata) using mesoporous NiMo-alumina composite catalysts. Bioresour Technol Reports 7:100288. https://doi.org/10.1016/j.biteb.2019.100288

17. Ameen M, Azizan MT, Ramli A, Yusup S, Alnarabiji MS (2018) Catalytic hydrodeoxygenation of rubber seed oil over sonochemically synthesized Ni–Mo/γ–Al$_2$O$_3$ catalyst for green diesel production. Ultrason Sonochem 51:90–102. https://doi.org/10.1016/j.ultsonch.2018.10.011

18. Ramesh A, Palanichamy K, Tamizhdurai P, Umasankar S, Sureshkumar K, Shanthi S (2019) Sulphated Zr–Al$_2$O$_3$ catalysts through jatropha oil to green-diesel production. Mater Lett 238:62–65. https://doi.org/10.1016/j.matlet.2018.11.158

19. Vázquez-Garrido I, López-Benítez A, Guevara-Lara A, Berhault G (2021) Synthesis of NiMo catalysts supported on Mn–Al2O3 for obtaining green diesel from waste soybean oil. Catal Today 365:327–340. https://doi.org/10.1016/j.cattod.2020.06.001

20. Nikolopoulos I, Kogkos G, Kordouli E, Bourikas K, Kordulis C, Lycourghiotis A (2020) Waste cooking oil transformation into third generation green diesel catalyzed by Nickel–Alumina catalysts. Mol Catal 482:110697. https://doi.org/10.1016/j.mcat.2019.110697

21. Sivaramakrishnan R, Incharoensakdi A (2018) Microalgae as feedstock for biodiesel production under ultrasound treatment—a review. Bioresour Technol 250:877–887. https://doi.org/10.1016/j.biortech.2017.11.095

22. Rosenberg JN, Oyler GA, Wilkinson L, Betenbaugh MJ (2008) A green light for engineered algae: redirecting metabolism to fuel a biotechnology revolution. Curr Opin Biotechnol 19:430–436. https://doi.org/10.1016/j.copbio.2008.07.008

23. Zhou L, Lawal A (2016) Hydrodeoxygenation of microalgae oil to green diesel over Pt, Rh and presulfided NiMo catalysts. Catal Sci Technol 6:1442–1454. https://doi.org/10.1039/c5cy01307k

24. Hongloi N, Prapainainar P, Prapainainar C (2021) Review of green diesel production from fatty acid deoxygenation over Ni-based catalysts. Msssol Catal (In Press) 111696. https://doi.org/10.1016/j.mcat.2021.111696

25. Di Visconte GS, Spicer A, Chuck CJ, Allen MJ (2019) The microalgae biorefinery: a perspective on the current status and future opportunities using genetic modification. Appl Sci 9:4793. https://doi.org/10.3390/app9224793

26. Alalwan HA, Alminshid AH, Aljaafari HAS (2019) Promising evolution of biofuel generations subject review. Renew Energy Focus 28:127–139. https://doi.org/10.1016/j.ref.2018.12.006

27. Bjelić A, Grilc M, Huš M, Likozar B (2018) Hydrogenation and hydrodeoxygenation of aromatic lignin monomers over Cu/C, Ni/C, Pd/C, Pt/C, Rh/C and Ru/C catalysts: mechanisms, reaction micro-kinetic modelling and quantitative structure-activity relationships. Chem Eng J 359:305–320. https://doi.org/10.1016/j.cej.2018.11.107

28. Liu Y, Sotelo-Boyás R, Murata K, Minowa T, Sakanishi K (2011) Hydrotreatment of vegetable oils to produce bio-hydrogenated diesel and liquefied petroleum gas fuel over catalysts containing sulfided Ni-Mo and solid acids. Energy Fuels 25:4675–4685. https://doi.org/10.1021/ef200889e

29. Toba M, Abe Y, Kuramochi H, Osako M, Mochizuki T, Yoshimura Y (2011) Hydrodeoxygenation of waste vegetable oil over sulfide catalysts. Catal Today 164:533–537. https://doi.org/10.1016/j.cattod.2010.11.049

30. Veriansyah B, Han JY, Kim SK, Hong SA, Kim YJ, Lim JS, Shu YW, Oh SG, Kim J (2012) Production of renewable diesel by hydroprocessing of soybean oil: effect of catalysts. Fuel 94:578–585. https://doi.org/10.1016/j.fuel.2011.10.057

31. Kumar R, Rana BS, Tiwari R, Verma D, Kumar R, Joshi RK, Garg MO, Sinha AK (2010) Hydroprocessing of jatropha and its mixtures with gas oil. Green Chem 12:2232–2239. https://doi.org/10.1039/c0gc00204f

32. Pimerzin AA, Nikulshin PA, Mozhaev AV, Pimerzin AA, Lyashenko AI (2015) Investigation of spillover effect in hydrotreating catalysts based on Co_2Mo10—heteropolyanion and cobalt sulphide species. Appl Catal B Environ 168–169:396–407. https://doi.org/10.1016/j.apcatb.2014.12.031

33. Phimsen S, Kiatkittipong W, Yamada H, Tagawa T, Kiatkittipong K, Laosiripojana N, Assabumrungrat S (2017) Nickel sulfide, nickel phosphide and nickel carbide catalysts for bio-hydrotreated fuel production. Energy Convers Manag 151:324–333. https://doi.org/10.1016/j.enconman.2017.08.089

34. Guzmán HJ, Vitale G, Carbognani-Ortega L, Scott CE, Pereira-Almao P (2020) Molybdenum sulfide nanoparticles prepared using starch as capping agent. Redispersion and activity in Athabasca Bitumen hydrotreating. Catal Today 377:38–49. https://doi.org/10.1016/j.cattod.2021.05.021

35. Silva LN, Fortes ICP, Sousa FPD, Pasa VMD (2016) Biokerosene and green diesel from macauba oils via catalytic deoxygenation over Pd/C. Fuel 164:329–338. https://doi.org/10.1016/j.fuel.2015.09.081

36. Camargo MDO, Pimenta JLCW, Camargo MDO, Arroyo PA (2020) Green diesel production by solvent-free deoxygenation of oleic acid over nickel phosphide bifunctional catalysts: effect of the support. Fuel 281:118719. https://doi.org/10.1016/j.fuel.2020.118719

37. Alvarez-Galvan MC, Campos-Martin JM, Fierro JLG (2019) Transition metal phosphides for the catalytic hydrodeoxygenation of waste oils into green diesel. Catalysts 9:3. https://doi.org/10.3390/catal9030293

38. Wang F, Jiang J, Wang K, Zhai Q, Sun H, Liu P, Feng J, Xia H, Ye J, Li Z, Li F, Xu J (2018) Activated carbon supported molybdenum and tungsten carbides for hydrotreatment of fatty acids into green diesel. Fuel 228:103–111. https://doi.org/10.1016/j.fuel.2018.04.150

39. Wang H, Yan S, Salley SO, Ng KYS (2013) Support effects on hydrotreating of soybean oil over NiMo carbide catalyst. Fuel 111:81–87. https://doi.org/10.1016/j.fuel.2013.04.066

40. Furimsky E (2003) Metal carbides and nitrides as potential catalysts for hydroprocessing. Appl Catal A Gen 240:1–28. https://doi.org/10.1016/S0926-860X(02)00428-3

41. Shu C, Sun T, Guo Q, Jia J, Lou Z (2014) Desulfurization of diesel fuel with nickel boride in situ generated in an ionic liquid. Green Chem 16:3881–3889. https://doi.org/10.1039/c4gc00695j

42. Kon K, Toyao T, Onodera W, Siddiki SMAH, Shimizu KI (2017) Hydrodeoxygenation of fatty acids, triglycerides, and ketones to liquid alkanes by a Pt–MoOx/TiO$_2$ catalyst. Chem Cat Chem 9:2822–2827. https://doi.org/10.1002/cctc.201700219

43. Serrano-Ruiz JC, Braden DJ, West RM, Dumesic JA (2010) Conversion of cellulose to hydrocarbon fuels by progressive removal of oxygen. Appl Catal B Environ 100:184–189. https://doi.org/10.1016/j.apcatb.2010.07.029

44. Janampelli S, Darbha S (2019) Highly efficient Pt–MoOx/ZrO$_2$ catalyst for green diesel production. Catal Commun 125:70–76. https://doi.org/10.1016/j.catcom.2019.03.027

45. Wang F, Xu J, Jiang J, Liu P, Li F, Ye J, Zhou M (2018) Hydrotreatment of vegetable oil for green diesel over activated carbon supported molybdenum carbide catalyst. Fuel 216:738–746. https://doi.org/10.1016/j.fuel.2017.12.059

46. Hongloi N, Prapainainar P, Seubsai A, Sudsakorn K, Prapainainar C (2019) Nickel catalyst with different supports for green diesel production. Energy 182:306–320. https://doi.org/10.1016/j.energy.2019.06.020

47. Kamaruzaman MF, Taufiq-Yap YH, Derawi D (2019) Green diesel production from palm fatty acid distillate over SBA-15-supported nickel, cobalt, and nickel/cobalt catalysts. Biomass Bioenergy 134:105476. https://doi.org/10.1016/j.biombioe.2020.105476

48. Jeong H, Shin M, Jeong B, Jang JH, Han GB, Suh YW (2020) Comparison of activity and stability of supported Ni2P and Pt catalysts in the hydroprocessing of palm oil into normal paraffins. J Ind Eng Chem 83:189–199. https://doi.org/10.1016/j.jiec.2019.11.027

49. Sousa FPD, Cardoso CC, Pasa VMD (2016) Producing hydrocarbons for green diesel and jet fuel formulation from palm kernel fat over Pd/C. Fuel Process Technol 143:35–42. https://doi.org/10.1016/j.fuproc.2015.10.024

50. Dujjanutat P, Kaewkannetra P (2020) Production of bio-hydrogenated kerosene by catalytic hydrocracking from refined bleached deodorised palm/ palm kernel oils. Renew Energy 147:464–472. https://doi.org/10.1016/j.renene.2019.09.015

51. Jeon KW, Shim JO, Cho JW, Jang WJ, Na HS, Kim HM, Lee YM, Jeon BJ, Bae JW, Roh HS (2019) Synthesis and characterization of Pt-, Pd-, and Ru-promoted Ni–Ce0.6Zr0.4O$_2$ catalysts for efficient biodiesel production by deoxygenation of oleic acid. Fuel 236:928–933. https://doi.org/10.1016/j.fuel.2018.09.078

52. Chen RX, Wang WC (2019) The production of renewable aviation fuel from waste cooking oil. Part I: Bio-alkane conversion through hydro-processing of oil. Renew Energy 135:819–835. https://doi.org/10.1016/j.renene.2018.12.048

53. Janampelli S, Darbha S (2018) Effect of support on the catalytic activity of WOx promoted Pt in green diesel production. Mol Catal 451:125–134. https://doi.org/10.1016/j.mcat.2017.11.029

54. Janampelli S, Darbha S (2021) Selective deoxygenation of fatty acids to fuel-range hydrocarbons over Pt–MOx/ZrO$_2$ (M = Mo and W) catalysts. Catal Today 375:174–180. https://doi.org/10.1016/j.cattod.2020.04.020

55. Chen Y, Li X, Liu S, Zhang W, Wang Q, Zi W (2020) Effects of metal promoters on one-step Pt/SAPO-11 catalytic hydrotreatment of castor oil to C_8–C_{16} alkanes. Ind Crops Prod 146:112182. https://doi.org/10.1016/j.indcrop.2020.112182

56. Ameen M, Azizan MT, Ramli A, Yusup S, Abdullah B (2019) The effect of metal loading over Ni/γ–Al_2O_3 and Mo/γ–Al_2O_3 catalysts on reaction routes of hydrodeoxygenation of rubber seed oil for green diesel production. Catal Today 355:51–64. https://doi.org/10.1016/j.cattod.2019.03.028

57. Lycourghiotis S, Kordouli E, Sygellou L, Bourikas K, Kordulis C (2019) Nickel catalysts supported on palygorskite for transformation of waste cooking oils into green diesel. Appl Catal B Environ 259:118059. https://doi.org/10.1016/j.apcatb.2019.118059

58. Gamal MS, Asikin-Mijan N, Khalit WNAW, Arumugam M, Izham SM, Taufiq-Yap YH (2020) Effective catalytic deoxygenation of palm fatty acid distillate for green diesel production under hydrogen-free atmosphere over bimetallic catalyst CoMo supported on activated carbon. Fuel Process Technol 208:106519. https://doi.org/10.1016/j.fuproc.2020.106519

59. Thongkumkoon S, Kiatkittipong W, Hartley UW, Laosiripojana N, Daorattanachai P (2019) Catalytic activity of trimetallic sulfided Re–Ni–Mo/Γ–Al_2O_3 toward deoxygenation of palm feedstocks. Renew Energy 140:111–123. https://doi.org/10.1016/j.renene.2019.03.039

60. Bezergianni S, Dimitriadis A (2013) Comparison between different types of renewable diesel. Renew Sustain Energy Rev 21:110–116. https://doi.org/10.1016/j.rser.2012.12.042

61. Oi LE, Choo MY, Lee HV, Taufiq-Yap YH, Cheng CK, Juan JC (2020) Catalytic deoxygenation of triolein to green fuel over mesoporous TiO_2 aided by in situ hydrogen production. Int J Hydrogen Energy 45:11605–11614. https://doi.org/10.1016/j.ijhydene.2019.07.172

62. Choo MY, Oi LE, Ling TC, Ng EP, Lin YC, Centi G, Juan JC (2020) Deoxygenation of triolein to green diesel in the H_2-free condition: effect of transition metal oxide supported on zeolite Y. J Anal Appl Pyrolysis 147:104797. https://doi.org/10.1016/j.jaap.2020.104797

63. Papanikolaou G, Lanzafame P, Giorgianni G, Abate S, Perathoner S, Centi G (2020) Highly selective bifunctional Ni zeo-type catalysts for hydroprocessing of methyl palmitate to green diesel. Catal Today 345:14–21. https://doi.org/10.1016/j.cattod.2019.12.009

64. Baharudin KB, Abdullah N, Taufiq-Yap YH, Derawi D (2020) Renewable diesel via solvent-less and hydrogen-free catalytic deoxygenation of palm fatty acid distillate. J Clean Prod 274:122850. https://doi.org/10.1016/j.jclepro.2020.122850

65. Liu Y, Zheng D, Yu H, Liu X, Yu S, Wang X, Li L, Pang J, Liu X, Yan Z (2020) Rapid and green synthesis of SAPO-11 for deoxygenation of stearic acid to produce bio-diesel fractions. Microporous Mesoporous Mater 303:110280. https://doi.org/10.1016/j.micromeso.2020.110280

66. Heriyanto H, Sumbogo SDM, Heriyanti SI, Sholehah I, Rahmawati A (2018) Synthesis of green diesel from waste cooking oil through hydrodeoxygenation technology with NiMo/γ-Al_2O_3 catalysts. MATEC Web Conference, Process for Energy Environ 156:03032. https://doi.org/10.1051/matecconf/201815603032

67. Moreira JDBD, Rezende DBD, Pasa VMD (2020) Deoxygenation of Macauba acid oil over Co-based catalyst supported on activated biochar from Macauba endocarp: a potential and sustainable route for green diesel and biokerosene production. Fuel 269:117253. https://doi.org/10.1016/j.fuel.2020.117253

68. Simacek P, Kubicka D, Sebor G, Pospíšil M (2010) Fuel properties of hydroprocessed rapeseed oil. Fuel 89:611–615. https://doi.org/10.1016/j.fuel.2009.09.017

69. Pelemo J, Inambao FL, Onuh EI (2020) Potential of used cooking oil as feedstock for hydroprocessing into hydrogenation derived renewable diesel: a review. Int J Eng Res Sci Technol 13:500–519. https://doi.org/10.37624/IJERT/13.3.2020.500-509

70. Vasquez MC, Silva EE, Castillo EF (2017) Hydrotreatment of vegetable oils: a review of the technologies and its developments for jet biofuel production. Biomass Bioenergy 105:197–206. https://doi.org/10.1016/j.biombioe.2017.07.008

71. Cremonez PA, Teleken JG, Meier TW (2020) Potential of green diesel to complement the Brazilian energy production: a review. Energy Fuels 35:176–186. https://doi.org/10.1021/acs.energyfuels.0c03805

72. Simakova I, Simakova O, Mäki-Arvela P, Simakov P, Estrada M, Murzin DY (2009) Deoxygenation of palmitic and stearic acid over supported Pd catalysts: effect of metal dispersion. Appl Catal A Gen 355:100–108. https://doi.org/10.1016/j.apcata.2008.12.001

73. Masoumi S, Dalai AK (2021) NiMo carbide supported on algal derived activated carbon for hydrodeoxygenation of algal biocrude oil. Energy Convers Manag 231:113834. https://doi.org/10.1016/j.enconman.2021.113834

74. Sankaranarayanan TM, Banu M, Pandurangan A, Sivasanker S (2011) Hydroprocessing of sunflower oil–gas oil blends over sulfided Ni–Mo–Al–zeolite beta composites. Bioresour Technol 102:10717–10723. https://doi.org/10.1016/j.biortech.2011.08.127

75. Oja S (2008) NExBTL–next generation renewable diesel, Neste Oil, Renewable Fuels. http://veranstaltungen.fnr.de/fileadmin/allgemein/pdf/veranstaltungen/NeueBiokraftstoffe/3_NExBTL.pdf. Accessed 26 Dec 2021

76. Kokossis A, Yang A (2009) Future system challenges in the design of renewable bio-energy systems and the synthesis of sustainable biorefineries. Des Energy Environ 107–123.https://doi.org/10.1201/9781439809136-c8

77. Asli H, Ahmadinia E, Zargar M, Karim MR (2012) Investigation on physical properties of waste cooking oil—rejuvenated bitumen binder. Constr Build Mater 37:398–405. https://doi.org/10.1016/j.conbuildmat.2012.07.042

78. Xu J, Long F, Jiang J, Li F, Zhai Q, Wang F, Liu P, Li J (2019) Integrated catalytic conversion of waste triglycerides to liquid hydrocarbons for aviation biofuels. J Clean Prod 222:784–792. https://doi.org/10.1016/j.jclepro.2019.03.094

79. Muradov NZ, Veziroğlu TN (2005) From hydrocarbon to hydrogen–carbon to hydrogen economy. Int J Hydrogen Energy 30:225–237. https://doi.org/10.1016/j.ijhydene.2004.03.033

80. Wang C, Lei H, Zhao Y, Qian M, Kong X, Mateo W, Zou R, Ruan R (2021) Integrated harvest of phenolic monomers and hydrogen through catalytic pyrolysis of biomass over nanocellulose derived biochar catalyst. Bioresour Technol 320:124352. https://doi.org/10.1016/j.biortech.2020.124352

81. Cao L, Yu IKM, Xiong X, Tsang DCW, Zhang S, Clark JH, Hu C, Ng YH, Shang J, Ok YS (2020) Biorenewable hydrogen production through biomass gasification: a review and future prospects. Environ Res 186:109547. https://doi.org/10.1016/j.envres.2020.109547

82. Bezergianni S, Dimitriadis A, Kalogianni A, Pilavachi PA (2010) Hydrotreating of waste cooking oil for biodiesel production. Part I: effect of temperature on product yields and heteroatom removal. Bioresour Technol 101:6651–6656. https://doi.org/10.1016/j.biortech.2010.03.081

83. Kalnes T, Marker T, Shonnard DR (2007) Green diesel: a second generation biofuel. Int J Chem React Eng 5:A48. https://doi.org/10.2202/1542-6580.1554

84. Pragya N, Pandey KK (2016) Life cycle assessment of green diesel production from microalgae. Renew Energy 86:623–632. https://doi.org/10.1016/j.renene.2015.08.064

85. Ramírez-Verduzco LF, Hernández-Sánchez MJ (2021) Blends of green diesel (Synthetized from Palm Oil) and petroleum diesel: a study on the density and viscosity. Bioenergy Res 14:1002–1013. https://doi.org/10.1007/s12155-020-10183-y

86. Wong A, Zhang H, Kumar A (2016) Life cycle assessment of renewable diesel production from lignocellulosic biomass. Int J Life Cycle Assess 21:1404–1424. https://doi.org/10.1007/s11367-016-1107-8

87. Nuss P (2012) Life cycle assessment handbook: a guide for environmentally sustainable products. In: Curran MA (ed), Chapter-4, Life cycle impact assessement. ISBN 9781118099728, pp 67–103

88. Arguelles-Arguelles A, Amezcua-Allieri MA, Ramírez-Verduzco LF (2021) Life cycle assessment of green diesel production by hydrodeoxygenation of palm oil. Frontiers Energy Res 9:690725. https://doi.org/10.3389/fenrg.2021.690725

89. Zaimes GG, Khanna V (2015) Chapter-8 Life cycle sustainability aspects of microalgal biofuels. In: Klemes JJ (ed) Assessing and measuring environmental impact and sustainability, Butterworth-Heinemann. ISBN 9780127999685. https://doi.org/10.1016/B978-0-12-799968-5.00008-7, pp 255–276

90. Luo Y, Ierapetritou M (2020) Comparison between different hybrid life cycle assessment methodologies: a review and case study of biomass-based p-Xylene production. Ind Eng Chem Res 59:22313–22329. https://doi.org/10.1021/acs.iecr.0c04709

91. Phukan MM, Chutia RS, Konwar BK, Kataki R (2011) Microalgae chlorella as a potential bio-energy feedstock. Appl Energy 88:3307–3312. https://doi.org/10.1016/j.apenergy.2010.11.026

92. Wu W, Wang PH, Lee DJ, Chang JS (2017) Global optimization of microalgae-to-biodiesel chains with integrated cogasification combined cycle systems based on greenhouse gas emissions reductions. Appl Energy 197:63–82. https://doi.org/10.1016/j.apenergy.2017.03.117

93. Kargbo H, Harris JS, Phan AN (2021) Drop-in" fuel production from biomass: critical review on techno-economic feasibility and sustainability. Renew Sustain Energy Rev 135:110168. https://doi.org/10.1016/j.rser.2020.110168

94. Patel M, Kumar A (2016) Production of renewable diesel through the hydroprocessing of lignocellulosic biomass-derived bio-oil: a review. Renew Sustain Energy Rev 58:1293–1307. https://doi.org/10.1016/j.rser.2015.12.146

95. Masoumi S, Borugadda VB, Dalai AK (2020) Chapter-11 Biocrude oil production via hydrothermal liquefaction of algae and upgradation techniques to liquid transportation fuels. In: Nanda S, N. Vo DV, Sarangi P (eds) Biorefinery of alternative resources: targeting green fuels and platform chemicals. Springer, Singapore. https://doi.org/10.1007/978-981-15-1804-1_11

96. Masoumi S, Dalai AK (2021) Techno-economic and life cycle analysis of biofuel production via hydrothermal liquefaction of microalgae in a methanol-water system and catalytic hydrotreatment using hydrochar as a catalyst support. Biomass Bioenergy 151:106168. https://doi.org/10.1016/j.biombioe.2021.106168

97. Zhang Y, Brown TR, Hu G, Brown RC (2013) Techno-economic analysis of two bio-oil upgrading pathways. Chem Eng J 225:895–904. https://doi.org/10.1016/j.cej.2013.01.030

98. Masoumi S, Borugadda, Dalai AK (2021) CHapter - 11 Techno-economic analysis of algal biorefineries. In: Dalai AK, Goud VV, Nanda S, Borugadda VB (eds) Algal biorefinery: developments, challenges and opportunities. eBook ISBN 9781003100317

99. Magdeldin M, Kohl T, Järvinen M (2017) Techno-economic assessment of the by-products contribution from non-catalytic hydrothermal liquefaction of lignocellulose residues. Energy 137:679–695. https://doi.org/10.1016/j.energy.2017.06.166

100. Wu L, Wang Y, Zheng L, Wang P, Han X (2019) Techno-economic analysis of bio-oil co-processing with vacuum gas oil to transportation fuels in an existing fluid catalytic cracker. Energy Convers Manag 197:111901. https://doi.org/10.1016/j.enconman.2019.111901

101. Rafati M, Wang L, Dayton DC, Schimmel K, Kabadi V, Shahbazi A (2017) Techno-economic analysis of production of Fischer–Tropsch liquids via biomass gasification: the effects of Fischer–Tropsch catalysts and natural gas co-feeding. Energy Convers Manag 133:153–166. https://doi.org/10.1016/j.enconman.2016.11.051

102. Meyer PA, Snowden-Swan LJ, Jones JB, Rappé KG, Hartley DS (2020) The effect of feedstock composition on fast pyrolysis and upgrading to transportation fuels: techno-economic analysis and greenhouse gas life cycle analysis. Fuel 259:116218. https://doi.org/10.1016/j.fuel.2019.116218

103. Leyva C, Romero-Galarza A, Castañón-Alonso SL, Farías-Cepeda L, Rosales-Marines L, Ríos AR (2021) Conversion of biomass to liquid transportation fuels through Fischer—Tropsch synthesis: a global perspective. In Handbook of research on bioenergy biomaterials. Apple Academic Press. eBook ISBN 9781003105053

104. Matayeva A, Basile F, Cavani F, Bianchi D, Chiaberge S (2019)Chapter-12 Development of upgraded bio-oil via liquefaction and pyrolysis. In: Albonetti S, Perathoner S, Quadrelli EA (eds) Studies in surface science and catalysis, Elsevier, 178:231–256. ISSN 0167-2991,ISBN 9780444641274.https://doi.org/10.1016/B978-0-444-64127-4.00012-4

105. Neste renewable diesel handbook (2020). https://www.neste.com/sites/default/files/attachments/neste_renewable_diesel_handbook.pdf. Accessed 26 Dec 2021

106. Amin A (2019) Review of diesel production from renewable resources: catalysis, process kinetics and technologies. Ain Shams Eng J 10:821–839. https://doi.org/10.1016/j.asej.2019.08.001

107. Bryan T (2021) Renewable diesel's rising tide. http://www.biodieselmagazine.com/articles/2517318/renewable-diesels-rising-tide. Accessed 26 Dec 2021

108. Honeywell introduces simplified technology to produce renewable diesel (2021) https://uop.honeywell.com/en/news-events/2021/january/honeywell-uop-ecofining-single-stage-process. Accessed 26 Dec 2021

109. Imperial to produce renewable diesel at Strathcona refinery (2021). https://news.imperialoil.ca/news-releases/news-releases/2021/Imperial-to-produce-renewable-diesel-at-Strathcona-refinery/default.aspx

Chapter 2
Feedstocks for Green Diesel

Sumit Sharma, Shikha Singh, Saurabh Jyoti Sarma, and Satinder Kaur Brar

Abstract Green diesel is an alternative fuel generated by the hydrotreating of oil or fat. Green diesel is a kind of next-generation diesel that can be derived from renewable feedstocks. A large range of feedstock is available for the production of green diesel. Various generations of feedstock such as 1st generation (Edible oil—sunflower, palm, corn, rapeseed, and soybean), 2nd generation (Non-edible oil—Jatropha and castor bean, plant waste biomass, and animal fat), 3rd generation (Microalgae) are being used for green diesel production. The chapter sheds light on the different feedstocks used in the production of green diesel along with their chemistry and classification. The use of different feedstock imparts distinct impacts on the production of green diesel. These impacts result in different fuel properties thereby leading to different effects on internal combustion engines. The type of production technology depends on the feedstock used. This chapter also discusses sources, distribution, properties, typical composition, comparative advantages and disadvantages, and current scenarios on commercial feasibility of each generation of feedstock. This is an insight for utilization of feedstocks for green diesel production along with generations.

Keywords Green diesel · Biomass feedstocks · 1st generation feedstocks · 2nd generation feedstocks · 3rd generation feedstocks

S. Sharma · S. Singh · S. J. Sarma (✉)
Department of Biotechnology, Bennett University, Greater Noida (UP) 201310, India
e-mail: saurabh.sarma@bennett.edu.in

S. K. Brar
Department of Civil Engineering, Lassonde School of Engineering, York University, North York (Toronto) M3J 1P3, Canada

© The Author(s), under exclusive license to Springer Nature Singapore Pte Ltd. 2022
M. Aslam et al. (eds.), *Green Diesel: An Alternative to Biodiesel and Petrodiesel*,
Advances in Sustainability Science and Technology,
https://doi.org/10.1007/978-981-19-2235-0_2

1 Introduction

Application of the green energy sources is an option to reduce the greenhouse gas emission. Green diesel is a fuel derived from renewable sources through a catalytic conversion of oil especially triglyceride through the hydrogenation process. The conventional biodiesel approaching towards an alternative to diesel is produced by transesterification of triglycerides in the presence of methanol and alkali catalysts such as caustic soda [1]. Unlike biodiesel, green diesel is produced by the deoxygenation of triglycerides in the presence of hydrogen [2] (Fig. 1). This hydroprocessing scheme is rebuilding the step towards sustainable fuel [3] production. The biomass-derived green diesel is similarly acceptable for currently running diesel engines and advantageous to have high cetane number than biodiesel [4]. There are a variety of renewable feedstocks that can be used directly or after pre-processing as an oil or fatty acid source.

Green diesel can be derived from three different generations of feedstock [5]. Dedicated food crop-based edible oil plant and seed feedstocks such as palm, canola, sunflower, soybean, mustard oils are generally named first-generation [6, 7]. The competitive crisis of the demand for food crops has extremely impacted the price increase. Therefore, the second-generation feedstocks such as lignocellulosic waste or other crop residues, for example—wheat straw, cotton stalk, sugar cane bagasse, rice husk, etc., and non-food crops like Castor, miscanthus, and jatropha are the most useful carbon sources [8]. The processing of lignocellulosic biomass needs pre-processing to convert into bio-oil before starting its hydrogenation. For omitting competition for food crops and their waste use, the third-generation feedstock such as microalgae [9]. Microalgae are an effective source of oil and have nearly half of the lipid content of their biomass [10–12]. Apart from yield and non-competitive cultivation, the microalgae need sunlight, multiple rich nutrients, air supply or medium mixing, major precautions, and larger area/land. Considering all facts, various generations of feedstocks 1st, 2nd, and 3rd are renewable feedstock supply directly reduce

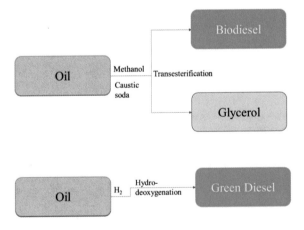

Fig. 1 Simplified scheme for biodiesel and green diesel production from oil or fat

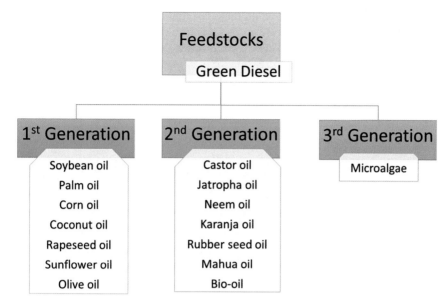

Fig. 2 Schematic classification of different generations of feedstocks proposed for green diesel production

the fossil-fuel reserve load. This chapter describes the classification of different generations of feedstock and their potential to be utilized for producing green diesel.

2 Renewable Feedstocks for Green Diesel

Green diesel is also known as "Renewable diesel" due to its renewable source. All sources are classified into three generations according to their source (Fig. 2) and availability. Renewable nature also enhances sustainability and process utilization at the commercial level. Most of the first-generation feedstocks are edible oils or fat sources, the second-generation includes non-edible oils and plant biomass, etc. The third-generation category is more specific to non-competitive feed like microalgae, waste oil, animal fat, etc. All kinds of feedstocks are further described in detail.

3 1st Generation Feedstocks

First-generation feedstocks are the ones started with the contribution from the native sources. The ago-industrial processed oil extracted from oilseeds is the primary feedstock used as a direct source of triglycerides. Edible oils are major sources of fats varied with different fatty acid profiles. The first sources such as sunflower oil, canola oil, mustard oil, and corn oil are only applicable to those geographical regions where the abundancy is high or these are cultivated dedicatedly without competing with the food demands. Therefore, these are a better substrate for making green diesel at a high scale. The efficient production of green diesel depends on the profile of the oil. The hydrocarbon chain length, percentage of saturated and unsaturated fatty acids, viscosity, density, cetane number, flash point, etc., are a few parameters that need to be considered before selecting a better feedstock. The availability of the feedstocks also affects the commercial feasibility of utilizing those feedstocks that are already in demand as feed.

3.1 Soybean Oil

Soybean contributes 18–20% of oil content and is the most promising feedstock in US and Brazil for diesel production [13]. Due to the increasing demand for green diesel, the soybean demand boosts up in the US to nearly 120 million acres of land for soybean, 11 billion pounds consumption and also need to double total edible oil cover by next 10 years for an efficient amount of green diesel production [14, 15]. As per the current report, the major four soybean oil consumption countries in the year 2020 are China, United States, Brazil, and India with 18,691, 10,433, 7700, 5150 TMT, respectively [16]. The United States is focusing more on green diesel production from soybean oil. 5 plants are running for green diesel production in the US are Diamon Green Diesel, Wyoming Renewable Diesel, Kern Oil and Refining, BP Cherry Point, and East Kansas Agri-Energy [14]. As more plants are being operating soon the demand for feedstock is seeking towards soybean oil. The five special fatty acids found in soybean oil are palmitic acid, oleic acid, stearic acid, linoleic acid, and linolenic acid. The major percentage of linoleic acid (~55%) causes oxygen instability that limits its industrial use that can be rectified by partial hydrogenation [17]. The hydrogenation approach is related to green diesel production that signifies the better processing as well as stability of the oil. The hydrogenation of soybean oil with niobium catalyst was studied to produce jet fuel and green diesel range hydrocarbons [18]. That showed the promising use of soybean feedstock for producing drop-in bio hydrocarbons matching Petro-fuel properties.

3.2 Palm Oil

Palm oil is extracted from reddish fruit pulp mesocarp of palm. The major producers are Indonesia and Malaysia contributing 58% and 26% of the world's palm oil production [19]. Unlike soybean oil, palm oil comprises saturated fatty acids 50% (especially palmitic acid), unsaturated fatty acid 40–44% (oleic acid), and polyunsaturated fatty acid ~10% (linoleic and linolenic acid) approximately [20, 21]. A stable composition of saturated and unsaturated fatty acids in palm oil made it convenient for making better diesel. A study showed that around 68% of green diesel can be yielded from crude palm oil using the catalytic hydrogenation method [22].

3.3 Corn Oil

Corn oil is considered edible oil. By its large-scale production, the inedible portion of it is being used for green diesel production. In the United States, Diamond Green Diesel with Valero company established the setup of corn-oil-based green diesel production [23]. The polyunsaturated fatty acids (~60%), especially linoleic acid are found in higher amounts in corn oil [24]. By country, United States is the major producer of corn in the world with around 360 million metric tons (MMT) capacity [25], China and Brazil also follow as leading corn producers.

3.4 Coconut Oil

Lauric acid (a kind of medium-chain fatty acid) contributes nearly half of the total fatty acid found in coconut oil [26]. It is the staple edible oil of the Philippines reaching 675,000 metric tons of production, the other following countries are EU-27 (645,000 metric tons), United States (497,000 metric tons), and India (470,000 metric tons) [27]. Currently, green diesel production from coconut oil is under consideration, and while the profound studies were focused on trans-esterified biodiesel from it.

3.5 Rapeseed Oil

Rapeseed oil is known for its consumption as a major cooking oil in house-hold. It has a very much competitive disadvantage if not being produced in large quantity that can balance the food inequality. Rapeseed oil is also known as canola oil in some countries like Canada and the higher producer of rapeseed oil is nearly 19 MMT. Apart from Canada, European Union, China, and India also utilized this as a staple oil source with a production capacity of 16.83, 13.1, and 7.7 MMT, respectively [28]. Oleic acid (56–64%) and linoleic acid (17–20%) are the major fatty acid comprised in the rapeseed oil [29]. The rapeseed oil was extensively studied for green diesel production using different hydrotreating processes like direct hydrotreating with Pt/zeolite [30] or subsequent hydrogenation and decarboxylation process [31].

3.6 Sunflower Oil

It is commonly used as frying oil and is a kind of ideal source of unsaturated fat. It is rich in unsaturated fatty acid just reverse to the rapeseed oil with 44–75% of linoleic acid and 14–43% of oleic acid [32]. Ukraine is the major producer of sunflower oil of 16.5 MMT and is also contributed by Russia (15.3 MMT) and European Union (9.6 MMT) [33]. The hydroconversion of sunflower oil was attempted with Renay nickel type catalyst to reduce the side reactions and to enhance the green diesel yield [34].

3.7 Olive Oil

Olive oil is rich in oleic acid of 55–83% and palmitic acid (7–20%) of the total fatty acids [35]. The major use of olive oil is medicine, soap-making, and lamps [36]. European Union is the highest producer of olive oil with 1595 thousand metric tons (TMT), the United States (386 TMT), Turkey (190 TMT), and China (60 TMT) come [37]. The studies reveal that the green diesel from olive oil is not being effectively processed than biodiesel. This feedstock is under-valued from green diesel production by its less production and consumption all over the world.

3.8 Other Edible Oils

There are many more edible oil sources such as Sesame, Cottonseed, and flaxseed grown in the world and are being studied for biodiesel production. However, these all also have the feasibility to be used as 1st generation feedstock for green diesel production.

4 2nd Generation Feedstocks

The use of edible sources for feed causes competition for its availability as food or commercial substrate. Therefore, those kinds of edible oils cannot be used as a direct source until they are not produced dedicatedly for commercial product formations. For example, in some countries, corn is produced dedicatedly for corn starch and corn oil production at a very large scale so that there is no limitation as food as well as commercial use of corn. Due to the demand for edible oil, much for food and practical possibility to fulfill demands is not possible in every region. That is why the second-generation feedstocks come into existence. The majorly defined second-generation feedstocks are non-edible oil sources such as Castor oil, Jatropha oil,

Neem oil, Rice bran oil, Rubber seed oil, Karanja oil, and Mahua oil. The non-edible parts of the crops or plant biomass such are wheat straw, rice straw, cotton stalk are also considered at second-generation feedstocks, while they need some initial pre-processing before use.

4.1 Castor Oil

Castor oil is produced from castor beans by pressing method. It is mostly used for medicinal and lubrication purposes. The castor oil is valued for ricinoleic acid and contributes around 90% of the total fatty acids [38]. India is the major producer and exporter of castor oil reaching the highest production of 1196 TMT in the Asia–Pacific region [39]. Green diesel can be produced from castor oil with high purity and yield using catalytic hydrotreatment [40].

4.2 Jatropha Oil

Jatropha oil has a vast potential of producing green diesel with more than 95% conversion yield using catalytic hydroprocessing [41, 42] and no further engine modification is also needed while using this hydrotreated oil as fuel [43]. Jatropha oil contains linoleic acid (50%) and palmitic acid (35%), and traces of oleic acid [44]. Apart from the high potential of Jatropha for diesel production, it is facing challenges related to high-yield, larger-area for cultivation, unexpended knowledge, and climatic instability. The fatty acid content of oil showed a maximum portion of saturated fatty acid with oleic acid (47%), linolenic acid (31%), and palmitic acid (14%) [45].

4.3 Other Food-Related Sources

Neem oil fatty acid profile showed the presence of linoleic acid 34% and oleic acid 21%, stearic acid 20%, and palmitic acid 18% [46]. Neem kernels when pressed in the machine releases nearly 50% of oil content [47]. Karanja oil with similar characteristics to neem oil is also a source of non-edible oil generated from *Pongamia pinnata* seeds [48, 49]. Rubber seed oil is generated from the rubber seeds consisting of oleic (22%), linoleic (37%), and linolenic (19%) [50, 51]. Mahua oil is also an alternative feedstock with oleic acid (40–50%), palmitic acid (16–28%), and linoleic acid (8–13%) [52]. The catalytic cracking of Mahua oil using coal fly ash catalyst showed the potential of producing hydrocarbon fuel [52]. Waste oil and animal fat are also included for feedstocks because of their low cost than fresh vegetable oil [53], but their limited availability is still a challenge.

4.4 Bio-Oil

Bio-oil is also a promising substrate that can be produced by the deconstruction of lignocellulosic biomass via pyrolysis treatment. The oleic and linoleic acid content jointly was found nearly 42% and palmitic and stearic acid content jointly was found nearly 33% and found effective with complete conversion into hydro-deoxygenated fuel [54]. Different varieties of biomass such as Wheat straw [55, 56], Rice husk [57], cotton stalk [58], and sugarcane bagasse [59].

5   3rd Generation Feedstocks

Third-generation feedstocks are the advanced resources that are not being a part of crop-type cultivation. The most common and majorly used third-generation feedstock is microalgae and also contributes to atmospheric CO_2 assimilation by [60] photosynthesis. These are beneficial over first- and second-generation feedstocks due to less greenhouse gas emission, cultivated on unimportant land, higher growth and productivity, no competition towards food, etc. [61]. Microalgae are a rich source of lipids [62] and are specifically grown in controlled in-situ or ex-situ environments as fatty sources. The fate of lipid extraction and preparing diesel out of it is very much feasible. However, microalgae have some cultivation challenges at a large scale such as the chance of contamination at an open area, high-throughput cost of photo-bioreactor supply and maintenance, pretreatment of biomass, and process optimization as well as downstream processing [62, 63]. As these feedstocks are not competing with the food supply, these can be grown in any habitat or land area [64]. The most common microalgae used for cultivation purposes are *Chlorella vulgaris* [65], *Tetradesmus obliquus* [66], *Nannochloropsis salina* [67], etc. The fatty acid content can be extracted through cell lysis and solvent purification method and the lipids can be directly used for further use without any specific separation.

6   Current Scenario of Feedstock Utilization for Green Diesel Production

There are many methods applied to advance the utilization of feedstock over a traditional way of hydrotreatment. The earlier process was dependent on processed oil for use as feedstock. However, second and third-generation feedstocks open the major gates for utilizing a variety of resources with different processing methods such as the deconstruction of biomass before processing, bio-oil formation through pyrolysis, hydrothermal treatment for decomposition of biomass, and direct catalytic conversion of biomass into hydrocarbon fuels [68]. Pyrolysis is a cost expensive pretreatment process that limits the use of biomass-derived oil generation subspecialty for green

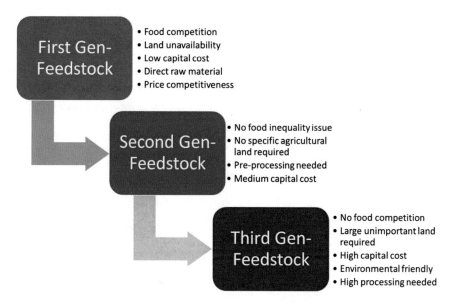

First Gen-Feedstock
- Food competition
- Land unavailability
- Low capital cost
- Direct raw material
- Price competitiveness

Second Gen-Feedstock
- No food inequality issue
- No specific agricultural land required
- Pre-processing needed
- Medium capital cost

Third Gen-Feedstock
- No food competition
- Large unimportant land required
- High capital cost
- Environmental friendly
- High processing needed

Fig. 3 Synergistic highlights of different generation feedstock utilization at commercial level

diesel production, the generated bio-oil may need more purification before using for the hydrogenation process. The shift from all dependable food and non-food-related sources, third-generation feedstock implies on availability of suitable environments and high-operation costs. A comparative scheme for the feasibility of different generations of feedstocks is shown in Fig. 3. There are some advancements in the stage of development to rectify this issue towards genetically modified renewable resources proposed as fourth-generation feedstocks in the future.

7 Conclusive Remarks

This chapter summarized different types of feedstock used for green diesel production. There have not been much differences between the feedstock of biodiesel and green diesel. This is because the initial substrate is approximately similar in both the cases. The method of production of biodiesel depends on alkali catalysts in the presence of alcohol, while the green diesel process needs H_2, like a catalyst. The feedstock generations are categorized based on their resources. First generation is food-related sources, second-generation feedstocks are non-food-related sources, and third-generation feedstocks are non-competitive and separately grown algal resources. This chapter gives insight to preferably identify a different kind of feedstock for green diesel production depending on their fatty acid content, availability, production area, and feasibility.

References

1. Ayoola AA, Fayomi OSI, Adegbite OA, Raji O (2021) Biodiesel fuel production processes: a short review. In: IOP conference series: materials science and engineering, vol 1107https://doi.org/10.1088/1757-899X/1107/1/012151, p 012151
2. Vignesh P, Kumar ARP, Ganesh NS, et al (2021) Biodiesel and green diesel generation: an overview. Oil & Gas Science and Technology—Revue d'IFP Energies nouvelles 76:6. https://doi.org/10.2516/OGST/2020088
3. Zhang Y, Ma L (2021) Application of high-throughput sequencing technology in HIV drug resistance detection. Biosaf Health 3:276–280. https://doi.org/10.1016/J.BSHEAL.2021.06.002
4. Green Diesel—Oil&Gas Portal. http://www.oil-gasportal.com/green-diesel/. Accessed 31 Oct 2021
5. Biofuel feedstocks. https://www.etipbioenergy.eu/everyone/biofuel-feedstocks. Accessed 31 Oct 2021
6. Singh D, Sharma D, Soni SL, et al (2020) A comprehensive review on 1st-generation biodiesel feedstock palm oil: production, engine performance, and exhaust emissions. BioEnergy Res 14:1 14:1–22. https://doi.org/10.1007/S12155-020-10171-2
7. Ng JH, Tan JA, Lim WL, et al (2017) On the economic feasibility of the first, second and third generations biodiesel feedstock. In: 3rd International conference on power generation systems and renewable energy technologies, PGSRET 2017 2018-January. https://doi.org/10.1109/PGSRET.2017.8251817, pp 142–147
8. Shaah MAH, Hossain MdS, Allafi FAS et al (2021) A review on non-edible oil as a potential feedstock for biodiesel: physicochemical properties and production technologies. RSC Adv 11:25018–25037. https://doi.org/10.1039/D1RA04311K
9. Saad MG, Dosoky NS, Zoromba MS, Shafik HM (2019) Algal biofuels: current status and key challenges. Energies 12:1920 12:1920. https://doi.org/10.3390/EN12101920
10. Mata TM, Martins AA, Caetano NS (2010) Microalgae for biodiesel production and other applications: a review. Renew Sustain Energy Rev 14:217–232. https://doi.org/10.1016/J.RSER.2009.07.020
11. Chisti Y (2007) Biodiesel from microalgae. Biotechnol Adv 25:294–306. https://doi.org/10.1016/J.BIOTECHADV.2007.02.001
12. Sun X-M, Ren L-J, Zhao Q-Y et al (2018) Microalgae for the production of lipid and carotenoids: a review with focus on stress regulation and adaptation. Biotechnol Biofuels 11:1–16. https://doi.org/10.1186/S13068-018-1275-9
13. McFarlane J (2011) Processing of soybean oil into fuels. Recent trends for enhancing the diversity and quality of soybean productshttps://doi.org/10.5772/18701
14. Fuel the Crush: Renewable Diesel Pumps Up Soybean Demand I AgWeb. https://www.agweb.com/news/crops/soybeans/fuel-crush-renewable-diesel-pumps-soybean-demand. Accessed 25 Nov 2021
15. Renewable diesel boom is wild card for U.S. soybeans I Successful Farming. https://www.agriculture.com/news/business/renewable-diesel-boom-is-wild-card-for-us-soybeans. Accessed 25 Nov 2021
16. Top countries of soybean oil consumption worldwide 2020 I Statista. https://www.statista.com/statistics/1199485/domestic-consumption-of-soybean-oil-worldwide-by-country/. Accessed 25 Nov 2021
17. Clemente TE, Cahoon EB (2009) Soybean oil: genetic approaches for modification of functionality and total content. Plant Physiol 151:1030. https://doi.org/10.1104/PP.109.146282
18. Scaldaferri CA, Pasa VMD (2019) Production of jet fuel and green diesel range biohydrocarbons by hydroprocessing of soybean oil over niobium phosphate catalyst. Fuel 245:458–466. https://doi.org/10.1016/J.FUEL.2019.01.179
19. Chart: Which Countries Produce The Most Palm Oil? I Statista. https://www.statista.com/chart/23097/amount-of-palm-oil-produced-in-selected-countries/. Accessed 25 Nov 2021

20. Sambanthamurthi R, Sundram K, Tan YA (2000) Chemistry and biochemistry of palm oil. Prog Lipid Res 39:507–558. https://doi.org/10.1016/S0163-7827(00)00015-1
21. Montoya C, Cochard B, Flori A, et al (2014) Genetic architecture of palm oil fatty acid composition in cultivated oil palm (Elaeis guineensis Jacq.) Compared to Its Wild Relative E. oleifera (H.B.K) Cortés. PLoS ONE 9. https://doi.org/10.1371/JOURNAL.PONE.0095412
22. Zikri A, Aznury M (2020) Green diesel production from Crude Palm Oil (CPO) using catalytic hydrogenation method. In: IOP conference series: materials science and engineering, vol 823.https://doi.org/10.1088/1757-899X/823/1/012026, p 012026
23. Renewable Diesel Production | Valero. https://www.valero.com/renewables/renewable-diesel. Accessed 26 Nov 2021
24. Dupont J, White PJ, Carpenter MP et al (1990) Food uses and health effects of corn oil. J Am Coll Nutr 9:438–470. https://doi.org/10.1080/07315724.1990.10720403
25. Corn production by country 2018/19 | Statista. https://www.statista.com/statistics/254292/glo bal-corn-production-by-country/. Accessed 26 Nov 2021
26. Boateng L, Ansong R, Owusu WB, Steiner-Asiedu M (2016) Coconut oil and palm oil's role in nutrition, health and national development: a review. Ghana Med J 50:189. https://doi.org/10.4314/gmj.v50i3.11
27. Top countries of coconut oil consumption worldwide 2020 | Statista. https://www.statista.com/statistics/1199479/domestic-consumption-of-coconut-oil-worldwide-by-country/. Accessed 25 Nov 2021
28. Production of rapeseed by main producing countries, 2019/2020 | Statista. https://www.statista.com/statistics/263930/worldwide-production-of-rapeseed-by-country/. Accessed 26 Nov 2021
29. Matthaus B, Özcan MM, Juhaimi F, al, (2016) Some rape/canola seed oils: fatty acid composition and tocopherols. Z Naturforsch [C] 71:73–77. https://doi.org/10.1515/ZNC-2016-0003
30. Sotelo-Boyás R, Liu Y, Minowa T (2010) Renewable diesel production from the hydrotreating of rapeseed oil with Pt/Zeolite and NiMo/Al$_2$O$_3$ catalysts. Ind Eng Chem Res 50:2791–2799. https://doi.org/10.1021/IE100824D
31. Sugami Y, Minami E, Saka S (2016) Renewable diesel production from rapeseed oil with hydrothermal hydrogenation and subsequent decarboxylation. Fuel 166:376–381. https://doi.org/10.1016/J.FUEL.2015.10.117
32. Akkaya MR (2018) Prediction of fatty acid composition of sunflower seeds by near-infrared reflectance spectroscopy. J Food Sci Technol 55:2318. https://doi.org/10.1007/S13197-018-3150-X
33. Major producer countries of sunflower seed 2019/2020 | Statista. https://www.statista.com/statistics/263928/production-of-sunflower-seed-since-2000-by-major-countries/. Accessed 26 Nov 2021
34. Onyestyák G, Harnos S, Szegedi Á, Kalló D (2012) Sunflower oil to green diesel over Raney-type Ni-catalyst. Fuel 102:282–288. https://doi.org/10.1016/J.FUEL.2012.05.001
35. Tsimidou M, Blekas G, Boskou D (2003) OLIVE OIL. Encyclopedia of Food Sciences and Nutrition. https://doi.org/10.1016/B0-12-227055-X/01347-X, pp 4252–4260
36. Olive oil—Wikipedia. https://en.wikipedia.org/wiki/Olive_oil. Accessed 26 Nov 2021
37. Leading consumers of olive oil worldwide 2020/21 | Statista. https://www.statista.com/statistics/940532/olive-oil-consumption-worldwide-by-leading-country/. Accessed 26 Nov 2021
38. Patel VR, Dumancas GG, Viswanath LCK et al (2016) Castor oil: properties, uses, and optimization of processing parameters in commercial production. Lipid Insights 9:1. https://doi.org/10.4137/LPI.S40233
39. APAC: castor oil seed production by country | Statista. https://www.statista.com/statistics/658666/asia-pacific-castor-oil-seed-production-by-country/. Accessed 26 Nov 2021
40. Orozco LM, Echeverri DA, Sánchez L, Rios LA (2017) Second-generation green diesel from castor oil: development of a new and efficient continuous-production process. Chem Eng J 322:149–156. https://doi.org/10.1016/J.CEJ.2017.04.027

41. Liu J, Lei J, He J, et al (2015) Hydroprocessing of jatropha oil for production of green diesel over non-sulfided Ni-PTA/Al$_2$O$_3$ catalyst. Sci Rep 5:1 5:1–13. https://doi.org/10.1038/srep11327
42. Liu C, Liu J, Zhou G et al (2013) Transformation of Jatropha oil into green diesel over a new heteropolyacid catalyst. Environ Prog Sustainable Energy 32:1240–1246. https://doi.org/10.1002/EP.11714
43. Hemanandh J, Narayanan KV (2016) Production of green diesel by hydrotreatment using jatropha oil: performance and emission analysis. Waste Biomass Valorization 8:6 8:1931–1939. https://doi.org/10.1007/S12649-016-9729-4
44. Verma R, Sharma DK, Bisen PS (2019) Determination of free fatty acid composition in jatropha crude oil and suitability as biodiesel feedstock. Curr Alternative Energy 3:59–64. https://doi.org/10.2174/2405463103666190722163037
45. Moniruzzaman M, Yaakob Z, Shahinuzzaman M et al (2017) Jatropha biofuel industry: the challenges. Front Bioenergy Biofuels. https://doi.org/10.5772/64979
46. Sandanasamy JD (2014) Fatty acid composition and antibacterial activity of neem (Azadirachta indica) seed oil. Open Conf Proc J 4:43–48. https://doi.org/10.2174/2210289201304020043
47. Neem NRC (US) P on (1992) Industrial Products
48. Karanjia Oil. https://manoramagroup.co.in/karanja-oil. Accessed 26 Nov 2021
49. Rahman MS, Islam MB, Rouf MA et al (2011) Extraction of alkaloids and oil from Karanja (*Pongamia pinnata*) seed. J Sci Res 3:669–675. https://doi.org/10.3329/JSR.V3I3.7227
50. Salimon J, Abdullah BM, Salih N (2012) Rubber (Hevea brasiliensis) seed oil toxicity effect and Linamarin compound analysis. Lipids Health Dis 11:74. https://doi.org/10.1186/1476-511X-11-74
51. Roschat W, Siritanon T, Yoosuk B et al (2017) Rubber seed oil as potential non-edible feedstock for biodiesel production using heterogeneous catalyst in Thailand. Renewable Energy 101:937–944. https://doi.org/10.1016/J.RENENE.2016.09.057
52. (6) (PDF) Cracked Madhuca Indica (Mahua) oil as a diesel engine application: an overview. https://www.researchgate.net/publication/344758252_Cracked_Madhuca_Indica_Mahua_Oil_as_a_Diesel_Engine_Application_An_Overview. Accessed 26 Nov 2021
53. Animal fats for biodiesel production—farm energy. https://farm-energy.extension.org/animal-fats-for-biodiesel-production/. Accessed 26 Nov 2021
54. Miao C, Marin-Flores O, Dong T et al (2018) Hydrothermal catalytic deoxygenation of fatty acid and bio-oil with in situ H$_2$. ACS Sustain Chem Eng 6:4521–4530. https://doi.org/10.1021/ACSSUSCHEMENG.7B02226/SUPPL_FILE/SC7B02226_SI_001.PDF
55. Xia H, Xu S, Yang L (2017) Efficient conversion of wheat straw into furan compounds, bio-oils, and phosphate fertilizers by a combination of hydrolysis and catalytic pyrolysis. RSC Adv 7:1200–1205. https://doi.org/10.1039/C6RA27072G
56. Tomás-Pejó E, Fermoso J, Herrador E et al (2017) Valorization of steam-exploded wheat straw through a biorefinery approach: bioethanol and bio-oil co-production. Fuel 199:403–412. https://doi.org/10.1016/J.FUEL.2017.03.006
57. Zhang XS, Yang GX, Jiang H, et al (2013) Mass production of chemicals from biomass-derived oil by directly atmospheric distillation coupled with co-pyrolysis. Sci Rep 3:1 3:1–7. https://doi.org/10.1038/srep01120
58. Xie Y, Zeng K, Flamant G et al (2019) Solar pyrolysis of cotton stalk in molten salt for bio-fuel production. Energy 179:1124–1132. https://doi.org/10.1016/J.ENERGY.2019.05.055
59. Teixeira Cardoso AR, Conrado NM, Krause MC et al (2019) Chemical characterization of the bio-oil obtained by catalytic pyrolysis of sugarcane bagasse (industrial waste) from the species Erianthus Arundinaceus. J Environ Chem Eng 7:102970. https://doi.org/10.1016/J.JECE.2019.102970
60. Tang Y, Rosenberg JN, Bohutskyi P, et al (2015) Microalgae as a feedstock for biofuel precursors and value-added products: Green fuels and golden opportunities. BioResources 11:2850–2885. https://doi.org/10.15376/BIORES.11.1.TANG
61. Singh D, Sharma D, Soni SL et al (2020) A review on feedstocks, production processes, and yield for different generations of biodiesel. Fuel 262:116553. https://doi.org/10.1016/J.FUEL.2019.116553

62. Khan MI, Shin JH, Kim JD (2018) The promising future of microalgae: current status, challenges, and optimization of a sustainable and renewable industry for biofuels, feed, and other products. Microbial Cell Factories 17:1 17:1–21. https://doi.org/10.1186/S12934-018-0879-X
63. Veeramuthu A, Ngamcharussrivichai C (2020) Potential of microalgal biodiesel: challenges and applications. Renew Energy Technol Appl.https://doi.org/10.5772/INTECHOPEN.91651
64. Medipally SR, Yusoff FM, Banerjee S, Shariff M (2015) Microalgae as sustainable renewable energy feedstock for biofuel production. BioMed Res Int.https://doi.org/10.1155/2015/519513
65. Pragya N, Pandey KK (2016) Life cycle assessment of green diesel production from microalgae. Renew Energy 86:623–632. https://doi.org/10.1016/J.RENENE.2015.08.064
66. Costa IG, Balmant W, Ramos LP, et al (2019) Green diesel from microalgae. In: 13th International conference on energy sustainability, ES 2019, collocated with the ASME 2019 Heat Transfer Summer Conference, ASME 2019. https://doi.org/10.1115/ES2019-3959
67. Zhou L, Lawal A (2016) Hydrodeoxygenation of microalgae oil to green diesel over Pt, Rh and presulfided NiMo catalysts. Catal Sci Technol 6:1442–1454. https://doi.org/10.1039/C5C Y01307K
68. Alternative fuels data center: renewable hydrocarbon biofuels. https://afdc.energy.gov/fuels/emerging_hydrocarbon.html. Accessed 24 Nov 2021

Chapter 3
Catalytic Materials for Green Diesel Production

Praveenkumar Ramprakash Upadhyay and Piyali Das

Abstract This chapter reviews green diesel alternately known as renewable diesel production pathways through deoxygenation routes with catalytic interventions. Deoxygenation processes of green diesel production involve milder reaction conditions than conventional biodiesel production thus appear more promising with respect to environmental and economic sustainability. The activity and selectivity of catalyst material depends on several factors among which reaction conditions (pressure, temperature, duration), type of catalysts, support (organic or inorganic or combined), and promoters are of maximum significance. In recent years, high activity and selectivity are being accomplished via deoxygenation with non-noble metal catalysts. Overall, present advancement/progress and forthcoming developments in green diesel production are thoroughly articulated in this chapter.

Keywords Green diesel · Renewable diesel · Triglyceride · Biocrude deoxygenation · Hydrocracking · Deoxyhydrogenation · Catalysts for green diesel

1 Introduction

Major technologies available today for green diesel production are (a) hydroprocessing; (b) catalytic upgrading of sugars, starches, and alcohols; (c) thermal conversion (pyrolysis) and upgrading of bio-oil; and (d) biomass to liquid (BTL) thermochemical processes, etc. [212]. Presently green diesel is produced exclusively through hydroprocessing route which utilizes triglycerides feed stocks like lipids, vegetable oils, and animal fats. The insufficient supply chain of such feedstock limits the scope of large-scale production of green diesel through this platform. Great opportunity lies ahead for other technology platforms like pyrolysis, biomass to liquid, and sugar

P. R. Upadhyay · P. Das (✉)
The Energy and Resources Institute (TERI), Darbari Seth Block, IHC Complex, Lodhi Road, New Delhi 110 003, India
e-mail: piyalid@teri.res.in

P. R. Upadhyay
e-mail: p.upadhyay@teri.res.in

© The Author(s), under exclusive license to Springer Nature Singapore Pte Ltd. 2022
M. Aslam et al. (eds.), *Green Diesel: An Alternative to Biodiesel and Petrodiesel*,
Advances in Sustainability Science and Technology,
https://doi.org/10.1007/978-981-19-2235-0_3

upgradation due to its feasibility of using more sustainable feedstocks like biomass, agricultural/forest residue [212].

Major pathways for green diesel production using Triglycerides and Biomass-based feedstocks are detailed below.

1.1 Green Diesel from Triglyceride-Based Feedstock

Possible reaction pathways of green diesel production from feedstocks like triglycerides, free fatty acids, and esters are illustrated in Fig. 1 [82] which is based on the reported results available so far.

Studies on model triglycerides like Triolein or tristearin have been instrumental in establishing the reaction mechanism of green diesel production from such feedstock [2, 43, 125, 156, 200]. This conversion is believed to proceed through hydrogenation of C = C to saturated triglyceride followed by hydrogenolysis or β-elimination to three molecules of saturated fatty acid and propane, C_3H_8, (step (b) → (c)) [2, 10, 43, 165, 200]. Oxygen removal from the carboxyl group takes place both in presence or absence of H2 environment. The latter one known as decarboxylation produces Heptadecane ($C_{17}H_{36}$) and CO_2 (step (m)). In presence of H_2, saturated fatty acid conversion follows another path to aldehyde intermediate (step (d)) which subsequently reject oxygen either via decarbonylation, i.e., CO removal (step (e)).

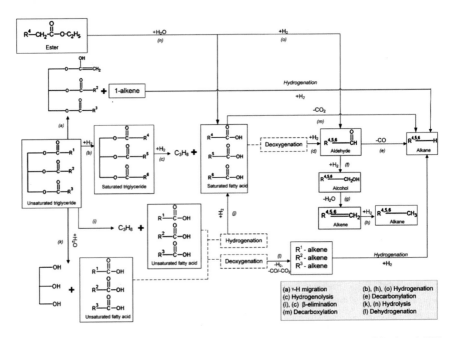

Fig. 1 Possible reaction pathways of green diesel production with several types of feedstock [82]

and/or hydrodeoxygenation route to first alcohols (step (f)) followed by H_2O removal (step (g) → (h)). Conversion of Macauba pulp oil with activated carbon-supported Co metal catalyst at 30 bars H_2, 350 °C, and 2 h reaction time to produce 88% saturated alkanes in the diesel range (C_{10}–C_{20} hydrocarbons) with 96.25% overall conversion is claimed to follow the above-mentioned pathways [47].

Alternate pathways to the β-elimination are perceived to progress through γ-H migration and β-γ-scission both resulting in alkene range of products as shown in step (a) in Fig. 1. The former route is followed during high-temperature decomposition, whereas the latter route progress through scission of C–C at the β and γ positions [184, 185, 246] these pathways are validated through production of green diesel from soybean oil over an $NbOPO_4$ catalyst at 10 bars H_2 and 350 °C, yielding hydrocarbons in the range C_9–C_{17}. Final products comprise of 62% bio-jet fuel, 40% green diesel, and 18% gasoline-range hydrocarbons [184, 185]. It is perceived that the initial reaction of triglyceride in soybean oil deoxygenation is γ-H migration, β-γ-scission, and β-elimination in parallel reactions followed by cracking and deoxygenation leading to hydrocarbons [184].

Another reaction pathway for cracking triglyceride to free fatty acid is proposed. The hydrolysis of triglyceride to produce fatty acid and glycerol is reported to occur over Ni–Mg supported on multi-walled carbon nanotubes catalysts at a reaction temperature of 350 ∘C for 2 h under nitrogen gas flow. This reaction route leads to free fatty acid as shown in step (k) [95].

The possible deoxygenation mechanism for Fatty acids-based feedstock is also shown in Fig. 1 which reveals that saturated and unsaturated fatty acids follow different pathways. The possible reactions of fatty acid are shown in Fig. 1. Palmitic acid (C_{16}:0) and stearic acid (C_{18}:0) are widely representative model compounds for saturated fatty acids [107, 114, 121, 124]. Hydrogenation of the carboxylic group in stearic acid initially form aldehyde intermediate, e.g., octadecanal (step (d)). Then, an aldehyde intermediate is converted via two pathways, decarbonylation (step (e)) and dehydration-hydrogenation (step (f) to (h)). The decarbonylation pathway forms n-heptadecane while, in the dehydration-hydrogenation pathway, the aldehyde octadecanal converts to respective alcohol, i.e., octadecanol which in turn by rejecting oxygen in the form of water leads to alkene compounds, i.e., octadecene. This is further hydrogenated to alkane compounds like n-octadecane. Complete deoxygenation of stearic acid with CuCo catalyst on carbon nanotubes support exhibiting 94.82% selectivity of long-chain hydrocarbon products validates the above-mentioned pathways [121]. Oleic acid is another commonly used model compound for green diesel production due to its presence in triglycerides, like palm oil, sunflower oil, and jatropha oil [140, 198]. The deoxygenation of unsaturated oleic acid is more complicated than that of saturated fatty acid due to the presence of double bonds. The general reaction pathway of oleic acid in the presence of external hydrogen is shown in Fig. 1.

It is reported that double bonds in unsaturated oleic acid can be hydrogenated to obtain saturated stearic acid, as shown in step (j), followed by decarboxylation, dehydrogenation, and cracking [19, 67, 89, 166, 243]. For hydrogen-free deoxygenation, direct decarboxylation or decarbonylation and dehydrogenation can occur to produce alkene hydrocarbons in step (1). Variety of alkene are obtained through oleic acid deoxygenation in hydrogen free environment, 1,8-heptadecadiene, 8-heptadecane, and polyunsaturated heptadecene. The alkene products still require a hydrogen atom for hydrogenation to produce alkane [1, 104, 179]. The hydrogen atoms used in the reaction can come not only from external hydrogen gas but can also be from gas-phase reaction, or a metal catalyst, solvent, or supports [18, 89, 156, 184].

1.2   Green Diesel from Pyrolytic and Hydrothermal Biocrude

Biomass pyrolysis, a thermochemical conversion routes, has emerged is as one of the most viable routes for making bio-oils/biocrude, bio-char, and non-condensable gases from wood and lignocellulosic biomass. Typically fast pyrolysis processes yield 60–75 wt. % of liquid bio-oil/biocrude, 15–25 wt. % of solid char, and 10–20 wt. % of non-condensable gases, depending on the feedstock used. Attributes of a fast pyrolysis process are high heating and heat transfer rates, carefully controlled pyrolysis temperature typically between 425 and 500 °C, short vapor residence times less than 2 s and rapid quenching and cooling of pyrolysis vapors and aerosols to give bio-oil/biocrude [27].

Pyrolytic biocrude is upgraded to transport fuel via full deoxygenating, which is conventionally carried out through hydrotreating and catalytic vapor cracking. Hydrotreating of bio-oil is carried out at high temperature, high hydrogen pressure (as high as 50 bar or higher), and in the presence of catalysts resulting in oxygen elimination as water. The catalysts used are typically sulphided Co-Mo or Ni-Mo supported on alumina where the process is similar to refining of petroleum cuts. Catalytic vapor cracking makes deoxygenation possible through simultaneous dehydration-decarboxylation over acidic zeolite catalysts and at around 450 °C and atmospheric pressure where oxygen is rejected as H_2O, CO_2, and CO. Up grading bio-oil directly to a quality transport liquid fuel. Stand-alone systems to convert biocrude to transport fuel is yet not proven economically viable due to high cost of Hydrogen upgrading agent and catalyst deactivation issues. Hydrotreating route has high Carbon retention but high H_2 requirement whereas, Catalytic upgrading has comparatively much lower H_2 requirement but it sacrifices carbon. Both routes need to be developed to a scale making them commercially viable.

In recent years, the importance of the enhancement of biocrude properties is recognized to fundamentally address the challenges related to bio-oil hydrotreating. Stabilization of the biocrude by mild hydrodeoxygenation (HDO) is now considered a necessary prerequisite for fruitful deep HDT upgrading [234]. Different preprocessing steps are adopted like physical modifications and chemical modifications like esterification, ion exchange, and azeotropic distillation to name a few [207]. As a better alternative to the conventional approach for the production of green transport fuel from pyrolytic biocrude, RTI International is developing an advanced technology integrating catalytic biomass pyrolysis and hydrotreating. The biocrude from catalytic fast pyrolysis has more desirable attributes like higher thermal stability, less acidity, and less oxygen content. Consequently, it is hypothesized that this biocrude could be hydrotreated fairly effectively without a preprocessing or stabilization step. the new process has the foreseeable advantage of potentially minimizing the overall hydrogen demand and increasing hydrocarbon product yield and quality. In fact, a recent techno-economic analysis found that the integration of mild catalytic biomass pyrolysis with hydroprocessing is a promising approach for the production of transportation fuels [144].

In recent years, co-refining of Biocrude made through from pyrolysis and Hydrothermal liquefaction process are being pilot tested by few refineries to make green diesel, gasoline, and other refinery products. Current petroleum catalysts are incompatible with biomass-derived crude. Pyrolytic biocrude cannot be readily co-processed with petroleum feeds as they typically contain up to 30% water and 40% oxygen and are thus not immiscible with a polar petroleum liquids [216].

With currently available technologies, the threshold oxygen content in biocrude is limited to <15wt.% for co-refining application [46]. There are several lab scale as well as Pilot scale studies reported in literature on biocrude co-refining. Concerns still persist on how assessment of the trade-offs for co-processing conventional petroleum feeds with biocrude in a petroleum refinery. Towards this very focused pilot level co-refining studies are conducted under DOE-BETO program largely between 2012 and 2017 to identify costs, opportunities, technical risks, information gaps, research needs, etc. These studies have provided some clarity on viable insertion points of biocrudes in petroleum oil refinery. Two favorable insertion points are identified as refinery's Fluid Catalytic Cracking (FCC) units or hydroprocessing (hydrotreating and hydrocracking) units [46].

FCC co-refining of biocrude/pyrolysis oils offers a complete trade-off approach where hydrogen use is minimum and the biocrude is mild pre-processed prior to co-processing with heavy vacuum gas oils to produce a low-value hydrocarbon intermediate and large amounts of renewable power. Recent studies under the European BIOCOUP program have shown the feasibility of bio-oil (upto 20 wt.% oxygen) insertion for co-refining with heavy fossil oils in FCCs unit. The second, potential insertion point of bio-oil is before the refinery's hydroprocessing unit. On the contrary Hydroprocessing requirs substantial hydrogen use (e.g., 800 L H_2 per kg of bio-oil processed, including pre-refinery hydrotreatment processing), expensive catalysts (e.g., ruthenium-based vs ZSM5-based), and heavily pre-processed bio-oils (e.g., de-oxygenated to 3–5 wt.% oxygen) [46]. Although biocrude Co-refining in hydroprocessing mode is more expensive option, it generates higher degree of middle distillates such as green diesel and jet fuels in comparison to FCC co-refining. Availability of low-cost green hydrogen is important in achieving favorable economics and realizing co-locating drop-in biofuel facilities with oil refineries synergistically beneficial.

A simplified schematic showing HDO bio-oil insertion points (red arrows) within a typical refinery is outlined in Fig. 2.

Biocrude upgradation to green transport fuel is presently under demonstration by Envergent Technology (Joint venture between Ensyn and Honeywell's UOP). GTI (Gas technology Institute, USA) has also completed proof of principal work

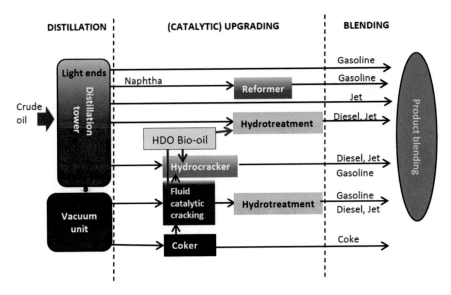

Fig. 2 Refinery Insertion Points (Red Arrows) for HDO bio-oils [46]

on Integrated two stage Hydropyrolysis and Hydroconversion (IH$_2$) technology for making biomass-based Gasoline and green Diesel from biomass. IH2 is carried out in two integrated stages. In the first stage catalytic hydro pyrolysis is conducted in a fluidized bed reactor under moderate hydrogen pressure. In the 2nd stage, vapors from the first stage pass through hydro-conversion step where a hydrodeoxygenation catalyst removes all remaining oxygen and produces gasoline and diesel boiling range material.

Another thermochemical conversion technology termed as Hydrothermal liquefaction (HTL) or direct liquefaction, also produces biomass to liquid products similar to pyrolytic biocrude. The HTL process usually takes place at temperatures of 200–400 °C and at elevated pressures of 50–200 bar with residence times of 10–60 min [212]. In HTL process, the organic biocrude is separated from the water; a distillate cut of green diesel is produced. Changing World Technologies (CWT) is commercializing this process to make ASTM D975 compliant green diesel and using the term "thermal depolymerization". HTL biocrude is upgraded by catalytic hydrotreatment similar to pyrolytic biocrude. HTL is considered the best option by many research groups for wet algal conversion to biocrude as it avoids the energy intensive drying of wet algae. After reaction, the organics are separated from the water; a distillate cut suitable for diesel use is produced. This technology is being commercialized in the United States by Changing World Technologies (CWT). CWT states that its product meets the requirements of ASTM D975 and uses the term "thermal depolymerization" to describe the process.

1.3   Green Diesel from Bio-Syngas

Another feasible route of Green or Renewable Diesel production is through indirect liquefaction of desired mixture of hydrogen and carbon monoxide called syngas. In this process biomass like agro residues are converted to syngas through gasification technology which after purification is catalytically converted to ultra-low sulfur diesel to meet ASTM D975 through Fischer–Tropsch (FT) synthesis Commercially FT Diesel is largely produced from natural gas though coal, and heavy crude oil are other potential resources.

1.4   Green Diesel from Olefins

Green diesel is also produced from lower alkenes such as propylene, butylene, and pentene to heavier alkenes through oligomerization using acidic catalysts like zeolites or supported Ni catalysts [212]. Mobil Olefins to Gasoline and Distillates (MOGD) and Conversion of Olefins to Diesel (COD) of Lurgi are two present commercial technologies in this space.

2   Catalysts for Green Diesel Production

2.1   Metal Catalysts

The role of catalysts in green diesel production are increasingly being under limelight since last few years. This section presents an exhaustive review of metal catalysts employed in deoxygenation processes of different resource materials relevant to green diesel research. Significant aspects of catalysts that discussed here are functionality, mechanisms, electronic interactions, and physicochemical properties of particular metals, etc., to name a few. Transition metals as well as few metals from other categories such as alkali and alkaline metals are also included.

2.1.1   Palladium Metal-Based Catalysts

Transition metal Palladium (Pd) is considered a noble metal and offers a high activity for hydrogenation for various chemical reactions. Pd metal-based catalysts are investigated in the deoxygenation processes of various types of model compounds like benzophenone, benzyl phenyl ether, furfurylideneacetone, guaiacol, lauric acid, m-cresol, palmitic acid, phenol, oleic acid, and stearic acid [22, 23, 54, 60, 71, 80, 81, 88, 98, 102, 133, 182, 195].

Several deoxygenation research studies are conducted to produce green diesel where Pd and Ni-based catalysts have exhibited superior activity than over other metal catalysts [43, 72, 92, 187]. Pd metal-based catalysts are successfully yielded over 90% green diesel through oxygen removal from >50% of different types of reactants such as aromatic compounds, fatty acids, and triglycerides [56, 183, 220]. However, due to inability of mono catalysts based on Pd in opening the aromatic ring the main products are found to comprise of cycloalkane ring in them [183]. The alkanes chain lengths largely depend on the number of carbon atoms in the reactant molecules.

It is worth mentioning that removal of oxygen is not facilitated by Pd metal-based catalysts due to their hydrogenation nature. The aromatic and furanic ring saturation frequently happens before the C-O bond scission as Pd surface exhibits a more robust surface-adsorbate synergy with the aromatic ring than carbon–oxygen single or double bonds [78]. The preferential aromatic rings hydrogenation with a Pd metal-based catalyst is confirmed in the phenol and benzyl phenyl ether deoxygenation [102, 241]. Carbonyl compounds deoxygenation with Pd metal-based catalysts usually hydrogenate the carbonyl group into corresponding alcohol and termed as a hydrogenation-hydrogenolytic mechanism [22, 170]. Pd metal-based catalysts often promote decarbonylation and decarboxylation in esters deoxygenation [23, 133]. Comparable reaction routes are suggested in deoxygenation studies using various raw feed such as castor oil, jatropha residue, macauba oil, and woody tar [96, 137, 139, 194].

Pd by virtue of its unique H_2 sticking coefficient properties facilitates H_2 spill over and simple H_2 activation at the surface in deoxygenation reactions [77]. Pd metal-based catalysts with acidic support sites are thus frequently bi-functionalized because of the facile hydrogenation and dehydration by palladium and acidic sites respectively, resulting in greater HDO activity [54, 58]. In addition, debenzylation with palladium is also effective, where deoxygenated product methylbenzene is recovered [39].

Pd metal-based catalysts in deoxygenation processes are studied in both gas–solid and liquid–solid phases with varying pressure ranges (1.01–60 bar) and a temperature range (403–573 K) (Table 1). Hydrogenation, hydrodeoxygenation, decarboxylation, and decarbonylation are dominant reaction pathways executed in the deoxygenation processes with Pd metal-based catalysts. Pd metal-based catalysts application in model compounds like benzaldehyde, benzophenone, benzyl phenyl ether, guaiacol, furfurylidene acetone, lauric acid, m-cresol, oleic acid, palmitic acid, and phenol are completely deoxygenated with 20–100% conversion and 22.2–100% selectivity. Pd/C and Pd/CNT catalysts are studied for deoxygenation of palmitic acid to obtain pentadecane [52, 195]. The results show that an increase in pressure had a negligible gain on the conversion rate and selectivity of the products [52, 195]. Though, Pd-Fe/OMC catalyst doped with Fe improves the toluene selectivity in the investigation of benzyl phenyl ether HDO. Pd metal is an outstanding hydrogenating metal, so additional Fe doping metal is required to reduce the unnecessary aromatic deoxygenated products during hydrogenation. In one study HY zeolite:γ-Al$_2$O$_3$ (20:80) supported catalyst is reported to have optimal pore size and number of active acid sites

Table 1 Pd metal-based catalysts

Entry	Metal catalysts	Preparation methods	Reactant	Product	Reaction pathways	Reaction phase	Reaction type	Temp (K)	Pressure (bar)	X (%)	S (%)	References
1	Pd/ZrO$_2$	Dry impregnation	Phenol	Benzene	Hydrodeoxygenation	G–S	Continuous	573	1.01	75	54	de Souza et al. [50]
2	Pd/OMC	Dry impregnation	Benzylphenyl ether	Toluene	Hydrodeoxygenation	L–S	Batch	523	10	93	17.3	Kim et al. [102]
3	Pd/C	Deposition	Palmiticacid	Pentadecane	Decarbonylation	L–S	Semi-batch	573	17.5	95	80	Simakova et al. [195]
4	Pd/C	Impregnation	Benzaldehyde	Toluene	Hydrodeoxygenation	L–S	Batch	403	60	99	28	Procházková et al. [170]
5	Pd/C	–	Lauricacid	Undecane	Decarboxylation	L–S	Continuous	543	10	100	100	Mäki-Arvela et al. [133]
6	Pd/C	Impregnation	Benzophenone	Diphenylmethane	Hydrogenation-hydrogenolytic mechanism	L–S	Batch	403	60	100	98.8	Bejblová et al. [22]
7	Pd/γ-Al$_2$O$_3$	Wet impregnation	Phenol	Cyclohexene Cyclohexane	Hydrodeoxygenation Hydrogenation	L–S	Continuous	573	15	35	98	Echeandia et al. [58]
8	Pd/Al$_2$O$_3$	–	Furfurylidene acetone	Octane	Hydrodeoxygenation	L–S	Batch	473	55	95	25	Faba et al. [60]
9	Pd/SiO$_2$	Dry impregnation	m-Cresol	Toluene	Hydrodeoxygenation	G–S	Continuous	523	1.01	99.8	68.4	Chen et al. [31]
10	Pd/Beta(35)	Impregnation	Benzaldehyde	Toluene	Hydrodeoxygenation	L–S	Batch	403	60	99	93	Procházková et al. [170]

(continued)

Table 1 (continued)

Entry	Metal catalysts	Preparation methods	Reactant	Product	Reaction pathways	Reaction phase	Reaction type	Temp (K)	Pressure (bar)	X (%)	S (%)	References
11	Pd/MB	Impregnation	Oleic acid	Heptadecane	Decarbonylation Decarboxylation	L-S	Batch	573	20.3	100	70.5	Dragu et al. [54]
12	Pd/C	–	Cashew nutshell liquid	Pentadecylcyclohexane	Deoxygenation Hydrogenation	L-S	Batch	573	40	98	89	Scaldaferri and Pasa [183]
13	Pd/ Ni-Ce$_{0.6}$Zr$_{0.4}$O$_2$	Co-precipitation	Oleic acid	C$_9$–C$_{17}$ Hydrocarbon	Deoxygenation	L-S	Batch	573	1	90.8	19	Jeon et al. [91]
14	Pd/C	–	Waste cooking oil	Alkanes Aromatics	Hydrodeoxygenation Decarbonylation Decarboxylation	G-S	Continuous	648	46	100	–	Chen and Wang [38]

delivering 57% phenol conversion and 93% selectivity for cyclohexene and cyclohexane [58]. Cashew nut shell liquid is reported as a raw material to make green diesel under different reaction conditions via deoxygenation, hydrogenation, and cracking over Pd metal-based catalyst supported on activated charcoal [183]. Diesel range hydrocarbons that are produced comprises of 6% oxygenated compounds, 3% aromatic compounds, and 89% saturated alkanes. Another research group has tested Pd/C catalyst for waste cooking oil and successfully achieved 100% conversion rate with product concentration 82 wt. % of C_{15}-C_{17}, and 18 wt. % of C_{16}-C_{18} [38].

2.1.2 Platinum Metal-Based Catalysts

Platinum has a high resistance to oxidation and is often classified as a noble metal. Pt-based catalyst offers an excellent hydrogenation activity and also deoxygenation of oxygenated compounds [3, 116, 128, 155, 235, 244]. The deoxygenation process with Pt-based catalysts often results in a saturation of $C = C$ bonds or aromatic rings prior to the expulsion of functional group-containing oxygen because of its high hydrogenating nature [235]. The interaction between platinum metal and an acid support like zeolites or γ-Al_2O_3 significantly improves performance of Pt-based catalyst in the deoxygenation process [63, 153]. Pt-based bifunctional catalyst has a high H_2 sticking potential with acidic support, which activates an oxygenated compound at the electrophilic acidic centers [81, 120]. The deoxygenation performance of the Pt-based catalyst is good, but it is still necessary to maximize the metal to acid sites ratio and acidic site strength in order to depreciate the critical coking of such metal–acid catalysts [236]. C=O bond chemo-selective hydrogenation or aromatic ring partial saturation is reported, where Pt metal-supported on oxophilic support (TiO_2 or ZrO_2) [152]. The Pt metal-catalyzed deoxygenation follows the tautomerization routes with an oxophilic support instead of the hydrogenation-dehydration routes. Pt metal-based catalysts promote decarboxylation and decarbonylation during the deoxygenation of aldehydes (-CHO), ketones (R_2C=O), alcohols (-OH), and carboxylic acids (-COOH) [128, 162]. Platinum metal catalysts promote C–C bond cleavage to an extent, however, still low as compared with Ni and Ru [31, 135]. The preferential catalyzing impact of Pt on hydrogenation is indicated by the order that proceeds as Pt > Ru > Pd > Ni > Cu [3, 31]. Pt-based catalyst provides exceptional deoxygenation performance but due to its high price, a more affordable non-noble metal may be advised as an alternative [36]. The conversion rates of Pt-based catalysts are reported approximately 23% at 220 °C reaction temperature, but an increase in temperature delivers a higher conversion [87, 89]. However, Pt-based metal catalysts under ideal reaction conditions offer more than 80% conversion, where alkane hydrocarbons are the main product [41, 87, 89, 92].

Pt metal-based catalyst assisted deoxygenation processes involve gas–solid or liquid–solid phases at temperature (373–673 K) and pressure (1–220 bar). During the deoxygenation process, Pt-based catalysts are used to catalyze dehydration, hydrodeoxygenation, and hydrogenation. The deoxygenation reaction of model compounds like anisole, acetophenone, 2- propanol, m-cresol, and methyl isobutyl ketone, using Pt metal-catalyzed in the vapor phase produces respective deoxy-genated products with a wide range of conversion (44–100%) and selectivity (40–100%).The extraordinary case of m-cresol HDO to toluene using Pt/SiO$_2$ resulted in 22% conversion and 40.9% selectivity, due to catalysts deactivation through coking and fouling. The Pt catalyst supported by H-ZSM-5 has achieved the highest methyl isobutyl ketone conversion and 2-methyl pentane selectivity. For example, H-ZSM-5 is the best zeolite support because it has the molecular sieve size and optimal acid sites number. The Pt metal loading effect over Pt/γ-Al$_2$O$_3$ catalyst is studied in the m-cresol HDO to toluene, where an increase in Pt metal loading (0.05–1.70 wt%) is found proficient in enhancing the toluene selectivity (46.4–86.11%). The effects of Pt on various metals are studied in the liquid phase to obtain propylbenzene via HDO of 4-propyl phenol. Pt metal-based bimetallic catalysts supported on zirconia had a fixed Pt and secondary metal ratio of 3. Pt/ZrO$_2$'s catalytic performances are increased by metals like Au, Ir, Re, and W, while catalytic performances are decreased by metals like Fe, Ga, Pd, In, Bi, and Mo. In another study, Pt/Ni-CZO catalyst is synthesized by co-precipitation method and used for deoxygenation of oleic acid [91]. Pt/Ni-CZO delivers 98.7% conversion of oleic acid with 54.7% selectivity and also demonstrates the absolute surface defect acidity and oxygen fraction [91]. 4Pt-8WO$_x$/ZrO$_2$ cata-lyst synthesized via wet impregnation method and studied for deoxygenation of oleic acid. A high catalytic oleic acid deoxygenation is observed with 100% conversion and 89.4% selectivity for hydrocarbon, due to adequate Pt metal dispersion and WO$_x$ reduced species [87] (Table 2).

2.1.3   Nickel Metal-Based Catalysts

One of the most widely studied transition metals in deoxygenation processes is nickel (Ni). Deoxygenation processes based on Ni metal catalysts exist in form of monometallic or bimetallic catalysts either supported or unsupported, however, bimetallic catalysts offer excellent deoxygenation activity due to their synergistic metallic effects. Ni metal-based catalysts for phenolic group's deoxygenation are shown high activity for aromatic rings hydrogenation [93]. Ni metal-based cata-lysts show high hydrogenation activity for hydrogenation of m-cresol to 3 methyl-cyclohexanol; levulinic acid to γ-valerolactone; 1,5 dinitronaphthalene to 1,5-diaminonaphthalene, and dimethyl oxalate to methyl glycolate [31, 32, 76, 151, 226]. In addition, C–C bond hydrogenolysis using nickel metal-based catalyst is reported, and the study shows furan ring-opening or formation of hydrocarbons with shorter chains like methane at a higher temperature [31]. Often decarboxylation, decarbony-lation, and hydrogenation reactions were observed in Ni metal-based catalysts. Ni metal's lower electrophilicity makes it unfavorable for direct scission of C–O and C = O bonds along with the activation as compared to other metals like molybdenum (Mo) [105, 192, 203]. In order to increase its deoxygenation activity, nickel is syner-gized with another metal with electrophilic properties like Mo [157]. Ni metal-based catalysts are shown to produce higher conversion under hydrogen-free and hydrogen-rich conditions. The activity of Ni metal-based catalysts is unaffected by the type of

Table 2 Pt metal-based catalysts

Entry	Metal catalysts	Preparation methods	Reactant	Product	Reaction pathways	Reaction phase	Reaction type	Temp (K)	Pressure (bar)	X (%)	S (%)	References
1	Pt/CsPW	Wet impregnation	Methyl isobutyl ketone	2-Methylpentane	Hydrogenation Dehydration	G-S	Continuous	373	1	100	100	Alharbi et al. [3]
2	Pt/CsPW	Wet impregnation	Acetophenone	Ethylcyclohexane	Hydrogenation Dehydration	G-S	Continuous	373	1	74	98	
3	Pt/Al₂O₃	Dry impregnation	2-Propanol	Propane	Hydrodeoxygenation	G-S	Continuous	523	1.01	99	98	Peng et al. [162]
4	Pt/C	–	Methyl isobutyl ketone	2-Methylpentane	Hydrogenation Dehydration	G-S	Continuous	473	1.01	47	40	Alotaibi et al. [5]
5	Pt/SiO₂	Wet impregnation	Methyl isobutyl ketone	2-Methylpentane	Hydrogenation Dehydration	G-S	Continuous	473	1.01	46	47	
6	Pt/H-ZSM-5	Ion exchange	Methyl isobutyl ketone	2-Methylpentane	Hydrogenation Dehydration	G-S	Continuous	473	1.01	100	83	
7	Pt/HY	Ion exchange	Methyl isobutyl ketone	2-Methylpentane	Hydrogenation Dehydration	G-S	Continuous	473	1.01	81	55	
8	Pt/H-Beta	Ion exchange	Methyl isobutyl ketone	2-Methylpentane	Hydrogenation Dehydration	G-S	Continuous	473	1.01	44	70	
9	Pt/SiO₂	Dry impregnation	m-Cresol	Toluene	Hydrodeoxygenation	G-S	Continuous	523	1.01	97.5	49.8	Chen et al. [31]
10	Pt/γ-Al₂O₃	Wet impregnation	m-Cresol	Toluene	Hydrodeoxygenation	G-S	Continuous	573	1.01	100	86.1	Zanuttini et al. [235]
11	Pt/SiO₂	Dry impregnation	m-Cresol	Toluene	Hydrodeoxygenation	G-S	Continuous	573	1.01	22	40.9	Nie and Resasco [152]

(continued)

Table 2 (continued)

Entry	Metal catalysts	Preparation methods	Reactant	Product	Reaction pathways	Reaction phase	Reaction type	Temp (K)	Pressure (bar)	X (%)	S (%)	References
12	Pt/H-MFI-90	Dry impregnation	Guaiacol	Cyclohexane	Hydrodeoxygenation	L-S	Batch	453	50	100	93	Hellinger et al. [75]
13	Pt/SAPO-11	Dry impregnation	Methyl palmitate	n-$C_{15,16}$ alkanes	Decarbonylation Decarboxylation	L-S	Continuous	648	30	81.3	50.4	Chen et al. [37]
14	Pt/C	–	Glycerol	Propylene glycol	Dehydration	L-S	Batch	473	40	13	79	Maris and Davis [135]
15	Pt/C with 0.8 M CaO	–	Glycerol	Propylene glycol	Dehydration	L-S	Batch	473	40	100	36	Maris and Davis [135]
16	Pt-W/ZrO$_2$	Co-precipitation	4-Propylphenol	Propylbenzene	Hydrodeoxygenation	L-S	Batch	573	20	66	50	Ohta et al.[155]
17	Pd/Ni-Ce$_{0.6}$Zr$_{0.4}$O$_2$	Co-precipitation	Oleic acid	C_9–C_{17} Hydrocarbon	Deoxygenation	L-S	Batch	573	1	98.7	54.2	Jeon et al. [91]
18	4Pt-8WO$_x$/ZrO$_2$	Wet impregnation	Oleic acid	Hydrocarbons	Deoxygenation	L-S	Batch	533	20	100	89.4	Janampelli and Darbha [87]
19	4Pt/ZrO$_2$	Hydrothermal Deposition	Oleic acid	Heptadecane	Decarboxylation Decarbonylation	L-S	Batch	493	20	23	58.9	Janampelli and Darbha [89]
20	1Pt/ Al$_2$O$_3$	Wet impregnation	Palm oil	C_{15} + C_{17}	Hydrodeoxygenation	G-S	Continuous	653	40	100	68	Jeong et al. [92]
21	Pt/SAPO-11	Wet impregnation	Castor oil	C_8-C_{16} alkanes	Deoxygenation	G-S	Continuous	673	40	84.82	49.96	Chen et al. [41]

reactant. Therefore, Ni metal-based catalysts are found to be suitable for different types of reactants like palm fatty acid, waste cooking oil, and triglyceride in the production of green diesel. Diesel range alkane hydrocarbons are produced apart from achieving a high conversion rate [21, 124, 129, 183]. NiO catalysts offer better conversion and alkane selectivity than the rest of the metal oxide catalysts [43, 88].

Ni metal-based catalysts deoxygenation reaction is performed in gas–solid or liquid–solid phases with a temperature (483–623 K) and a pressure (1.01–170 bar). Ni metal-based catalysts follow hydrogenation, dehydrogenation, hydrodeoxygenation, decarbonylation, and decarboxylation reaction routes during deoxygenation processes. Ni metal-based catalysts application in deoxygenation of compounds such as anisole, dibenzofuran, guaiacol, m-cresol, 2-methyl tetrahydrofuran, phenol, palmitic acid, ethyl heptanoate, stearic acid, and methyl laurate are potentially viable to deoxygenate absolutely with 21.7–100% conversion rate and 27.9–99% selectivity. Ni-based catalysts for anisole have good deoxygenation activity. Ni/SiO$_2$ catalyst is preferred for anisole deoxygenation to obtain cyclohexane in the liquid phase at 30 bar pressure and 483 K temperature due to its high efficiency. In the case of aromatics products, e.g., benzene, toluene, and xylene, supported Ni/CeO$_2$ catalysts are used for anisole deoxygenation in the vapor phase at 3 bar pressure and 563 K temperature. Ni-based (Ni$_2$P/SiO$_2$) catalyst was chosen due to 99% conversion rate and 99% selectivity for methyl laurate deoxygenation to n-dodecane and n-undecane. Ni metal in oxide form supported on zeolite Y (NiO$_x$/zeolite Y catalyst)) are efficient to make green diesel through deoxygenation process without H$_2$ and solvent [43]. NiO$_x$/zeolite Y catalyst exhibits excellent deoxygenation activity and selectivity for desired product (diesel 93% selectivity). A low-cost palm oil feedstock obtained from refining process was converted to advanced biofuel by deoxygenation process, where Ni/SBA-15 catalyts produced 85.8% of liquid hydrocarbons with more than 85% selectivity [98]. A semi-batch reactor is used for the catalytical analysis of NiO–ZnO catalysts for the deoxygenation process of palm fatty acid distillate. The catalytic activity of the NiO–ZnO catalyst for the substrate deoxygenation process is significantly improved due to the NiO and ZnO synergistic effect [21]. Linear hydrocarbons in the form of green diesel compounds with 83.4% yield and 86.0% selectivity for C$_{11}$–C$_{17}$ were achieved [21] (Table 3).

2.1.4 Cobalt Metal-Based Catalyst

Inexpensive metal-based catalysts include metals like Co, Ni, and Mo. Co metal-based catalysts in the presence of a hydrogen environment provides greater than 90% conversion and are found highly efficient in removing oxygen atoms from Oleic acid and Macauba oil [47]. The hydrocarbon yield is found less than 40% with less than 60% conversion rate without use of any external hydrogen gas [43, 67, 98]. Co metal-based catalysts with multi-metal compositions like CoMoS catalyst are shown to have good catalytic activity for oxygen-containing compounds hydrodeoxygenation and sulfur-containing compounds hydrodesulfurization [65]. CoMoS catalyst is studied for p-cresol, 2-ethylphenol, phenol, guaiacol, canola oil, and pyrolysis fuel

Table 3 Ni metal-based catalysts

Entry	Metal catalysts	Preparation methods	Reactant	Product	Reaction Pathways	Reaction phase	Reaction type	Temp (K)	Pressure (bar)	X (%)	S (%)	References
1	Ni/C	Wet impregnation	Anisole	Aromatics	Hydrodeoxygenation	G-S	Continuous	563	3	98	50	Yang et al. [228]
2	Ni/Al-SBA-15	Wet impregnation	Anisole	Cyclohexane	Hydrodeoxygenation Hydrogenation	G-S	Continuous	563	3	100	60	
3	Ni/SBA-15	Wet impregnation	Anisole	Cyclohexane	Hydrodeoxygenation Hydrogenation	G-S	Continuous	563	3	100	70	
4	Ni/SiO$_2$	Dry impregnation	m-Cresol	Toluene	Hydrodeoxygenation	G-S	Continuous	523	1.01	100	71.5	Chen et al. [31]
5	Ni-Mo-S	Hydrothermal	Phenol	Cyclohexane	Hydrodeoxygenation	L-S	Batch	623	28	96.2	61.8	Yoosuk et al. [231]
6	Ni-Mo-S	Thermal decomposition	Ethylheptanoate	Hexane	Decarboxylation	G-S	Continuous	523	15	21.7	27.9	Ruinart de Brimont et al. [177]
7	Ni$_2$P/SiO$_2$	Dry impregnation	Dibenzofuran	Bicyclohexane	Hydrodeoxygenation	L-S	Continuous	573	30	90	80	Cecilia et al. [29]
8	Ni/HZSM	Dry impregnation	Stearic acid	Heptadecane	Dehydrogenation Decarbonylation	L-S	Batch	543	13.5	40	39	Kumar et al. [108]
9	Ni/AC	Dry impregnation	Anisole	Cyclohexylmethyl ether	Hydrogenation	L-S	Batch	483	30	100	74	Jin et al. [93]
10	Ni$_2$P/SiO$_2$	Dry impregnation	Methyl laurate	n-C$_{11,12}$ alkanes	Hydrodeoxygenation Decarbonylation	L-S	Continuous	613	20	99	99	Shi et al. [192]
11	Ni/γ-Al$_2$O$_3$	Dry impregnation	Anisole	Cyclohexylmethyl ether	Hydrogenation	L-S	Batch	483	30	100	57	Jin et al. [93]
12	Ni/CeO$_2$	Wet impregnation	Anisole	Aromatics	Hydrodeoxygenation	G-S	Continuous	563	3	90	55	Yang et al. [228]

(continued)

Table 3 (continued)

Entry	Metal catalysts	Preparation methods	Reactant	Product	Reaction Pathways	Reaction phase	Reaction type	Temp (K)	Pressure (bar)	X (%)	S (%)	References
13	NiO$_x$/zeolite Y	Wet impregnation	Triolein	Hydrocarbons	Deoxygenation	L-S	Batch	653	–	76.21	93	Choo et al. [43]
14	Ni/D$_S$D$_A$-β-zeolite	Deposition	Methyl Palmitate	Alkanes	Hydroprocessing	L-S	Batch	513	25	77	61.9	Papanikolaou et al. [159]
15	Ni/SBA-15	Wet impregnation	Palm Fatty acid distillate	Hydrocarbons	Decarbonylation Decarboxylation	L-S	Semi-batch	623	–	88.1	85	Kamaruzaman et al. [98]
16	Ni/γ-Al$_2$O$_3$	Wet impregnation	Rubber seed oil	Hydrocarbons	Hydrodeoxygenation Decarboxylation	G-S	Continuous	623	35	99	45	Ameen et al. [10]
17	NiO/ZnO	Wet impregnation	Palm Fatty acid distillate	Hydrocarbons	Deoxygenation	L-S	Semi-batch	623	–	83.5	77.8	Baharudin et al. [21]
18	Ni/palygorskite	Deposition–precipitation	Waste cooking oil	Hydrocarbons	Deoxygenation	L-S	Semi-batch	583	40	100	82	Lycourghiotis et al. [129]
19	Ni/SAPO-11	Incipient impregnation	Stearic acid	Hydrocarbons	Deoxygenation	L-S	Batch	563	40	97	91	Liu et al. [124]

oil hydrodeoxygenation [20, 139, 213, 214, 221, 227, 239]. Many studies reveal that Co metal affects MoS_2 catalyst properties significantly. For example, Co metal 4.2 wt% doping on MoS_2 catalyst significantly increases overall HDO activity of 2-ethylphenol and phenol, apart from that higher selectivity is also noted for aromatics [20]. p-cresol HDO results show an increase in the DDO activity and enhanced selectivity for toluene with the addition of Co metal [221]. Co metal promoting effects on MoS_2 catalyst cause metal-sulfur deficient bond which enables the facile sulfur vacancies formation [20]. Saturated sulfur sites are activated in the CoMoS catalyst by Co metal to promote hydrogen adsorption and scission of C–C bonds for the stearic acid decarbonylation into formic acid, n-heptadecene, and n-heptadecane [239]. However, excessive Co metal addition causes CoS_2 species aggregation on MoS_2 active sites, resulting in a reduced catalytic activity [221].

Monometallic Co metal catalysts like Co/Al-MCM-41, Co/SiO_2, Co/γ-Al_2O_3 significantly increase deoxygenation activity [139, 203, 211]. C_{18} fatty acids deoxygenation over Co/γ-Al_2O_3 follows decarbonylation and hydrodeoxygenation resulting in 43.0% and 45.8% forming hydrocarbons n-C_{15} to n-C_{18} as primary products [203]. Guaiacol deoxygenation is observed superior with Co/SiO_2 catalysts than CoMo/Al_2O_3 catalysts in terms of conversion of guaiacol and HDO selectivity [139]. Co metal-based catalysts deoxygenation processes are performed in gas–solid and liquid–solid phases with temperature (548–723 K) and pressure (1.0–70 bar). The deoxygenation processes over Co metal-based catalysts follow decarboxylation, hydrodeoxygenation, and trans alkylation as the main reaction routes. Co metal-based catalysts are found suitable for deoxygenation of phenolic compounds like p-cresol, guaiacol, 2-ethylphenol, and phenol with high conversion (84–100%) and excellent selectivity (20–92.2%). Nevertheless, Co metal-based monometallic catalysts like Co/SiO_2 and Co/γ-Al_2O_3 are able to produce deoxygenated products with high yield. Endocarp activated carbon as support was used to synthesize Co metal-based catalyst (Co/endocarp AC) and further applied to evaluate the Macauba acid oil deoxygenation process [47]. The deoxygenation process with highest degree of 98.4% wt. of deoxygenated products like jet fuel, gasoline, and mainly diesel was achieved [47]. Palm fatty acid distillate is used for the production of green diesel by catalytic deoxygenation in an H_2 free environment using a mesoporous activated carbon-supported Co-Mo catalyst [67]. The bimetallic catalyst ($Co_{10}Mo_{10}$/AC) has displayed excellent catalytic performance in terms of hydrocarbon production with 92% C_8–C_{20} yield and 89% n-($C_{15} + C_{17}$) selectivity [67]. Moreover, $Co_{10}Mo_{10}$/AC catalysts are shown to maintain excellent stability until the sixth run during the deoxygenation process with >80% n-($C_{15} + C_{17}$) yield and selectivity [67] (Table 4).

2.1.5 Molybdenum Metal-Based Catalyst

Molybdenum (Mo) metal-based catalysts are utilized in the production and processing of fuels. Mo metal-based catalysts like Mo_2C, MoO_3, MoS_2, Co–Mo–S, and Ni–Mo–S are reported for high activity in the deoxygenation process. Mo_2C catalysts are studied on the deoxygenation of several organic compounds like anisole,

Table 4 Co metal-based catalysts

Entry	Metal catalysts	Preparation methods	Reactant	Product	Reaction pathways	Reaction phase	Reaction type	Temp (K)	Pressure (bar)	X (%)	S (%)	References
1	Co/SiO$_2$	Dry impregnation	Guaiacol	Benzene	Hydrodeoxygenation	L-S	Batch	573	10	100	53.1	Mochizuki et al. [139]
2	Co/γ-Al$_2$O$_3$	Dry impregnation	Palm oil	C$_{15}$–C$_{18}$ alkanes	Hydrodeoxygenation Decarboxylation	L-S	Continuous	573	50	92.2	45.8/ 43	Srifa et al. [203]
3	Co–Mo–S	Hydrothermal	p-Cresol	Toluene	Hydrodeoxygenation	L-S	Batch	548	40	100	92.2	Wang et al. [214]
4	CoMoP/MgO	Dry impregnation	Phenol	Benzene	Hydrodeoxygenation	L-S	Batch	723	50	89.4	74.8	Yang et al. [227]
5	CoMo/TiO$_2$	Equilibrium-deposition-filtration	Phenol	Cyclohexane	Hydrodeoxygenation	L-S	Continuous	623	15	84	–	Plataniitis et al. [168]
6	Co/Al-MCM-41	Wet impregnation	Guaiacol	Benzene	Hydrodeoxygenation Transalkylation	G-S	Continuous	673	1.01	100	20	Tran et al. [211]
7	Co/ endocarp AC	Impregnation	Oleic acid	Hydrocarbons	Deoxygenation	L-S	Batch	523	30	97.38	96.68	de Barros Dias Moreira et al. [47]
8	Co/ endocarp AC	Impregnation	Macauba acid oil	Hydrocarbons	Deoxygenation	L-S	Batch	523	30	96.25	96.68	de Barros Dias Moreira et al. [47]
9	CoO$_x$/ zeolite Y	Wet impregnation	Triolein	Hydrocarbons	Deoxygenation	L-S	Batch	653	–	57.75	76	Choo [43]
10	Co/SBA-15	Wet impregnation	Palm Fatty acid distillate	Hydrocarbons	Deoxygenation	L-S	Semi-batch	623	–	16.3	48	Kamaruzaman et al. [98]
11	Co/AC	Impregnation	Palm Fatty acid distillate	Hydrocarbons	Deoxygenation	L-S	Semi-batch	623	–	38	45	Gamal et al. [67]

ethyl formate, furfural, glycolaldehyde, phenol, propanol, propanal, 1-octanol, and lignocellulosic biomass [26, 40, 69, 117, 136, 141, 225]. Selective deoxygenation is accomplished with Mo_2C catalyst without the hydrogenation of aromatic or furanic rings. Mo_2C catalyst is shown to facilitate oxygenated compounds $\eta_2(C, O)$ adsorption, causing an apparent segmentation of the intermediate C–O or C=O bond [136, 225]. The preferential electronegative oxygen adsorption and C= O bond activation are attributed to the electrons inferior density and the high electrophilicity nature of Mo [157]. Moreover, this enables facile esters hydrogenolysis like ethyl formate [40]. Decarboxylation and decarbonylation are the minor routes since the C–C bond cleavage is averse due to the deficiency of acidic sites within the Mo_2C catalyst [117]. However, Mo metal-based catalyst supported on carbide gets deactivated in the proximity of water as water oxidizes Mo_2C to MoO_2 causing inferior activity [141]. Co–Mo–S, Ni–Mo–S, and MoS_2 types of catalysts have sulfur vacancy sites and MoS_2 active phase, causing C–O bond direct scission in the deoxygenation process. A high yield of deoxygenated products is reported for organic compounds like ethyl heptanoate, 2-ethylphenol, methyl stearate methyl heptanoate, and phenol in the deoxygenated process with sulfided Mo metal-based catalysts [20, 172, 177, 186, 231]. HDO activities is enhanced by reforming the property of the active sites using promoters like cobalt and nickel [20, 231]. The Ni-Mo synergy offers overall higher deoxygenation activity when a sulfided bimetallic (NiMo) catalyst is involved. However, a Mo monometallic sulfided catalyst delivers higher selectivity for hydrodeoxygenation reaction instead of decarbonylation and decarboxylation [105]. Mo metal-based sulfided catalysts are popular for uninterrupted sulfur replenishment via external sulfiding agents like CS_2 or H_2S to preserve catalytical activity due to the sulfur absence in bio-oils [139].

Mo oxide catalysts are studied for compounds like anisole, acrolein, phenol, palmitic acid, and tetrahydro furfuryl alcohol in the deoxygenation process [26, 52, 138, 197, 222]. Mo oxide catalysts often found in the MoO_3 phase can be converted into the MoO_2 phase by reduction which causes the lower activity [169]. The deoxygenation performance is affected by the strength of the metal oxide bond. It is observed that weak metal–oxygen bonds in metals are unapt to abstract oxygen effectively from the oxygenated products while metal with sturdy metal–oxygen bond properties strongly binds oxygen to the surface of the catalyst and prevents the emergence of vacant sites of oxygen [138]. Mo metal-based catalysts deliver a high conversion rate but suffer from lower selectivity of alkane hydrocarbons as compared to Ni-based catalysts [10, 67, 88, 208]. Moreover, metals in the zero oxidation state give excellent production in green diesel production than metal oxide. Table 5, deoxygenation processes were performed in gas–solid and liquid–solid phases with pressure (25–100 bar) and temperature (408–673 K) over Mo metal-based catalysts. Dehydration, hydrodeoxygenation, hydrogenation, hydrogenolysis, and demethoxylation were the major reaction routes catalyzed with Mo metal-based catalysts in the deoxygenation activity. Mo metal-based catalysts are used for organic compounds like anisole, methyl stearate, octanol, phenol, and palmitic acid in the deoxygenation process. The respective products were achieved with a conversion rate (42–100%) and selectivity (45.2–93%). Some Mo metal-based catalysts like Mo_2C and Mo_2N which

Table 5 Mo metal-based catalysts

Entry	Metal catalysts	Preparation methods	Reactant	Product	Reaction pathways	Reaction phase	Reaction type	Temp (K)	Pressure (bar)	X (%)	S (%)	References
1	NiMoO$_x$-SiO$_2$	Sol–gel	Anisole	Cyclohexane	Demethoxylation Hydrogenation	L–S	Batch	573	60	100	90	Smirnov et al. [197]
2	Mo$_2$C/ZrO$_2$	Dry impregnation, Carburization	Phenol	Benzene	Hydrodeoxygenation	L–S	Continuous	633	100	42	45.2	Mortensen et al. [141]
3	Mo$_2$C/TiO$_2$	Dry impregnation	Phenol	Benzene	Hydrodeoxygenation	L–S	Continuous	673	25	90	90	Boullosa-Eiras et al. [26]
4	Mo$_2$N	Temperature programmed reaction	Ethyl formate	Ethanol	Hydrogenolysis	L–S	Batch	408	30	–	51	Chen et al. [40]
5	NiMo/SAPO-11	Impregnation	Methyl stearate	n-Octadecane	Hydrodeoxygenation	L–S	Continuous	573	30	100	55.25	Qian et al. [172]
6	NiMo/Al$_2$O$_3$	Impregnation	Methyl stearate	n-Octadecane	Hydrodeoxygenation	L–S	Continuous	573	30	100	70.37	
7	NiMo/SBA-15	Impregnation	Methyl stearate	n-Octadecane	Hydrodeoxygenation	L–S	Continuous	573	30	100	70.08	
8	NiMo/Al-SAPO-11	Impregnation	Methyl stearate	n-Octadecane	Hydrodeoxygenation	L–S	Continuous	573	30	100	71.75	
9	MoO$_2$/CNTs	Dry impregnation	Palmitic acid	Hexadecane	Hydrodeoxygenation	L–S	Batch	493	40	100	92.2	Ding et al. [52]
10	Mo/γ-Al$_2$O$_3$	Wet impregnation	Rubber seed oil	Hydrocarbons	Deoxygenation Hydrodeoxygenation	G–S	Continuous	623	35	98	48	Ameen et al. [10]

(continued)

Table 5 (continued)

Entry	Metal catalysts	Preparation methods	Reactant	Product	Reaction pathways	Reaction phase	Reaction type	Temp (K)	Pressure (bar)	X (%)	S (%)	References
11	Mo/γ-Al$_2$O$_3$	Impregnation	Oleic acid	Hydrocarbons	Deoxygenation	L-S	Batch	573	40	91.39	42.22	Thongkumkoon et al. [208]
12	Mo/AC	Impregnation	Palm Fatty acid distillate	Hydrocarbons	Deoxygenation	L-S	Semi-batch	623	–	41	39	Gamal et al. [67]
13	MoO$_x$/ZrO$_2$	Wet impregnation	Oleic acid	Hydrocarbon	Deoxygenation	L-S	Batch	493	20	12	10	Janampelli and Darbha [88]

were synthesized via a temperature-programmed technique had good selectivity to deliver fractional deoxygenated products. Hydrodeoxygenation of rubber seed oil into hydrocarbons of diesel range over Mo/γ-Al$_2$O$_3$ catalysts was studied [10]. Mo metal-based catalysts produced 15.7 wt.% of C$_{16}$ and 30.7 wt.% of C$_{18}$ hydrocarbons as compared to Ni metal-based catalysts. Mo/γ-Al$_2$O$_3$ catalysts exhibit notable hydrodeoxygenation activity with less oxygenated components concentration in the mixture, however, Mo-supported catalysts suffer from high coke deposition [149]. In another study over Mo/γ-Al$_2$O$_3$ catalysts for oleic acid deoxygenation, 91.39% conversion and 42.22% selectivity for C$_{13}$–C$_{18}$ is achieved [208].

2.1.6 Copper Metal-Based Catalyst

Several Cu metal-based catalysts are reported in the domain of deoxygenation reactions. Cu metal-based catalysts applied in the glycerol hydrogenolysis are Ru-Cu/ZrO$_2$, Ru-Cu/Al$_2$O$_3$, and Cu–ZrO$_2$ [57, 199]. Cu metal crystallites offer active sites in the glycerol hydrogenolysis for the selective synthesis of 1,2-propanediol but in the absence of Cu, ZrO$_2$ yielded a negligible 1,2-propanediol [57]. Cu metal-based bimetallic catalysts show synergy with other metal elements in the glycerol hydrogenolysis. The small Cu crystallites formation is beneficial to grow hydrogenolysis activity and ZrO$_2$ presence is favored. Transition metal Cu has a totally filled and parochial d-band that is possible to be used as a stabilizer in the Cu-Ru catalyst for the reduced metallic particles [199]. The 1,2 propanediol synthesis from glycerol without C–C bonds cleavage requires stabilized reduced particles of metal.

Cu metal-based catalysts like Cu/ZnO/TiO and Cu/ZnO/ZrO$_2$ are studied for the catalytic transformation of maleic anhydride into tetrahydrofuran and γ-butyrolactone [83, 238, 240]. The maleic anhydride selective hydrogenation to γ-butyrolactone is favored by ZnO whereas the γ-butyrolactone deep hydrogenation to tetrahydrofuran is favored by Cu and ZrO$_2$ species [238]. Cu metal-based catalysts loaded with external phases like TiO, ZnO, and ZrO$_2$ enable the formation of smaller crystallites of copper [83]. The high dispersion of the Cu phase is found feasible with ZnO species in Cu–Zn/Al$_2$O$_3$ catalyst for ester hydrogenation to alcohols [232]. Similarly, cerium addition is found to enhance the Cu metal dispersion, reduce Cu metal aggregation, and improve the Cu$^+$ species surface enrichment for methyl acetate hydrogenation to alcohol [230]. Synergic effects play an important role in the hydrogenation activity of Cu metal, which increases with the reduction of crystallite size of Cu metal and increase in CuO structure lattice distortion degree [240]. Typically, Cu^0 and Cu$^+$ species are crucial phases of Cu metal-based catalysts for hydrogenation activity [51, 173]. Moreover, Cu species hydrogenation activity is increased via keeping a synergistic metal like Fe to interact with Cu electronically [73]. Co doping promotes micropores genesis in β-zeolite, however, an excess of doped copper causes Cu aggregates formation on the zeolite surface and blocks zeolite pores [223].

Plenty of Cu metal-based catalysts are reported for the deoxygenation process yielding congruent result. Bicyclohexyl-1-ene synthesis via hydrodeoxygenation of dibenzofuran over NiCu/CeO$_2$, NiCu/ZrO$_2$, and NiCu/TiO$_2$ catalysts are studied with product selectivity 83%, 95%, and 61% respectively [9]. Aromatic ketones like 9,10-anthraquinone, 9-fluorenone, benzhydrol, benzophenone, and tetralone were promptly deoxygenated over Cu/SiO$_2$ catalyst in presence of toluene at 90 °C temperature and 1 atm pressure of H$_2$ into deoxygenated polyaromatics with the conversion of >84%. Generally, 1,3-propanediol deoxygenation over Cu/ZnO/Al$_2$O$_3$ occurs via decarbonylation, decarboxylation, and hydrogenolysis of an equilibrium mixture of propanol-propanal through Cu [178]. Table 6, Cu metal-based catalysts deoxygenation processes are performed in gas–solid and liquid–solid phases with temperature (363–563 K) and pressure (0.8–100 bar). Generally, decarboxylation, decarbonylation, dehydrogenation, hydrogenation, hydrodeoxygenation, and hydrogenolysis are reaction pathways during deoxygenation processes catalyzed by Cu metal-based catalysts. Cu metal-based catalysts tested for organic compounds like dibenzofuran, glycerol, furfural, 1,3-propanediol, aromatic ketones, methyl esters, and maleic anhydride leading to respective deoxygenated products with conversion rate (11.5–100%) and selectivity (10–100%). Cu/SiO$_2$ is an ergastic deoxygenation catalyst for aromatic ketones with more than 80% conversion and selectivity [233].

2.1.7 Ruthenium Metal-Based Catalyst

Like other transition metals, ruthenium (Ru) is well-known for its hydrogenation activity. The lower price of Ru makes it more economical to be exploited in large-scale processes. Hydrogenation based on Ru metal-catalysts is employed as a precursor step in the reduction of complex oxygenated compounds derived from biomass into simple deoxygenation. Hydrogenations of compounds like methyl oleate, methyl laurate, glucose, levulinic acid, lactic acid, and succinic acid have been studied over Ru metal-based catalysts [44, 59, 70, 90, 99, 109, 176, 215]. Deep hydrogenation products from levulinic acid are achieved with the help of Ru supported on graphene oxide [215]. Ru-based catalysts are not much studied as compared to Pt and Pd catalysts, although it offers a conversion rate of more than 80%. However, Ru-based catalysts underperformed in oxygen removal as compared to Pt and Pd catalysts, which causes oxygenated products (alkane hydrocarbon) with lower selectivity [4, 91].

Usually, Ru-catalyzed hydrogenation is perceived together with hydrogenolysis [62]. For example, sugar compounds like glycerol deoxygenation are often required Ru metal-based catalysts. Ru metal-based catalysts incorporate crystalline nanoparticles of Ru^0 and the amorphous phase of Ru(O)$_x{}^{H+}$ which is believed to be essential for the protonic acidic sites formation to catalyze hydrogenolysis [113]. However, Ru metal-based catalyst is effective in C–C bond cleavage which causes a significant carbon length reduction in hydrocarbons during the deoxygenation activity [135]. Hence, metals like Cu that suppress cleavage of C–C bond are doped into Ru metal-based catalysts to enhance its selectivity for desired deoxygenated products [199]. Deoxygenation routes like dehydroxylation and decarbonylation are reported for Ru metal-catalyzed phenolics hydrodeoxygenation [127]. DFT study on Ru metal-based catalysts reveals that the high CAR-O bond energy is the reason why the direct erosion

Table 6 Cu metal-based catalysts

Entry	Metal catalysts	Preparation methods	Reactant	Product	Reaction pathways	Reaction phase	Reaction type	Temp (K)	Pressure (bar)	X (%)	S (%)	References
1	Cu–ZrO$_2$	Co-precipitation	Glycerol	1,2-Propanediol	Hydrogenolysis	L-S	Batch	473	40	11.5	95.7	Durán-Martín et al. [57]
2	Cu/SiO$_2$	Dry impregnation	Furfural	2-Methylfuran	Hydrodeoxygenation Hydrogenation	G-S	Continuous	563	1.01	70	14.2	Sitthisa et al. [196]
3	Cu/SiO$_2$	Deposition	2-Acetonaphthone	2-Ethylnaphthalene	Deoxygenation	L-S	Batch	363	1.01	100	100	Zaccheria et al. [233]
4	Cu/SiO$_2$	Deposition	9,10-Anthraquinone	9,10-Dihydroanthracene	Deoxygenation	L-S	Batch	363	1.01	100	100	
5	Cu/SiO$_2$	Deposition	9-Fluorenone	Fluorene	Deoxygenation	L-S	Batch	363	1.01	100	100	
6	NiCu/ZrO$_2$	Wet impregnation	Dibenzofuran	Bicyclohexyl	Hydrodeoxygenation	L-S	Batch	523	100	65	10	Ambursa et al. [9]
7	NiCu/TiO$_2$	Wet impregnation	Dibenzofuran	Bicyclohexyl	Hydrodeoxygenation	L-S	Batch	523	100	96	38	
8	CuPd/rGO	In-situ reduction	Glycerol	Lactic acid	Dehydrogenation	L-S	Batch	413	14	56.2	88.1	Jin et al. [94]
9	Cu/ZnO/Al$_2$O$_3$	Co-precipitation	1,3-Propanediol	Propanol	Decarbonylation Decarboxylation	G-S	Continuous	503	0.8	100	83	Sad et al. [178]
10	Cu–Zn/Al$_2$O$_3$	Co-precipitation	Methyl esters	Alcohols	Hydrogenolysis Hydrogenation	L-S	Batch	513	100	84	–	Yuan et al. [232]

of the C–O bond is very unlikely [127]. However, the Ru addition onto TiO$_2$ support (amphoteric) is capable of reducing C–O bond activation energy, which makes the DDO routes to be more congruent as compared to the HYD routes [149]. Table 7, Ru metal-based catalysts are studied in the deoxygenation processes with temperature (373–473 K) and pressure (1–64 bar) in gas–solid and liquid–solid phases. The major reaction routes using Ru metal-based catalysts in the deoxygenation process were hydrogenolysis, decarbonylation, dehydration, and dehydrogenation. Methyl isobutyl ketone vapor-phase deoxygenation using Ru/CsPW-I catalyst leads to 96% methyl isobutyl ketone conversion and 100% 2-methyl pentane selectivity at 373 K temperature and 1 bar pressure.

2.1.8   Rhenium Metal-Based Catalyst

Rhenium (Re) is an expensive transition metal and thus unlikely to be viable for industrial processes, and consequently got very little attention in deoxygenation studies so far [40]. However, Re-based bimetallic catalysts have also few favorable aspects like isomerization of intermediates and reduction of metal crystallite size, which is beneficial in facile conversion [55, 155]. Hydrogenation activity is reported in its synergistic application with other metals like Cu and Ru for the succinic acid hydrogenation to 1,4-butanediol [99, 100]. Various reports indicate that Re metal-based catalysts led deoxygenation often stimulate the hydrogenolysis of oxygenated compounds like tetrahydropyran-2-methanol, tetrahydrofurfuryl alcohol, and glycerol [33, 34, 148]. The active sites for deoxygenation in Re metal-based catalysts are ReS$_2$ or ReO$_x$ [118]. Principally ReS$_2$ sites follow demethoxylation pathway which is essential for methoxy group-containing compound deoxygenation, whereas ReO$_x$ sites lead to the C–O bond hydrogenolysis of an adsorbed substrate [188]. Adding acids to Re-based catalysts is reported to increase the glycerol activity on the ReO$_x$ surface cluster [147]. The acid protonation effect assists to increase hydroxorhenium sites number that is required for the glycerol activation. Usually, other hydrogenating metals can be used with ReO$_x$ species to enhance hydride attack synergistically [8]. Table 8, Re metal-based catalysts are tested for the deoxygenation processes in liquid–solid phases at different temperatures (393–573 K) and pressure (50–80 bar). The deoxygenation process over Re metal-based catalyst is believed to follow hydrogenation, hydrogenolysis, hydrodeoxygenation, and demethoxylation as major reaction pathways. Guaiacol hydrodeoxygenation over ReO$_x$-based catalyst produces a full deoxygenation product (cyclohexane). ReO$_x$ catalysts modified with iridium deliver partial deoxygenation product with high selectivity via hydrogenolysis, however suffer from low conversion rate.

2.1.9   Cesium Metal-Based Catalyst

Alkali metal Cesium (Cs) is very soft in nature and hard to find in a stable elemental form. The Cs metal-based catalysts application is limited for deoxygenation process.

Table 7 Ru metal-based catalysts

Entry	Metal catalysts	Preparation methods	Reactant	Product	Reaction pathways	Reaction phase	Reaction type	Temp (K)	Pressure (bar)	X (%)	S (%)	References
1	Ru/ZrO$_2$	Wet impregnation	Propionic acid	Ethane	Dehydrogenation Decarbonylation	L-S	Continuous	463	64	80.3	44.2	Chen et al. [35]
2	Ru/HUSY	Wet impregnation	Stearic acid	Hydrocarbons Esters	Hydrodeoxygenation	L-S	Batch	433	50	100	91	Ali and Zhao [4]
3	Ru/AC	Wet impregnation	Stearic acid	Hydrocarbons	Hydrodeoxygenation	L-S	Batch	433	50	100	95	
4	Ru/SiO$_2$	Wet impregnation	Stearic acid	Hydrocarbons	Hydrodeoxygenation	L-S	Batch	433	50	100	92	
5	Ru/Ni–Ce$_{0.6}$Zr$_{0.4}$O$_2$	Co-precipitation	Oleic acid	Hydrocarbons	Deoxygenation	L-S	Batch	573	1	88.3	19.9	Jeon et al. [91]

Table 8 Re metal-based catalysts

Entry	Metal catalysts	Preparation methods	Reactant	Product	Reaction pathways	Reaction phase	Reaction type	Temp (K)	Pressure (bar)	X (%)	S (%)	References
1	Ir–ReO_x/SiO_2	Impregnation	1-Propanol	Propane	Hydrogenolysis	L-S	Batch	393	80	14.9	99	Nakagawa et al. [148]
2	ReS_2/C	Dry impregnation	Guaiacol	Phenol	Demethoxylation	L-S	Batch	573	50	75	86.7	Sepúlveda et al. [188]
3	ReO_x/SiO_2	Dry impregnation	Guaiacol	Cyclohexane	Hydrodeoxygenation Hydrogenation	L-S	Batch	573	50	81	54.3	Leiva et al. [118]
4	ReO_x/SiO_2	Impregnation	1,2-Propane diol	1-Propanol	Hydrogenolysis	L-S	Batch	393	80	71.7	85	Nakagawa et al. [148]
5	ReO_x/SiO_2	Impregnation	Glycerol	1,3-Propane diol	Hydrogenolysis	L-S	Batch	393	80	62.8	49	
6	ReO_x/SiO_2	Impregnation	2,3-Butane diol	2-Butanol	Hydrogenolysis	L-S	Batch	393	80	18.2	81	

Hence, Cs metal-based monometallic catalyst is unlikely to be utilized in the deoxygenation process. However, Cs metal supported on zeolite (CsNaX) has been studied for the methyl octanoate and benzaldehyde deoxygenation process [163, 201]. Cs metal supported on NaX zeolite is synthesized via the cationic exchange between $CsNO_3$ and NaX zeolite solution [45]. Cs species function as the active sites for both methyl octanoate and benzaldehyde deoxygenation, which further increases catalysts basicity [163, 201]. A significant drop in yield of the desired product is observed over NaX zeolite catalyst without Cs element, leading to severe coking. Benzaldehyde deoxygenation over Cs metal-based catalysts to benzene involves Cs cation interaction with aromatic ring electrons to debilitate the benzylic C–C bond of adsorbed intermediate (benzoate) [163]. Hence, the benzylic C–C bond weakening causes a direct decarbonylation to produce CO and benzene. Benzoate intermediates coupling reaction is noticed as a side reaction that becomes cardinal without Cs metal presence to form condensation products. In case of methyl octanoate deoxygenation, n-hept-1-ene is synthesized via elimination of β-hydrogen and decarbonylation whereas n-hex-1-ene is produced via structural rearrangements and deacetalation. Cs metal presence increases both deacetalation and decarbonylation activity, indicating the importance of Cs species in catalytic reactions [45, 201]. Cs metal-based catalysts deoxygenation processes are performed in gas–solid phases with temperatures from 698 to 748 K at atmospheric pressure. Deacetalation and decarbonylation are principal reaction pathways during the deoxygenation process catalyzed by Cs metal-based catalysts. CsNaX zeolite-supported types of catalyst synthesized via the ion exchange method are found capable of completely deoxygenating ideal compounds like methyl octanoate and benzaldehyde to deoxygenated products with 62.3%–100% conversion rate and 41.9%–80% selectivity, respectively (Table 9).

2.1.10 Magnesium Metal-Based Catalyst

Alkaline earth metal Magnesium (Mg) often creates basic oxide like magnesium oxide and Mg metal-based catalyst utilized in heterogeneous catalysis. The study on biodiesel production via transesterification over Mg metal-based catalysts (MgO and Mg–Al hydrotalcite) was found to be effective. Also, several studies for deoxygenation processes were done to investigate Mg metal-based catalysts catalytic performances. Generally, acid-supported catalysts have a higher tendency to coking; therefore, solid basic catalysts resistive to coking were utilized in the deoxygenation process [58]. However, acidic site's absence may cause hydroxyl groups dehydration during the deoxygenation process [3]. A comparative study is performed with CsNaX, MgO, and NaX catalysts for the methyl octanoate deoxygenation process. Methyl octanoate decarbonylation over MgO catalyst delivers an 8% methyl octanoate conversion and 8% hexene selectivity. MgO catalyst has no polarized environment and very low surface area, which causes poor performance as compared to other catalysts. Highly polarized environment absence in MgO catalysts makes it unable for methyl octanoate adsorption and decomposition [201]. However, MgO catalyst is capable to activate coupling reactions (aldol condensation) [175, 201].

Table 9 Cs metal-based catalysts

Entry	Metal catalysts	Preparation methods	Reactant	Product	Reaction pathways	Reaction phase	Reaction type	Temp (K)	Pressure (bar)	X (%)	S (%)	References
1	CsNaX zeolite	Ion exchange	Methyl octanoate	Heptene	Decarbonylation Deacetalation	G-S	Continuous	698	1.01	62.3	41.9	Sooknoi et al. [201]
2	CsNaX zeolite	Ion exchange	Benzaldehyde	Benzene	Decarbonylation	G-S	Continuous	748	1.01	100	80	Peralta et al. [163]

Many research studies are performed on basic catalysts like $MgO-ZrO_2$, MgO/NaY, and MgO/USY for furfural aldol condensation with acetone during deoxygenation [84, 171, 191]. Aldol condensation involves a feasible C–C coupling step in the furanic compounds deoxygenation, which increases the chain length of carbon in the deoxygenated hydrocarbons [189]. Likewise, MgO catalyst activates the C = O group, leading to simple carbonyl group hydrogenation with metals like Pd and Cu [119, 173]. MgO catalyst supported on CoMo offers more resistance to coking during the hydrodeoxygenation process [227].

3 Factors Affecting Green Diesel Productions

Some of the factors that play a pivotal role in green diesel productions, e.g., catalysts synthesis methods, catalysts morphology, metal loading in supports, reaction conditions (time, temperature, and pressure), and phosphides content in catalysts are elaborated in this section.

3.1 Effects of Catalyst Synthesis Methods

The importance of catalyst synthesis processes and its subsequent effect on physicochemical properties as well as on deoxygenation are discussed in many literatures [101, 130]. Among different synthesis routes to synthesize deoxygenation metal support catalysts, such as impregnation, precipitation, sol–gel, hydrothermal, ion exchange and temperature programmed reactions, the first two are most widely practised.

In Impregnation method, metal precursor solution comes in contact with porous support material in order to facilitate metal deposition in support pore [115, 134]. Under two categories namely wet and dry impregnation, the former is accomplished with addition of excess volume of metal solution and the latter with volume of metal solution equivalent to total pore volume of support. The terms incipient wetness impregnation or pore volume impregnation are also interchangeably used for wet impregnation. This tends to result in better metal deposition in support pores as the uptake of metal solution into the pore is driven by capillary action. Co-impregnation method is applied for bi or tri metallic catalyst preparation [101]. $Pd-Fe_2O_3$ catalyst synthesis via DIM at 300 °C yielded 56% and 92% selectivity for green diesel [77]. DIM synthesized Cu/SiO_2 catalyst at 290 °C yielded 70% and 14.2% selectivity for green diesel [242]. Another, DIM synthesized MoS/Al_2O_3 catalyst achieved a 70% yield and 7.9% selectivity for green diesel [217]. Contrarily, WIM synthesized Co/MCM-41 zeolite yielded 100% green diesel at 400 °C and WIM synthesized Ru/CsPW-1 catalyst achieved 100% selectivity for green diesel at 100 °C [3, 211].

Catalyst drying conditions are as important as the impregnation process itself as rate of drying and temperature critically affect metal redistribution inside the pores. Disadvantage of high drying rate causing metal deposition on support surface instead of the pores is reported [145]. To achieve highly dispersed supported metal catalyst with high metal loading, precipitation method is commonly employed [48]. Co-precipitation and deposition–precipitation are the two most common methods of precipitation. Co-precipitation is chosen in order to synthesize catalysts with excessive high metal weight to volume ratio. High metal loading of 80 wt. % with small particle size is reported [126]. In this process salts of metal and support are co-dissolved and mixed in order to make combined precursor particles which are primarily metal carbonates or hydroxides of active metal and support. In deposition–precipitation method, precipitation of active metal from metal precursor solution is achieved with preformed support material. Precipitation method is known to be far more complex compared to impregnation method. One of the challenges is localized precipitation of metal salts leading to agglomeration of catalyst [145].

Sol–gel processing is another widely practiced method for preparation of supported deoxygenation catalysts which employs condensation of colloidal particles (miscelles) in the stable colloidal solution into an integrated network of polymers or discrete particles (gel) [25, 111, 115]. Highly active Ni-based catalysts made through Sol–gel method are reported for guaiacol HDO [28]. To overcome the limitation of aggregation and agglomeration issues of sol–gel and precipitation methods, Hydrothermal synthesis methods are resorted [115]. This synthesis method involves/synonym crystallization within metal salt solution at elevated temperature and pressure where growth of crystal size is well controlled. This method is prevalent/ substantially used in the preparation of highly active Mo-based sulfided catalyst including nano-sized particles for hydrodeoxygenation reactions [221, 231].

Ion exchange method is well known for making metal-doped zeolite catalyst, whereas, temperature programmed reaction is preferred practise for conversion of active elemental metals into metal compounds like metal carbides, sulfides, nitrides, phosphides, and oxides to demonstrate desired catalytic properties [5, 40]. To generalize one preparation method as best for making deoxygenation catalyst for all pathways is unrealistic/implausible as the desired catalyst attributes vary widely for different pathways. Another interesting study is reported on the HDO of guaiacol over Pt/SiO_2 catalysts with different preparation processes like incipient wetness impregnation (IWI), wetness impregnation (W), flame-spray method (FSP) along with sol–gel [236] method. The guaiacol conversion is reported to increase in the following order SG (7%). $< Pt/SiO_2 <$ FSP (10%) $<$ W (35%) $<$ IWI (86%). Although SG method has produced catalyst with higher surface area, large Pt particles of low dispersion finally led to lower conversion [75].

Methods employed for catalyst synthesis and its downstream purification significantly influence the deoxygenation processes [11, 15]. Y-zeolite of 65 nm size is synthesized via two competing methods like deposition precipitation (DP-Y65) and the impregnation (IM-Y65) for the triolein deoxygenation process. The surface areas reported are 43 m^2/g 62 m^2/g, 124 m^2/g and for, IM-Y65, Y-zeolite and DP-Y65 respectively. The conversion of triolein is observed in the decreasing trend as 74% (DP-Y65) > 46.2% (IM-Y65) > 20.8% (Y-zeolites) at 380 °C in 30 min. While used for green diesel production with addition of Ni content, the selectivity too followed the expected trend as 88.9%(DP-Y65) > 78.2% (IM-Y65) [42]. The highest deoxygenation by DP-Y65 is achieved at 380 °C in 1 h with 80% conversion and 93.7% selectivity for green diesel [42]. High Brönsted-Lewis acidity ratio and Ni content (NiO particles with 3.57 ± 0.40 nm size) are believed to be responsible in improving the reactants accessibility and deoxygenation catalytic activity, causing effective triolein diffusion over active surface sites leading to its high conversion into green diesel [79]. Y-zeolites synthesized by impregnation method on contrary shown a poor deoxygenation process due to the low strength of acid sites and diminutive external surface area [42].

The above research studied shows that catalysts synthesis plays an important role in the catalytic deoxygenation process and decides its capability. Catalyst needs to be treated carefully to broaden the active surface sites and to expel impurities. The simplicity of the process has led to much wider use of impregnation method for catalyst treatment over other methods.

3.2 Effects of Catalyst Morphology

Catalysts with nanoparticles size perform efficiently in the deoxygenation process, as they provide high surface area, accessibility for interaction, as well as offers reactants and products rapid diffusion over the surface of the catalysts [205]. Deoxygenation has been studied using nanoparticles catalysts like multi-walled carbon nanotubes coated with NiO-ZnO and NiO-Fe$_2$O$_3$ catalysts [19, 122]. The researchers found that the high metal oxide and catalyst content was unable to affect deoxygenation activity, whereas carbon nanoparticles significantly contributed to achieving catalyst stability with 89% hydrocarbon and 79% n-(C_{15} + C_{17}) selectivity due to strong Brønsted-Lewis acid sites. Green diesel production over the NiO–Fe$_2$O$_3$ nanocatalysts offers low oxygen content and high heating value, along with more acceptable physicochemical properties as compared to standard biodiesel and commercial diesel fuels. Nanocatalysts NiO–Fe$_2$O$_3$ can be recycled up to 6 cycles and deliver the strongest basic active sites (9046 mol/g) and acid active sites (5313 mol/g) [180, 181]. NiO–ZnO nanocatalysts are studied to re-produce green diesel using palm fatty acid distillate (PFAD) at 350 °C [21]. The co-precipitation method is used to synthesize meso-macro-sized ZnO particles of crystalline nature and later NiO content of nanosize added. NiO–ZnO nanocatalysts have strong Brønsted-Lewis acidity properties and provide a synergetic effect between catalyst active sites and support due to the large size of crystallites. NiO-ZnO nanocatalyst yielded 83.4% with 86% selectivity for green diesel successfully [21]. Cu/Ni–Al$_2$O$_3$ catalyst of nanosize (3–5 nm) with 192–285 m^2/g an active surface area was synthesized via co-precipitation method and later tested for deoxygenation process [68]. Cu/Ni-Al$_2$O$_3$ nanosize catalyst delivered the highest catalytic activity due to stronger acidity [67]. Nanocatalysts have many advantages such as enhanced deoxygenation activity, catalyst recyclability, high green diesel selectivity, and green diesel quality, due to their enlarged crystallite sizes, large active sites, and high Brønsted-Lewis acid-basic sites.

3.3   Effects of Metal Loading in Supports

Metal oxides broadly categorized under noble metals like Rh, Ru, Ag, Au, Pt, Pd, Ir, Os, etc., and transition metals like Ni, Mo, Mn, Co, Cu, Cr, Fe, Sc, Ti, V, Zn, etc., are employed to enhance catalytic activity [6, 229]. Metal oxides impregnated catalysts exhibit high activity, selectivity, stability and are found to produce green diesel with high yields [16, 17]. Green diesel production over two separate catalysts (Pd/Al_2O_3 and $Fe–Pd/Al_2O_3$) are investigated using palm oil [204]. Fe metal loading into Pd/Al_2O_3 catalyst is found to significantly increase active sites, pore size, and porosity of the catalyst ($Fe–Pd/Al_2O_3$) thereby producing 94% biofuels with 50–62.02% green diesel proportion as compared to Pd/Al_2O_3 catalyst (86%) [204]. A noble metal and a transition metal combination are examined using $Pt-MoO_2$ catalyst with a Pt (0.5–3%) content in the palm oil deoxygenation process [61]. The catalyst acidity is remarkably enhanced, leading to outstanding decarbonylation and decarboxylation processes with a high conversion rate and yield [87, 88]. However, excessive Pt loading causes loss in catalytic activity due to coagulated Pt particles formation leading to pore blockage on the active sites of the catalyst [61]. 1% Pt doped MoO_2 catalyst yielded 56.4% green diesel and an increase in Pt metal doping causes a drop in green diesel yield (55.2% at 2% Pt) and (54.5% at 3% Pt) [61]. Although noble metals incorporated catalysts usually deliver excellent results, however, noble metals high price makes it unsustainable on a large-scale application. Transition metals are considerably cheaper than noble metals, hence once an equivalent activity is achieved, their application are expected to be more attractive [7]. A comparative study of Ru/ZSM-5 and NiMo catalysts is done to examine catalytic activity for green diesel production, where both catalysts Ru/ZSM-5 and NiMo deliver similar yield, i.e., 20–29% green diesel and 16% bio-jet fuel [219]. Thus, it is observed that catalysts based on transition metals are continuously researched for comparative outcomes and cost-efficiency.

The effect of transition metals doping into catalysts is studied using Y-zeolite support with different metals such as Co, Cu, Ni, Mn, and Zn [43]. Ni metal supported on Y-zeolites delivers high deoxygenation activity with 76.21% conversion and 92.61% green diesel selectivity as compared to other doped catalysts at 380 °C for 2 h. Ni-doped catalysts showed excellent performance when doped on other support SBA-15. The deoxygenation of palm fatty acid distillate is carried out for the production of green diesel using 5 wt % doped metal (Ni, Co, and Ni-Co) on 10 wt% SBA-15 [98]. Ni-assisted SBA-15 catalyst showed much better catalytic performance as compared to a Co-assisted SBA-15 catalyst. Ni metal-doped SBA-15 catalyst yielded 85.8% hydrocarbon with 97.2% selectivity for green diesel and recycled up to 5 cycles with 16.3% hydrocarbon yield [98]. The poor performance of the Co-assisted SBA-15 catalyst was due to the large particle size of Co (5.45 nm), carbon, and coke formation on the active surface sites. Another study of Co metal-doped catalysts showed deactivation effects, where Co/Al_2O_3 catalyst was applied for palm oil deoxygenation process [203]. Ni metal-doped MoO_2/Mo_2CT_x catalyst showed strong catalytic activity with 100% conversion and high selectivity for the product due to Ni metal and catalyst active sites synergistic effects [121]. Ni metal doping delivers many advantageous features such as small particle size, large crystallite size, more active sites, less coke formation, and lower reduction [10]. Overall transition metals-based catalysts are highly advantageous as compared to noble metal-based catalysts, due to their low cost as well as comparable activity for green diesel production. Metal oxide with a larger size than oils kinetic diameter can deliver reactants accessibility over catalyst pores and surface. Ni metal is recognized and studied as the best metal for green diesel and other biofuel production as well. Optimization of metal loading in catalyst is undoubtedly instrumental in order to maintain its physicochemical properties favorable for high catalytic activity without compromising the active metal sites for reaction.

3.4 Effects of Metal Support Interaction

The value addition through deposition of active metal sites on support surfaces or pores is well known in order to achieve higher catalytic surface area, higher metal dispersion, and reduced metal agglomeration [101]. Supports with wide diversity like carbon, metal oxides (Al_2O_3, TiO_2, SiO_2, CeO_2, ZrO_2), zeolites (HY, H-Beta, HZSM-5) silica with mesoporosity (e.g., SBA-15, MCM-41, Al-SBA-15, Al-MCM-41) are extensively tested for deoxygenation reaction [14]. Al_2O_3 and zeolites support with acidity are shown to catalyze hydrogenolysis of C–O bond in deoxygenation reactions [30, 65, 235]. Comparative acidity of supports ranging from zeolites to metal oxides are found to be in ascending order as H-ZSM-5 > H-Beta > HY > Al_2O_3 > TiO_2 > ZrO_2 > CeO_2 > SiO_2 with zeolites reporting higher acidity over metal oxide supports [5, 30, 142]. Supports with greater acidity are believed to enhance C–O hydrogenolysis leading to increased selectivity of HDO products such as benzene, toluene, and xylene. In addition to acidity, pore size and porosity of supports also influence deoxygenation of bulkier molecules such as phenolics via diffusion limitations.

Supports with high acidity like zeolite reportedly undergo faster deactivation due to enhanced coking which results from condensation products acting as coke precursors [58, 209, 210, 235]. A study shows the degree of acidity and coke content having the same order on zeolite-supported Pd catalyst as Pd/Al < Pd/10%HY-Al < Pd/20%HY-Al < Pd/HY [58]. In addition, supports with high acidity also caused cracking of alkanes with the possible influence of the selectivity of long-chained hydrocarbons results from deoxygenation process [66]. Basic support like MgO are studied for phenol HDO over sulfided CoMo/MgO catalyst and are reported to perform better in terms of sintering and coking resistance although exhibiting significantly lower HDO activity against its acidic counterparts. In view of this mild to moderate acidic supports like C, TiO_2, ZrO_2, Al_2O_3, and CeO_2 are predominately selected for HDO process due to their ability to catalyze C–O bond hydrogenolysis with least coking effect. The use of C, TiO_2, and ZrO_2 supports are considered owing to their oxophilicity which is perceived to increase aromatics selectivity in HDO processes.

The use of supports for catalysts is also known to exhibit metal-support interactions and physicochemical properties which were not present when metals were merely physically mixed with support materials. Metal-support interactions generally arise due to the electronic perturbation of metallic atoms by surrounding foreign atoms by supports which would result in the change of electronic properties of the metal catalyst [206]. However, certain metal-support interactions are expected not to be conducive to those specifically possessing high bond strength between metal and the oxide of supports. Metal–oxygen bond strength is reported to be in descending order of Mg > Al > Zr > Ti > W > Cr > Zn > V > Sn > Fe > Ge > Mn > Ni > Bi > Cu > Pb [164]. The strength of metal-support interactions is a factor which ought to be considered during the selection of an appropriate supported metal catalyst. In accordance to Sabatier rule, moderate metal–oxygen bond strength is the optimal while choosing an oxide support [74]. This is attributed to inhibition executed by high metal–oxygen bond strength in formation of oxygen vacancy for O containing compound adsorption while low metal–oxygen bond strength makes the catalyst unable to abstract oxygen effectively [14].

Thus, the preference of support for deoxygenation catalyst is crucial for superior deoxygenation activity with least deactivation effects.

3.5 Effects of Reaction Conditions

The catalysts activity and stability are widely affected by reaction conditions such as pressures, temperatures, time, catalyst content, gas hourly space velocities (GHSV), and liquid hourly space velocities (LHSV), therefore, these parameters are widely studied to optimize the production of green diesel [53, 103, 158, 161]. A palm oil deoxygenation study was done to understand the reaction condition effect with Ni metal-doped catalyst supported on Al_2O_3, SiO_2, and ZrO_2. It is observed that temperature at specific points delivered favorable effects for the different catalysts, while higher pressure has similar impacts on all catalysts [158]. The best activity was observed at 300 ^{0}C for Ni/ZrO$_2$, 350 °C for Ni/SiO$_2$, and 375 °C for Ni/Al$_2$O$_3$ catalyst, while catalytic activity loss was observed after 6 h of reaction time [158]. The reaction temperature contribution is not always adaptable, however, temperature increase can improve catalyst activation energy at specific temperatures [146]. It is also observed that after reaching above desired temperatures causes to reaction overheat and the catalyst framework gets ruptured, hence leading to reduced catalytic activity [131, 132]. An increase in yield and selectivity was observed at 350 °C and decreased at 400 °C temperature, due to an overheated reaction causing light hydrocarbons formation [13, 15]. Another study on the temperature effect in the deoxygenation process over CaO catalyst reported a two-fold yield (27–54%) at higher reaction temperatures conditions with 100% conversions [17]. However, the selectivity and yield significantly declined to 45% at 400 °C reaction temperatures due to cracking during deoxygenation reaction [97]. A range of aromatic compounds is strongly affected by the reaction temperatures, apart from that concentration of free fatty acid have no effect on the efficiency of deoxygenation or catalytic activity reaction rate [167].

A specific reaction condition is directly influenced by increasing pressure, although not always positive for the reactions [24]. A study on the palm oil deoxygenation process was performed over Ni_2P/Na-MOR catalyst with pressure range 1–50 bar and increased pressure delivered higher green diesel yield [174]. Another catalyst ($NiMo/\gamma$-Al_2O_3) is used to study the pressure effect at 15, 30, 50, and 80 bar in the palm oil deoxygenation process, and 100% conversion with enhanced product yield was achieved at higher pressures [202]. Overall 95.25 product yield was obtained at 80 bar pressure although a slight stall in the range of 30–50 bar pressure was experienced. An improvement in yield (78.1%) is also seen in the hydrodeoxygenation process at increased reaction pressures [158]. It is assumed that hydrogen absorption onto active sites of the catalyst surface at higher pressures causes enriched hydrogen solubility within oils and thus a higher product yield [123]. The positive and negative effects of long reaction times is observed in terms of product selectivity and yield. The catalytic deoxygenation process is conducted to study the impact of reaction time (0.5–2 h) and an excellent selectivity achieved at 1 h, which is further decreased as reaction times proceed [15]. The deoxygenated hydrocarbons cracking creates lighter molecules (butane, propane, and other fractions) causing reduced yield [15]. However, another study shows a contrasting development, where the Macauba acid oil conversion (97.38–98.80%) and deoxygenated product yield (96.68–98.40%) are improved with reaction times [47]. Hence, the reaction time effect is not straightforward and precise, but it strongly depends on the particular condition and catalyst properties.

Catalyst content plays a major role in the activity, stability, catalytic performances, and reaction speed [68, 129]. A deoxygenation study was performed over catalyst doped with metal different concentrations (1–9%), and catalyst with 1–5% significantly increased deoxygenation activity due to large active sites on the catalyst surface [15, 110]. Catalyst with 5% metal loading yielded 80% of the desired product. However, catalysts with 6–9% metal content showed poor deoxygenation activity and low product yield due to polymerization reactions [110, 143]. Another study also confirms catalysts with 5 wt.% show maximum activity in the deoxygenation reaction over Ag–Ni/AC catalyst [18]. Catalysts with 5 wt.% content yielded 78–95% hydrocarbons and up to 83% selectivity for green diesel under the best operating condition, apart from that catalyst recycled up to 5 times [18]. The catalysts and reactants contact time is controlled by LHSV, which is the most important parameter in the deoxygenation process especially to avoid cracking and isomerization type reactions [24, 202, 237]. The impact of LHSV was studied over CoP/PC catalyst in the palm oil deoxygenation by varying LHSV from 0.5 to 1.5 per hour, and 100% conversion occurs at 0.5 and 1.0/h LHSV while it was reduced to 85.9% at 1.5/h LHSV [97]. LHSV operation at 1.0/h yielded 75.9% hydrocarbon with a 50.4% yield of green diesel, which reduced at 1.5/h LHSV to less than 70%. Excellent selectivity of 90% achieved at 0.5/h LHSV, which was reduced at 1.5/h LHSV to 50% for green diesel.

In another study, palm oil conversion was unaffected at higher LHSV, but with an increase in LHSV from 0.25 to 5/h product yield reduced to 84.3% from 95% [202]. Therefore, lower operations of LHSV favor the deoxygenation process, leading to high selectivity and yield of green diesel [36]. Cracking and isomerization reactions are caused by higher LHSV, which contribute to reactor plugging due to esters and free fatty acids formation [12, 85, 106, 160]. The best result and operating condition are more affected by the catalyst content of the reaction than any other factor.

3.6   Effects of Phosphide Content in Catalysts

Role of phosphides in green diesel production is well documented in recent literature [130]. Doping of phosphides on metal oxide is proved to play substantial roles in green diesel production by providing H-active surface sites and enabling reactants' easy accessibility into catalyst pores [86]. Phosphide doped catalysts (Ni_2P and Co_2P) are widely studied in the vegetable oils deoxygenation process [6, 211, 237]. Ni_2P catalysts have been used in biofuels and green diesel productions [29, 64, 112, 193, 218].

A study on the addition of Ni_2P on different support such as Al-SBA-15, USY, and HZSM-5 is considered for oleic acid deoxygenation processes in a batch reactor at different temperatures (e.g., 260, 280, and 300 °C) and at 50 bar pressure [49]. Noble metals doped Al-SBA-15 show inferior activity for the deoxygenation process as compared to Ni_2P doped Al-SBA-15. Ni_2P/Al-SBA-15 catalysts with large micropore volume delivers 42% yield, while Ni_2P/USY and Ni_2P/HZSM-5 catalysts yield 24% and 29% respectively [49]. Another study concludes that Ni_2P assisted Na-MOR zeolite significantly improves palm oil conversion and green diesel yield in the hydrodeoxygenation process due to large active surface sites and high Lewis-Brønsted acidity [174, 193]. Successful recycle of Ni_2P doped Y-zeolite catalyst is also reported [218]. Cobalt phosphide is another metal phosphide that is studied in the deoxygenation process for green diesel production. Another study is conducted on palm oil deoxygenation using Co_2P impregnated porous carbon (PC) catalyst at temperature range 340–420 °C and 0.5–1.5 /h LHSV, where 100% conversion and maximum 67% green diesel selectivity are achieved at 340 °C [97]. Niobium metals integrated with niobium phosphate are also being recognized as promising catalysts for vegetable oils deoxygenation and green diesel production [150, 154, 184, 185, 190, 224, 245]. A deoxygenation study is conducted over niobium phosphate catalysts (0–25%) with temperatures ranging from 300 to 350 °C, at 10 bar pressure for 3–5 h [185] with reported 40% green diesel and 62% bio-jet fuels yields.

4 Conclusions

Green diesel has emerged as one of the sustainable alternatives to diesel fuel as it is derived from renewable resources like waste oils, triglyceride, algal lipid, and agro residues. In spite of its compositional resemblance to fossil diesel green diesel has advanced properties compared to the later. Green diesel production occurs via the deoxygenation process in the gas-phase and liquid-phase. The gas-phase reactions involve the Boudouard reaction, methanation, and water–gas shift reaction, whereas the liquid-phase progress through three major pathways decarbonylation, decarboxylation, and hydrodeoxygenation.

Optimized use of low cost and green hydrogen holds the key to large-scale commercial production of green diesel through hydrogenation route. In recent years, several research leads of green diesel production through catalytic intervention under inert gas conditions too show future potential of its production at industrial scale. Use of Ni metal catalyst is increasingly being found as preferred choice owing to its cost effectiveness and catalytic performance in terms of yield and selectivity. Activated carbon, alumina, zeolite, and zirconia are universally being explored as catalyst supports due to their acidity, oxygen vacancy properties, high surface area, and other synergistic effects. Additionally, deoxygenation processes are also affected by pressure, temperature, reaction time (GHSV and LHSV), and promoter, etc. Ni_2P as catalytic support offers excellent performance when modified with other catalysts and this strategy can help achieving twin target of meeting environmental as well as economic sustainability. More such studies on the comparative performance analyses of deoxygenation catalysts prepared through different processes for same deoxygenation reaction pathway will help establishing direct correlation of preparation method to its efficiency for green diesel preparation.

Pilot scale Environmental and economic sustainability studies through comprehensive life cycle as well as techno-economic analyses of the promising routes both hydrogenation and catalytic pathways are of extreme importance to move forward in right direction towards large scale commercial deployment.

References

1. Abdulkareem-Alsultan G, Asikin-Mijan N, Mansir N, Lee HV, Zainal Z, Islam A, Taufiq-Yap YH (2019) Pyro-lytic de-oxygenation of waste cooking oil for green diesel production over Ag_2O_3-La_2O_3/AC nano-catalyst. J Anal Appl Pyrol 137:171–184. https://doi.org/10.1016/j.jaap.2018.11.023
2. Afshar Taromi A, Kaliaguine S (2018) Hydrodeoxygenation of triglycerides over reduced mesostructured Ni/γ-alumina catalysts prepared via one-pot sol-gel route for green diesel production. Appl Catal A: General 558:140–149. https://doi.org/10.1016/j.apcata.2018.03.030
3. Alharbi K, Kozhevnikova EF, Kozhevnikov IV (2015) Hydrogenation of ketones over bifunctional Pt-heteropoly acid catalyst in the gas phase. Appl Catal A: General 504:457–462. https://doi.org/10.1016/j.apcata.2014.10.032

4. Ali A, Zhao C (2020) Ru nanoparticles supported on hydrophilic mesoporous carbon catalyzed low-temperature hydrodeoxygenation of microalgae oil to alkanes at aqueous-phase. Chin J Catal 41(8):1174–1185. https://doi.org/10.1016/S1872-2067(20)63539-2

5. Alotaibi MA, Kozhevnikova EF, Kozhevnikov IV (2012) Hydrogenation of methyl isobutyl ketone over bifunctional Pt–zeolite catalyst. J Catal 293:141–144. https://doi.org/10.1016/j. jcat.2012.06.013

6. Alvarez-Galvan MC, Blanco-Brieva G, Capel-Sanchez M, Morales-delaRosa S, Campos-Martin JM, Fierro JLG (2018) Metal phosphide catalysts for the hydrotreatment of non-edible vegetable oils. Catal Today 302:242–249. https://doi.org/10.1016/j.cattod.2017.03.031

7. Alvarez-Galvan MC, Campos-Martin JM, Fierro JLG (2019) Transition metal phosphides for the catalytic hydrodeoxygenation of waste oils into green diesel. Catalysts 9(3):293. https:// www.mdpi.com/2073-4344/9/3/293

8. Amada Y, Shinmi Y, Koso S, Kubota T, Nakagawa Y, Tomishige K (2011) Reaction mechanism of the glycerol hydrogenolysis to 1,3-propanediol over Ir–ReOx/SiO2 catalyst. Appl Catal B: Environ 105(1):117–127. https://doi.org/10.1016/j.apcatb.2011.04.001

9. Ambursa MM, Ali TH, Lee HV, Sudarsanam P, Bhargava SK, Hamid SBA (2016) Hydrodeoxygenation of dibenzofuran to bicyclic hydrocarbons using bimetallic Cu–Ni catalysts supported on metal oxides. Fuel 180:767–776. https://doi.org/10.1016/j.fuel.2016. 04.045

10. Ameen M, Azizan MT, Ramli A, Yusup S, Abdullah B (2020) The effect of metal loading over Ni/γ-Al2O3 and Mo/γ-Al2O3 catalysts on reaction routes of hydrodeoxygenation of rubber seed oil for green diesel production. Catal Today 355:51–64. https://doi.org/10.1016/j.cattod. 2019.03.028

11. Ameen M, Azizan MT, Yusup S, Ramli A, Shahbaz M, Aqsha A, Kaur H, Wai CK (2020) Parametric studies on hydrodeoxygenation of rubber seed oil for diesel range hydrocarbon production. Energy Fuels 34(4):4603–4617. https://doi.org/10.1021/acs.energyfuels.9b03692

12. Anand M, Sinha AK (2012) Temperature-dependent reaction pathways for the anomalous hydrocracking of triglycerides in the presence of sulfided Co–Mo-catalyst. Bioresour Technol 126:148–155. https://doi.org/10.1016/j.biortech.2012.08.105

13. Arend M, Nonnen T, Hoelderich WF, Fischer J, Groos J (2011) Catalytic deoxygenation of oleic acid in continuous gas flow for the production of diesel-like hydrocarbons. Appl Catal A: General 399(1):198–204. https://doi.org/10.1016/j.apcata.2011.04.004

14. Arun N, Sharma RV, Dalai AK (2015) Green diesel synthesis by hydrodeoxygenation of bio-based feedstocks: strategies for catalyst design and development. Renew Sustain Energy Rev 48:240–255. https://doi.org/10.1016/j.rser.2015.03.074

15. Asikin-Mijan N, Lee HV, Abdulkareem-Alsultan G, Afandi A, Taufiq-Yap YH (2017) Production of green diesel via cleaner catalytic deoxygenation of Jatropha curcas oil. J Clean Prod 167:1048–1059. https://doi.org/10.1016/j.jclepro.2016.10.023

16. Asikin-Mijan N, Lee HV, Juan JC, Noorsaadah AR, Abdulkareem-Alsultan G, Arumugam M, Taufiq-Yap YH (2016) Waste clamshell-derived CaO supported Co and W catalysts for renewable fuels production via cracking-deoxygenation of triolein. J Anal Appl Pyrol 120:110–120. https://doi.org/10.1016/j.jaap.2016.04.015

17. Asikin-Mijan N, Lee HV, Juan JC, Noorsaadah AR, Taufiq-Yap YH (2017) Catalytic deoxygenation of triglycerides to green diesel over modified CaO-based catalysts [https://doi.org/10. 1039/C7RA08061A]. RSC Adv 7(73):46445–46460. https://doi.org/10.1039/C7RA08061A

18. Asikin-Mijan N, Ooi JM, AbdulKareem-Alsultan G, Lee HV, Mastuli MS, Mansir N, Alharthi FA, Alghamdi AA, Taufiq-Yap YH (2020) Free-H2 deoxygenation of Jatropha curcas oil into cleaner diesel-grade biofuel over coconut residue-derived activated carbon catalyst. J Clean Prod 249:119381. https://doi.org/10.1016/j.jclepro.2019.119381

19. Asikin-Mijan N, Rosman NA, AbdulKareem-Alsultan G, Mastuli MS, Lee HV, Nabihah-Fauzi N, Lokman IM, Alharthi FA, Alghamdi AA, Aisyahi AA, Taufiq-Yap YH (2020) Production of renewable diesel from Jatropha curcas oil via pyrolytic-deoxygenation over various multi-wall carbon nanotube-based catalysts. Process Saf Environ Protection 142:336–349. https:// doi.org/10.1016/j.psep.2020.06.034

20. Badawi M, Paul J-F, Payen E, Romero Y, Richard F, Brunet S, Popov A, Kondratieva E, Gilson J-P, Mariey L, Travert A, Maugé F (2013) Hydrodeoxygenation of Phenolic Compounds by Sulfided (Co)Mo/Al2O3 Catalysts, a combined experimental and theoretical study. Oil Gas Sci Technol—Rev IFP Energies nouvelles 68(5):829–840. https://doi.org/10.2516/ogst/201 2041

21. Baharudin KB, Abdullah N, Taufiq-Yap YH, Derawi D (2020) Renewable diesel via solvent-less and hydrogen-free catalytic deoxygenation of palm fatty acid distillate. J Clean Prod 274:122850. https://doi.org/10.1016/j.jclepro.2020.122850

22. Bejblová M, Zámostný P, Červený L, Čejka J (2005) Hydrodeoxygenation of benzophenone on Pd catalysts. Appl Catal A: General 296(2):169–175. https://doi.org/10.1016/j.apcata.2005. 07.061

23. Bernas H, Eränen K, Simakova I, Leino A-R, Kordás K, Myllyoja J, Mäki-Arvela P, Salmi T, Murzin DY (2010) Deoxygenation of dodecanoic acid under inert atmosphere. Fuel 89(8):2033–2039. https://doi.org/10.1016/j.fuel.2009.11.006

24. Bezergianni S, Dimitriadis A, Kalogianni A, Knudsen KG (2011) Toward hydrotreating of waste cooking oil for biodiesel production. Effect of pressure, H_2/Oil ratio, and liquid hourly space velocity. Indus Eng Chem Res 50(7):3874–3879. https://doi.org/10.1021/ie200251a

25. Boda L, Onyestyák G, Solt H, Lónyi F, Valyon J, Thernesz A (2010) Catalytic hydroconversion of tricaprylin and caprylic acid as model reaction for biofuel production from triglycerides. Appl Catal A: General 374(1):158–169. https://doi.org/10.1016/j.apcata.2009.12.005

26. Boullosa-Eiras S, Lødeng R, Bergem H, Stöcker M, Hannevold L, Blekkan EA (2014) Catalytic hydrodeoxygenation (HDO) of phenol over supported molybdenum carbide, nitride, phosphide and oxide catalysts. Catal Today 223:44–53. https://doi.org/10.1016/j.cattod.2013. 09.044

27. Bridgwater AV (2017) Biomass conversion technologies: fast pyrolysis liquids from biomass: quality and upgrading. In: Rabaçal M, Ferreira AF, Silva CAM, Costa M (eds) Biore-fineries: targeting energy, high value products and waste valorisation. Springer International Publishing, pp 55–98. https://doi.org/10.1007/978-3-319-48288-0_3

28. Bykova MV, Ermakov DY, Kaichev VV, Bulavchenko OA, Saraev AA, Lebedev MY, Yakovlev VA (2012) Ni-based sol–gel catalysts as promising systems for crude bio-oil upgrading: guaiacol hydrodeoxygenation study. Appl Catal B: Environ 113–114:296–307. https://doi. org/10.1016/j.apcatb.2011.11.051

29. Cecilia JA, Infantes-Molina A, Rodríguez-Castellón E, Jiménez-López A, Oyama ST (2013) Oxygen-removal of dibenzofuran as a model compound in biomass derived bio-oil on nickel phosphide catalysts: role of phosphorus. Appl Catal B: Environ 136–137:140–149. https:// doi.org/10.1016/j.apcatb.2013.01.047

30. Centi G, Perathoner S (2009) Catalysis: role and challenges for a sustainable energy. Top Catal 52(8):948–961. https://doi.org/10.1007/s11244-009-9245-x

31. Chen C, Chen G, Yang F, Wang H, Han J, Ge Q, Zhu X (2015) Vapor phase hydrodeoxygena-tion and hydrogenation of m-cresol on silica supported Ni, Pd and Pt catalysts. Chem Eng Sci 135:145–154. https://doi.org/10.1016/j.ces.2015.04.054

32. Chen H, Tan J, Zhu Y, Li Y (2016) An effective and stable $Ni2P/TiO_2$ catalyst for the hydro-genation of dimethyl oxalate to methyl glycolate. Catal Commun 73:46–49. https://doi.org/ 10.1016/j.catcom.2015.10.010

33. Chen K, Koso S, Kubota T, Nakagawa Y, Tomishige K (2010) Chemoselective hydrogenolysis of tetrahydropyran-2-methanol to 1,6-hexanediol over rhenium-modified carbon-supported rhodium catalysts. ChemCatChem 2(5):547–555. https://doi.org/10.1002/cctc.201000018

34. Chen K, Mori K, Watanabe H, Nakagawa Y, Tomishige K (2012) C–O bond hydrogenolysis of cyclic ethers with OH groups over rhenium-modified supported iridium catalysts. J Catal 294:171–183. https://doi.org/10.1016/j.jcat.2012.07.015

35. Chen L, Zhu Y, Zheng H, Zhang C, Li Y (2012) Aqueous-phase hydrodeoxygenation of propanoic acid over the Ru/ZrO2 and Ru–Mo/ZrO2 catalysts. Appl Catal A: General 411–412:95–104. https://doi.org/10.1016/j.apcata.2011.10.026

36. Chen N, Gong S, Shirai H, Watanabe T, Qian EW (2013) Effects of Si/Al ratio and Pt loading on Pt/SAPO-11 catalysts in hydroconversion of Jatropha oil. Appl Catal A: General 466:105–115. https://doi.org/10.1016/j.apcata.2013.06.034

37. Chen N, Ren Y, Qian EW (2016) Elucidation of the active phase in PtSn/SAPO-11 for hydrodeoxygenation of methyl palmitate. J Catal 334:79–88. https://doi.org/10.1016/j.jcat.2015.11.001

38. Chen R-X, Wang W-C (2019) The production of renewable aviation fuel from waste cooking oil. Part I: bio-alkane conversion through hydro-processing of oil. Renew Energy 135:819–835. https://doi.org/10.1016/j.renene.2018.12.048

39. Chen W, Bao H, Wang D, Wang X, Li Y, Hu Y (2015) Chemoselective hydrogenation of nitrobenzyl ethers to aminobenzyl ethers catalyzed by palladium–nickel bimetallic nanoparticles. Tetrahedron 71(49):9240–9244. https://doi.org/10.1016/j.tet.2015.10.037

40. Chen Y, Choi S, Thompson LT (2016) Ethyl formate hydrogenolysis over Mo2C-based catalysts: towards low temperature CO and CO2 hydrogenation to methanol. Catal Today 259:285–291. https://doi.org/10.1016/j.cattod.2015.08.021

41. Chen Y, Li X, Liu S, Zhang W, Wang Q, Zi W (2020) Effects of metal promoters on one-step Pt/SAPO-11 catalytic hydrotreatment of castor oil to C8-C16 alkanes. Indus Crops Prod 146:112182. https://doi.org/10.1016/j.indcrop.2020.112182

42. Choo M-Y, Oi LE, Daou TJ, Ling TC, Lin Y-C, Centi G, Ng E-P, Juan JC (2020) Deposition of NiO nanoparticles on nanosized zeolite NaY for production of biofuel via hydrogen-free deoxygenation. Materials 13(14):3104. https://www.mdpi.com/1996-1944/13/14/3104

43. Choo M-Y, Oi LE, Ling TC, Ng E-P, Lin Y-C, Centi G, Juan JC (2020) Deoxygenation of triolein to green diesel in the H2-free condition: effect of transition metal oxide supported on Zeolite Y. J Anal Appl Pyrol 147:104797. https://doi.org/10.1016/j.jaap.2020.104797

44. Corradini SA, Gonçalves G, Lenzi M, Soares C, Santos O (2008) Characterization and hydrogenation of methyl oleate over Ru/TiO2, Ru–Sn/TiO2 catalysts. J Non-Cryst Solids 354:4865–4870. https://doi.org/10.1016/j.jnoncrysol.2008.04.040

45. Danuthai T, Sooknoi T, Jongpatiwut S, Rirksomboon T, Osuwan S, Resasco DE (2011) Effect of extra-framework cesium on the deoxygenation of methylester over CsNaX zeolites. Appl Catal A: General 409–410:74–81. https://doi.org/10.1016/j.apcata.2011.09.029

46. Das P (2022) Chapter 11—Pyrolytic bio-oil—production and applications. In: Tuli D, Kasture S, Kuila A (eds) Advanced biofuel technologies. Elsevier, pp 243–304. https://doi.org/10.1016/B978-0-323-88427-3.00016-7

47. de Barros Dias Moreira J, Bastos de Rezende D, Márcia Duarte Pasa V (2020) Deoxygenation of Macauba acid oil over Co-based catalyst supported on activated biochar from Macauba endocarp: a potential and sustainable route for green diesel and biokerosene production. Fuel 269:117253. https://doi.org/10.1016/j.fuel.2020.117253

48. de Jong KP (2009) Deposition precipitation. In: Synthesis of solid catalysts, pp 111–134. https://doi.org/10.1002/9783527626854.ch6

49. de Oliveira Camargo M, Castagnari Willimann Pimenta JL, de Oliveira Camargo M, Arroyo PA (2020) Green diesel production by solvent-free deoxygenation of oleic acid over nickel phosphide bifunctional catalysts: effect of the support. Fuel 281:118719. https://doi.org/10.1016/j.fuel.2020.118719

50. de Souza PM, Rabelo-Neto RC, Borges LEP, Jacobs G, Davis BH, Sooknoi T, Resasco DE, Noronha FB (2015) Role of keto intermediates in the hydrodeoxygenation of phenol over Pd on oxophilic supports. ACS Catal 5(2):1318–1329. https://doi.org/10.1021/cs501853t

51. Di W, Cheng J, Tian S, Li J, Chen J, Sun Q (2016) Synthesis and characterization of supported copper phyllosilicate catalysts for acetic ester hydrogenation to ethanol. Appl Catal A: General 510:244–259. https://doi.org/10.1016/j.apcata.2015.10.026

52. Ding R, Wu Y, Chen Y, Liang J, Liu J, Yang M (2015) Effective hydrodeoxygenation of palmitic acid to diesel-like hydrocarbons over MoO2/CNTs catalyst. Chem Eng Sci 135:517–525. https://doi.org/10.1016/j.ces.2014.10.024

53. Douvartzides SL, Charisiou ND, Papageridis KN, Goula MA (2019) Green diesel: biomass feedstocks, production technologies, catalytic research, fuel properties and performance in

compression ignition internal combustion engines. Energies 12(5):809. https://www.mdpi.com/1996-1073/12/5/809

54. Dragu A, Kinayyigit S, García-Suárez EJ, Florea M, Stepan E, Velea S, Tanase L, Collière V, Philippot K, Granger P, Parvulescu VI (2015) Deoxygenation of oleic acid: influence of the synthesis route of Pd/mesoporous carbon nanocatalysts onto their activity and selectivity. Appl Catal A: General 504:81–91. https://doi.org/10.1016/j.apcata.2015.01.008

55. Duan J, Kim YT, Lou H, Huber GW (2014) Hydrothermally stable regenerable catalytic supports for aqueous-phase conversion of biomass. Catal Today 234:66–74. https://doi.org/10.1016/j.cattod.2014.03.009

56. Dujjanutat P, Kaewkannetra P (2020) Production of bio-hydrogenated kerosene by catalytic hydrocracking from refined bleached deodorised palm/palm kernel oils. Renew Energy 147:464–472. https://doi.org/10.1016/j.renene.2019.09.015

57. Durán-Martín D, Ojeda M, Granados ML, Fierro JLG, Mariscal R (2013) Stability and regeneration of Cu–ZrO$_2$ catalysts used in glycerol hydrogenolysis to 1,2-propanediol. Catal Today 210:98–105. https://doi.org/10.1016/j.cattod.2012.11.013

58. Echeandia S, Pawelec B, Barrio VL, Arias PL, Cambra JF, Loricera CV, Fierro JLG (2014) Enhancement of phenol hydrodeoxygenation over Pd catalysts supported on mixed HY zeolite and Al$_2$O$_3$. An approach to O-removal from bio-oils. Fuel 117:1061–1073. https://doi.org/10.1016/j.fuel.2013.10.011

59. Echeverri DA, Marín JM, Restrepo GM, Rios LA (2009) Characterization and carbonylic hydrogenation of methyl oleate over Ru-Sn/Al$_2$O$_3$: effects of metal precursor and chlorine removal. Appl Catal A: General 366(2):342–347. https://doi.org/10.1016/j.apcata.2009.07.029

60. Faba L, Díaz E, Ordóñez S (2014) Hydrodeoxygenation of acetone–furfural condensation adducts over alumina-supported noble metal catalysts. Appl Catal B: Environ 160–161:436–444. https://doi.org/10.1016/j.apcatb.2014.05.053

61. Fangkoch S, Boonkum S, Ratchahat S, Koo-amornpattana W, Eiad-Ua A, Kiatkittipong W, Klysubun W, Srifa A, Faungnawakij K, Assabumrungrat S (2020) Solvent-free hydrodeoxygenation of triglycerides to diesel-like hydrocarbons over Pt-decorated MoO$_2$ catalysts. ACS Omega 5(12):6956–6966. https://doi.org/10.1021/acsomega.0c00326

62. Feng J, Fu H, Wang J, Li R, Chen H, Li X (2008) Hydrogenolysis of glycerol to glycols over ruthenium catalysts: effect of support and catalyst reduction temperature. Catal Commun 9(6):1458–1464. https://doi.org/10.1016/j.catcom.2007.12.011

63. Foster AJ, Do PTM, Lobo RF (2012) The synergy of the support acid function and the metal function in the catalytic hydrodeoxygenation of m-cresol. Top Catal 55(3):118–128. https://doi.org/10.1007/s11244-012-9781-7

64. Fujita S, Nakajima K, Yamasaki J, Mizugaki T, Jitsukawa K, Mitsudome T (2020) Unique catalysis of nickel phosphide nanoparticles to promote the selective transformation of biofuranic aldehydes into diketones in water. ACS Catal 10(7):4261–4267. https://doi.org/10.1021/acscatal.9b05120

65. Furimsky E (2000) Catalytic hydrodeoxygenation. Appl Catal A: General 199(2):147–190. https://doi.org/10.1016/S0926-860X(99)00555-4

66. Galadima A, Muraza O (2015) Catalytic upgrading of bioethanol to fuel grade biobutanol: a review. Ind Eng Chem Res 54(29):7181–7194. https://doi.org/10.1021/acs.iecr.5b01443

67. Gamal MS, Asikin-Mijan N, Khalit WNAW, Arumugam M, Izham SM, Taufiq-Yap YH (2020) Effective catalytic deoxygenation of palm fatty acid distillate for green diesel production under hydrogen-free atmosphere over bimetallic catalyst CoMo supported on activated carbon. Fuel Process Technol 208:106519. https://doi.org/10.1016/j.fuproc.2020.106519

68. Gousi M, Kordouli E, Bourikas K, Symianakis E, Ladas S, Kordulis C, Lycourghiotis A (2020) Green diesel production over nickel-alumina nanostructured catalysts promoted by copper. Energies 13(14):3707. https://www.mdpi.com/1996-1073/13/14/3707

69. Grilc M, Veryasov G, Likozar B, Jesih A, Levec J (2015) Hydrodeoxygenation of solvolysed lignocellulosic biomass by unsupported MoS2, MoO2, Mo2C and WS2 catalysts. Appl Catal B: Environ 163:467–477. https://doi.org/10.1016/j.apcatb.2014.08.032

70. Guo X, Wang X, Guan J, Chen X, Qin Z, Mu X, Xian M (2014) Selective hydrogenation of D-glucose to D-sorbitol over Ru/ZSM-5 catalysts. Chin J Catal 35(5):733–740. https://doi.org/10.1016/S1872-2067(14)60077-2
71. Gutierrez A, Kaila RK, Honkela ML, Slioor R, Krause AOI (2009) Hydrodeoxygenation of guaiacol on noble metal catalysts. Catal Today 147(3):239–246. https://doi.org/10.1016/j.cattod.2008.10.037
72. Hachemi I, Kumar N, Mäki-Arvela P, Roine J, Peurla M, Hemming J, Salonen J, Murzin DY (2017) Sulfur-free Ni catalyst for production of green diesel by hydrodeoxygenation. J Catal 347:205–221. https://doi.org/10.1016/j.jcat.2016.12.009
73. He L, Li X, Lin W, Li W, Cheng H, Yu Y, Fujita S-I, Arai M, Zhao F (2014) The selective hydrogenation of ethyl stearate to stearyl alcohol over Cu/Fe bimetallic catalysts. J Mol Catal A: Chem 392:143–149. https://doi.org/10.1016/j.molcata.2014.05.009
74. He Z, Wang X (2012) Hydrodeoxygenation of model compounds and catalytic systems for pyrolysis bio-oils upgrading. Catal Sustain Energy 1(2013):28–52. https://doi.org/10.2478/cse-2012-0004
75. Hellinger M, Carvalho HWP, Baier S, Wang D, Kleist W, Grunwaldt J-D (2015) Catalytic hydrodeoxygenation of guaiacol over platinum supported on metal oxides and zeolites. Appl Catal A: General 490:181–192. https://doi.org/10.1016/j.apcata.2014.10.043
76. Hengst K, Schubert M, Carvalho HWP, Lu C, Kleist W, Grunwaldt J-D (2015) Synthesis of γ-valerolactone by hydrogenation of levulinic acid over supported nickel catalysts. Appl Catal A: General 502:18–26. https://doi.org/10.1016/j.apcata.2015.05.007
77. Hensley AJR, Hong Y, Zhang R, Zhang H, Sun J, Wang Y, McEwen J-S (2014) Enhanced Fe2O3 reducibility via surface modification with Pd: characterizing the synergy within Pd/Fe catalysts for hydrodeoxygenation reactions. ACS Catal 4(10):3381–3392. https://doi.org/10.1021/cs500565e
78. Hensley AJR, Wang Y, McEwen J-S (2014) Adsorption of phenol on Fe (110) and Pd (111) from first principles. Surf Sci 630:244–253. https://doi.org/10.1016/j.susc.2014.08.003
79. Hermida L, Amani H, Saeidi S, Abdullah AZ, Mohamed AR (2018) Selective acid-functionalized mesoporous silica catalyst for conversion of glycerol to monoglycerides: state of the art and future prospects. Rev Chem Eng 34(2):239–265. https://doi.org/10.1515/revce-2016-0039
80. Hong Y-K, Lee D-W, Eom H-J, Lee K-Y (2014) The catalytic activity of Pd/WOx/γ-Al₂O₃ for hydrodeoxygenation of guaiacol. Appl Catal B: Environ 150–151:438–445. https://doi.org/10.1016/j.apcatb.2013.12.045
81. Hong Y, Zhang H, Sun J, Ayman KM, Hensley AJR, Gu M, Engelhard MH, McEwen J-S, Wang Y (2014) Synergistic catalysis between Pd and Fe in gas phase hydrodeoxygenation of m-cresol. ACS Catal 4(10):3335–3345. https://doi.org/10.1021/cs500578g
82. Hongloi N, Prapainainar P, Prapainainar C (2021) Review of green diesel production from fatty acid deoxygenation over Ni-based catalysts. Mol Catal 111696. https://doi.org/10.1016/j.mcat.2021.111696
83. Hu T, Yin H, Zhang R, Wu H, Jiang T, Wada Y (2007) Gas phase hydrogenation of maleic anhydride to γ-butyrolactone by Cu–Zn–Ti catalysts. Catal Commun 8(2):193–199. https://doi.org/10.1016/j.catcom.2006.06.009
84. Huang X-M, Zhang Q, Wang T-J, Liu Q-Y, Ma L-l, Zhang Q (2012) Production of jet fuel intermediates from furfural and acetone by aldol condensation over MgO/NaY. J Fuel Chem Technol 40(8):973–978. https://doi.org/10.1016/S1872-5813(12)60035-8
85. Huber GW, O'Connor P, Corma A (2007) Processing biomass in conventional oil refineries: production of high quality diesel by hydrotreating vegetable oils in heavy vacuum oil mixtures. Appl Catal A: General 329:120–129. https://doi.org/10.1016/j.apcata.2007.07.002
86. Ivars-Barceló F, Asedegbega-Nieto E, Aguado ER, Cecilia JA, Molina AI, Rodríguez-Castellón E (2020) 6. Advances in the application of transition metal phosphide catalysts for hydrodeoxygenation reactions of bio-oil from biomass pyrolysis. In: Alina MB, Araceli García N (eds) Biomass and biowaste: new chemical products from old. De Gruyter, pp 145–166. https://doi.org/10.1515/9783110538151-006

87. Janampelli S, Darbha S (2018) Effect of support on the catalytic activity of WOx promoted Pt in green diesel production. Mol Catal 451:125–134. https://doi.org/10.1016/j.mcat.2017.11.029

88. Janampelli S, Darbha S (2019) Highly efficient Pt-MoOx/ZrO$_2$ catalyst for green diesel production. Catal Commun 125:70–76. https://doi.org/10.1016/j.catcom.2019.03.027

89. Janampelli S, Darbha S (2021) Selective deoxygenation of fatty acids to fuel-range hydrocarbons over Pt-MOx/ZrO$_2$ (M=Mo and W) catalysts. Catal Today 375:174–180. https://doi.org/10.1016/j.cattod.2020.04.020

90. Jang H, Kim S-H, Lee D, Shim SE, Baeck S-H, Kim BS, Chang TS (2013) Hydrogenation of lactic acid to propylene glycol over a carbon-supported ruthenium catalyst. J Mol Catal A: Chem 380:57–60. https://doi.org/10.1016/j.molcata.2013.09.006

91. Jeon K-W, Shim J-O, Cho J-W, Jang W-J, Na H-S, Kim H-M, Lee Y-L, Jeon B-H, Bae JW, Roh H-S (2019) Synthesis and characterization of Pt-, Pd-, and Ru-promoted Ni–Ce0.6Zr0.4O$_2$ catalysts for efficient biodiesel production by deoxygenation of oleic acid. Fuel 236:928–933. https://doi.org/10.1016/j.fuel.2018.09.078

92. Jeong H, Shin M, Jeong B, Jang JH, Han GB, Suh Y-W (2020) Comparison of activity and stability of supported Ni2P and Pt catalysts in the hydroprocessing of palm oil into normal paraffins. J Indus Eng Chem 83:189–199. https://doi.org/10.1016/j.jiec.2019.11.027

93. Jin S, Xiao Z, Li C, Chen X, Wang L, Xing J, Li W, Liang C (2014) Catalytic hydrodeoxygenation of anisole as lignin model compound over supported nickel catalysts. Catal Today 234:125–132. https://doi.org/10.1016/j.cattod.2014.02.014

94. Jin X, Dang L, Lohrman J, Subramaniam B, Ren S, Chaudhari RV (2013) Lattice-matched bimetallic CuPd-graphene nanocatalysts for facile conversion of biomass-derived polyols to chemicals. ACS Nano 7(2):1309–1316. https://doi.org/10.1021/nn304820v

95. Kaewmeesri R, Srifa A, Itthibenchapong V, Faungnawakij K (2015) Deoxygenation of waste chicken fats to green diesel over Ni/Al$_2$O$_3$: effect of water and free fatty acid content. Energy Fuels 29(2):833–840. https://doi.org/10.1021/ef5023362

96. Kaewpengkrow P, Atong D, Sricharoenchaikul V (2014) Effect of Pd, Ru, Ni and ceramic supports on selective deoxygenation and hydrogenation of fast pyrolysis Jatropha residue vapors. Renew Energy 65:92–101. https://doi.org/10.1016/j.renene.2013.07.026

97. Kaewtrakulchai N, Kaewmeesri R, Itthibenchapong V, Eiad-Ua A, Faungnawakij K (2020) Palm oil conversion to bio-jet and green diesel fuels over cobalt phosphide on porous carbons derived from palm male flowers. Catalysts 10(6):694. https://www.mdpi.com/2073-4344/10/6/694

98. Kamaruzaman MF, Taufiq-Yap YH, Derawi D (2020) Green diesel production from palm fatty acid distillate over SBA-15-supported nickel, cobalt, and nickel/cobalt catalysts. Biomass Bioenergy 134:105476. https://doi.org/10.1016/j.biombioe.2020.105476

99. Kang KH, Hong UG, Bang Y, Choi JH, Kim JK, Lee JK, Han SJ, Song IK (2015) Hydrogenation of succinic acid to 1,4-butanediol over Re–Ru bimetallic catalysts supported on mesoporous carbon. Appl Catal A: General 490:153–162. https://doi.org/10.1016/j.apcata.2014.11.029

100. Kang KH, Hong UG, Jun JO, Song JH, Bang Y, Choi JH, Han SJ, Song IK (2014) Hydrogenation of succinic acid to γ-butyrolactone and 1,4-butanediol over mesoporous rhenium–copper–carbon composite catalyst. J Mol Catal A: Chem 395:234–242. https://doi.org/10.1016/j.molcata.2014.08.032

101. Kay Lup AN, Abnisa F, Wan Daud WMA, Aroua MK (2017) A review on reactivity and stability of heterogeneous metal catalysts for deoxygenation of bio-oil model compounds. J Indus Eng Chem 56:1–34. https://doi.org/10.1016/j.jiec.2017.06.049

102. Kim JK, Lee JK, Kang KH, Song JC, Song IK (2015) Selective cleavage of CO bond in benzyl phenyl ether to aromatics over Pd–Fe bimetallic catalyst supported on ordered mesoporous carbon. Appl Catal A: General 498:142–149. https://doi.org/10.1016/j.apcata.2015.03.034

103. Kim SK, Han JY, Lee H-S, Yum T, Kim Y, Kim J (2014) Production of renewable diesel via catalytic deoxygenation of natural triglycerides: comprehensive understanding of reaction intermediates and hydrocarbons. Appl Energy 116:199–205. https://doi.org/10.1016/j.apenergy.2013.11.062

104. Krobkrong N, Itthibenchapong V, Khongpracha P, Faungnawakij K (2018) Deoxygenation of oleic acid under an inert atmosphere using molybdenum oxide-based catalysts. Energy Convers Manage 167:1–8. https://doi.org/10.1016/j.enconman.2018.04.079

105. Kubička D, Kaluža L (2010) Deoxygenation of vegetable oils over sulfided Ni, Mo and NiMo catalysts. Appl Catal A: General 372(2):199–208. https://doi.org/10.1016/j.apcata.2009.10.034

106. Kubička D, Šimáček P, Žilková N (2009) Transformation of vegetable oils into hydrocarbons over mesoporous-alumina-supported CoMo catalysts. Top Catal 52(1):161–168. https://doi.org/10.1007/s11244-008-9145-5

107. Kumar P, Maity SK, Shee D (2020) Hydrodeoxygenation of stearic acid using Mo modified Ni and Co/alumina catalysts: effect of calcination temperature. Chem Eng Commun 207(7):904–919. https://doi.org/10.1080/00986445.2019.1630396

108. Kumar P, Yenumala SR, Maity SK, Shee D (2014) Kinetics of hydrodeoxygenation of stearic acid using supported nickel catalysts: effects of supports. Appl Catal A: General 471:28–38. https://doi.org/10.1016/j.apcata.2013.11.021

109. Kuwahara Y, Magatani Y, Yamashita H (2015) Ru nanoparticles confined in Zr-containing spherical mesoporous silica containers for hydrogenation of levulinic acid and its esters into γ-valerolactone at ambient conditions. Catal Today 258:262–269. https://doi.org/10.1016/j.cattod.2015.01.015

110. Kwon KC, Mayfield H, Marolla T, Nichols B, Mashburn M (2011) Catalytic deoxygenation of liquid biomass for hydrocarbon fuels. Renew Energy 36(3):907–915. https://doi.org/10.1016/j.renene.2010.09.004

111. Landau MV (2009) Sol-gel processing. In: Synthesis of solid catalysts, pp 83–109. https://doi.org/10.1002/9783527626854.ch5

112. Landau MV, Herskowitz M, Hoffman T, Fuks D, Liverts E, Vingurt D, Froumin N (2009) Ultradeep hydrodesulfurization and adsorptive desulfurization of diesel fuel on metal-rich nickel phosphides. Ind Eng Chem Res 48(11):5239–5249. https://doi.org/10.1021/ie9000579

113. Lazaridis PA, Karakoulia S, Delimitis A, Coman SM, Parvulescu VI, Triantafyllidis KS (2015) d-Glucose hydrogenation/hydrogenolysis reactions on noble metal (Ru, Pt)/activated carbon supported catalysts. Catal Today 257:281–290. https://doi.org/10.1016/j.cattod.2014.12.006

114. Lee C-W, Lin P-Y, Chen B-H, Kukushkin RG, Yakovlev VA (2021) Hydrodeoxygenation of palmitic acid over zeolite-supported nickel catalysts. Catal Today 379:124–131. https://doi.org/10.1016/j.cattod.2020.05.013

115. Lee DW, Yoo BR (2014) Advanced metal oxide (supported) catalysts: synthesis and applications. J Indus Eng Chem 20(6):3947–3959. https://doi.org/10.1016/j.jiec.2014.08.004

116. Lee EH, Park R-S, Kim H, Park SH, Jung S-C, Jeon J-K, Kim SC, Park Y-K (2016) Hydrodeoxygenation of guaiacol over Pt loaded zeolitic materials. J Indus Eng Chem 37:18–21. https://doi.org/10.1016/j.jiec.2016.03.019

117. Lee W-S, Wang Z, Wu RJ, Bhan A (2014) Selective vapor-phase hydrodeoxygenation of anisole to benzene on molybdenum carbide catalysts. J Catal 319:44–53. https://doi.org/10.1016/j.jcat.2014.07.025

118. Leiva K, Sepulveda C, García R, Laurenti D, Vrinat M, Geantet C, Escalona N (2015) Kinetic study of the conversion of 2-methoxyphenol over supported Re catalysts: sulfide and oxide state. Appl Catal A: General 505:302–308. https://doi.org/10.1016/j.apcata.2015.08.010

119. Li X, Su H, Ren G, Wang S (2016) The role of MgO in the performance of Pd/SiO$_2$/cordierite monolith catalyst for the hydrogenation of 2-ethyl-anthraquinone. Appl Catal A: General 517:168–175. https://doi.org/10.1016/j.apcata.2016.01.011

120. Li Y, Huang X, Zhang Q, Chen L, Zhang X, Wang T, Ma L (2015) Hydrogenation and hydrodeoxygenation of difurfurylidene acetone to liquid alkanes over Raney Ni and the supported Pt catalysts. Appl Energy 160:990–998. https://doi.org/10.1016/j.apenergy.2015.02.077

121. Liang J, Chen T, Liu J, Zhang Q, Peng W, Li Y, Zhang F, Fan X (2020) Chemoselective hydrodeoxygenation of palmitic acid to diesel-like hydrocarbons over Ni/MoO2@Mo2CTx catalyst with extraordinary synergic effect. Chem Eng J 391:123472. https://doi.org/10.1016/j.cej.2019.123472

122. Liu Q, Zuo H, Wang T, Ma L, Zhang Q (2013) One-step hydrodeoxygenation of palm oil to isomerized hydrocarbon fuels over Ni supported on nano-sized SAPO-11 catalysts. Appl Catal A: General 468:68–74. https://doi.org/10.1016/j.apcata.2013.08.009

123. Liu Y, Sotelo-Boyás R, Murata K, Minowa T, Sakanishi K (2011) Hydrotreatment of vegetable oils to produce bio-hydrogenated diesel and liquefied petroleum gas fuel over catalysts containing sulfided Ni–Mo and solid acids. Energy Fuels 25(10):4675–4685. https://doi.org/10.1021/ef200889e

124. Liu Y, Zheng D, Yu H, Liu X, Yu S, Wang X, Li L, Pang J, Liu X, Yan Z (2020) Rapid and green synthesis of SAPO-11 for deoxygenation of stearic acid to produce bio-diesel fractions. Microporous Mesoporous Mater 303:110280. https://doi.org/10.1016/j.micromeso.2020.110280

125. Loe R, Huff K, Walli M, Morgan T, Qian D, Pace R, Song Y, Isaacs M, Santillan-Jimenez E, Crocker M (2019) Effect of Pt promotion on the Ni-catalyzed deoxygenation of tristearin to fuel-like hydrocarbons. Catalysts 9(2):200. https://www.mdpi.com/2073-4344/9/2/200

126. Lok M (2009) Coprecipitation. In: Synthesis of solid catalysts, pp 135–151. https://doi.org/10.1002/9783527626854.ch7

127. Lu J, Heyden A (2015) Theoretical investigation of the reaction mechanism of the hydrodeoxygenation of guaiacol over a Ru(0001) model surface. J Catal 321:39–50. https://doi.org/10.1016/j.jcat.2014.11.003

128. Lugo-José YK, Monnier JR, Williams CT (2014) Gas-phase, catalytic hydrodeoxygenation of propanoic acid, over supported group VIII noble metals: metal and support effects. Appl Catal A: General 469:410–418. https://doi.org/10.1016/j.apcata.2013.10.025

129. Lycourghiotis S, Kordouli E, Sygellou L, Bourikas K, Kordulis C (2019) Nickel catalysts supported on palygorskite for transformation of waste cooking oils into green diesel. Appl Catal B: Environ 259:118059. https://doi.org/10.1016/j.apcatb.2019.118059

130. Mahdi HI, Bazargan A, McKay G, Azelee NIW, Meili L (2021) Catalytic deoxygenation of palm oil and its residue in green diesel production: a current technological review. Chem Eng Res Des 174:158–187. https://doi.org/10.1016/j.cherd.2021.07.009

131. Mahdi HI, Muraza O (2016) Conversion of isobutylene to octane-booster compounds after methyl tert-butyl ether phaseout: the role of heterogeneous catalysis. Ind Eng Chem Res 55(43):11193–11210. https://doi.org/10.1021/acs.iecr.6b02533

132. Mahdi HI, Muraza O (2019) An exciting opportunity for zeolite adsorbent design in separation of C4 olefins through adsorptive separation. Separ Purification Technol 221:126–151. https://doi.org/10.1016/j.seppur.2018.12.004

133. Mäki-Arvela P, Snåre M, Eränen K, Myllyoja J, Murzin DY (2008) Continuous decarboxylation of lauric acid over Pd/C catalyst. Fuel 87(17):3543–3549. https://doi.org/10.1016/j.fuel.2008.07.004

134. Marceau E, Carrier X, Che M (2009) Impregnation and drying. In: Synthesis of solid catalysts, pp 59–82. https://doi.org/10.1002/9783527626854.ch4

135. Maris EP, Davis RJ (2007) Hydrogenolysis of glycerol over carbon-supported Ru and Pt catalysts. J Catal 249(2):328–337. https://doi.org/10.1016/j.jcat.2007.05.008

136. McManus JR, Vohs JM (2014) Deoxygenation of glycolaldehyde and furfural on Mo2C/Mo(100). Surf Sci 630:16–21. https://doi.org/10.1016/j.susc.2014.06.019

137. Meller E, Green U, Aizenshtat Z, Sasson Y (2014) Catalytic deoxygenation of castor oil over Pd/C for the production of cost effective biofuel. Fuel 133:89–95. https://doi.org/10.1016/j.fuel.2014.04.094

138. Moberg DR, Thibodeau TJ, Amar FG, Frederick BG (2010) Mechanism of hydrodeoxygenation of acrolein on a cluster model of MoO3. The J Phys Chem C 114(32):13782–13795. https://doi.org/10.1021/jp104421a

139. Mochizuki T, Chen S-Y, Toba M, Yoshimura Y (2014) Deoxygenation of guaiacol and woody tar over reduced catalysts. Appl Catal B: Environ 146:237–243. https://doi.org/10.1016/j.apcatb.2013.05.040

140. Moradi-kheibari N, Ahmadzadeh H, Murry MA, Liang HY, Hosseini M (2019) Chapter 13—Fatty acid profiling of biofuels produced from microalgae, vegetable oil, and waste vegetable

oil. In: Hosseini M (ed) Advances in feedstock conversion technologies for alternative fuels and bioproducts. Woodhead Publishing, pp 239–254. https://doi.org/10.1016/B978-0-12-817 937-6.00013-8

141. Mortensen PM, de Carvalho HWP, Grunwaldt J-D, Jensen PA, Jensen AD (2015) Activity and stability of Mo2C/ZrO2 as catalyst for hydrodeoxygenation of mixtures of phenol and 1-octanol. J Catal 328:208–215. https://doi.org/10.1016/j.jcat.2015.02.002

142. Mortensen PM, Grunwaldt J-D, Jensen PA, Jensen AD (2013) Screening of catalysts for hydrodeoxygenation of phenol as a model compound for bio-oil. ACS Catal 3(8):1774–1785. https://doi.org/10.1021/cs400266e

143. Mortensen PM, Grunwaldt JD, Jensen PA, Knudsen KG, Jensen AD (2011) A review of catalytic upgrading of bio-oil to engine fuels. Appl Catal A: General 407(1):1–19. https://doi.org/10.1016/j.apcata.2011.08.046

144. Mu W, Ben H, Du X, Zhang X, Hu F, Liu W, Ragauskas AJ, Deng Y (2014) Noble metal catalyzed aqueous phase hydrogenation and hydrodeoxygenation of lignin-derived pyrolysis oil and related model compounds. Bioresour Technol 173:6–10. https://doi.org/10.1016/j.biortech.2014.09.067

145. Munnik P, de Jongh PE, de Jong KP (2015) Recent developments in the synthesis of supported catalysts. Chem Rev 115(14):6687–6718. https://doi.org/10.1021/cr500486u

146. Murti SDS, Yanti FM, Sholihah A, Juwita AR, Prasetyo J, Thebora ME, Pramana E, Saputra H (2020) Synthesis of green diesel through hydrodeoxygenation reaction of used cooking oil over NiMo/Al2O3 catalyst. AIP Conf Proc 2217(1):030178.https://doi.org/10.1063/5.0000604

147. Nakagawa Y, Ning X, Amada Y, Tomishige K (2012) Solid acid co-catalyst for the hydrogenolysis of glycerol to 1,3-propanediol over Ir-ReOx/SiO2. Appl Catal A: General 433–434:128–134. https://doi.org/10.1016/j.apcata.2012.05.009

148. Nakagawa Y, Shinmi Y, Koso S, Tomishige K (2010) Direct hydrogenolysis of glycerol into 1,3-propanediol over rhenium-modified iridium catalyst. J Catal 272(2):191–194. https://doi.org/10.1016/j.jcat.2010.04.009

149. Nelson RC, Baek B, Ruiz P, Goundie B, Brooks A, Wheeler MC, Frederick BG, Grabow LC, Austin RN (2015) Experimental and theoretical insights into the hydrogen-efficient direct hydrodeoxygenation mechanism of phenol over Ru/TiO2. ACS Catal 5(11):6509–6523. https://doi.org/10.1021/acscatal.5b01554

150. Nico C, Monteiro T, Graça MPF (2016) Niobium oxides and niobates physical properties: review and prospects. Progress Mater Sci 80:1–37. https://doi.org/10.1016/j.pmatsci.2016.02.001

151. Nie L, de Souza PM, Noronha FB, An W, Sooknoi T, Resasco DE (2014) Selective conversion of m-cresol to toluene over bimetallic Ni–Fe catalysts. J Mol Catal A: Chem 388–389:47–55. https://doi.org/10.1016/j.molcata.2013.09.029

152. Nie L, Resasco DE (2014) Kinetics and mechanism of m-cresol hydrodeoxygenation on a Pt/SiO2 catalyst. J Catal 317:22–29. https://doi.org/10.1016/j.jcat.2014.05.024

153. Nimmanwudipong T, Runnebaum RC, Block DE, Gates BC (2011) Catalytic conversion of guaiacol catalyzed by platinum supported on alumina: reaction network including hydrodeoxygenation reactions. Energy Fuels 25(8):3417–3427. https://doi.org/10.1021/ef200803d

154. Nowak I, Ziolek M (1999) Niobium compounds: preparation, characterization, and application in heterogeneous catalysis. Chem Rev 99(12):3603–3624. https://doi.org/10.1021/cr9800208

155. Ohta H, Feng B, Kobayashi H, Hara K, Fukuoka A (2014) Selective hydrodeoxygenation of lignin-related 4-propylphenol into n-propylbenzene in water by Pt-Re/ZrO2 catalysts. Catal Today 234:139–144. https://doi.org/10.1016/j.cattod.2014.01.022

156. Oi LE, Choo M-Y, Lee HV, Taufiq-Yap YH, Cheng CK, Juan JC (2020) Catalytic deoxygenation of triolein to green fuel over mesoporous TiO2 aided by in situ hydrogen production. Int J Hydrogen Energy 45(20):11605–11614. https://doi.org/10.1016/j.ijhydene.2019.07.172

157. Pan Z, Wang R, Li M, Chu Y, Chen J (2015) Deoxygenation of methyl laurate to hydrocarbons on silica-supported Ni-Mo phosphides: effect of calcination temperatures of precursor. J Energy Chem 24(1):77–86. https://doi.org/10.1016/S2095-4956(15)60287-X

158. Papageridis KN, Charisiou ND, Douvartzides SL, Sebastian V, Hinder SJ, Baker MA, AlKhoori S, Polychronopoulou K, Goula MA (2020) Effect of operating parameters on the selective catalytic deoxygenation of palm oil to produce renewable diesel over Ni supported on Al_2O_3, ZrO_2 and SiO_2 catalysts. Fuel Process Technol 209:106547. https://doi.org/10.1016/j.fuproc.2020.106547

159. Papanikolaou G, Lanzafame P, Giorgianni G, Abate S, Perathoner S, Centi G (2020) Highly selective bifunctional Ni zeo-type catalysts for hydroprocessing of methyl palmitate to green diesel. Catal Today 345:14–21. https://doi.org/10.1016/j.cattod.2019.12.009

160. Patil SJ, Vaidya PD (2018) On the production of bio-hydrogenated diesel over hydrotalcite-like supported palladium and ruthenium catalysts. Fuel Process Technol 169:142–149. https://doi.org/10.1016/j.fuproc.2017.09.026

161. Pattanaik BP, Misra RD (2017) Effect of reaction pathway and operating parameters on the deoxygenation of vegetable oils to produce diesel range hydrocarbon fuels: a review. Renew Sustain Energy Rev 73:545–557. https://doi.org/10.1016/j.rser.2017.01.018

162. Peng B, Zhao C, Mejía-Centeno I, Fuentes GA, Jentys A, Lercher JA (2012) Comparison of kinetics and reaction pathways for hydrodeoxygenation of C3 alcohols on Pt/Al_2O_3. Catal Today 183(1):3–9. https://doi.org/10.1016/j.cattod.2011.10.022

163. Peralta MA, Sooknoi T, Danuthai T, Resasco DE (2009) Deoxygenation of benzaldehyde over CsNaX zeolites. J Mol Catal A: Chem 312(1):78–86. https://doi.org/10.1016/j.molcata.2009.07.008

164. Pestman R, Koster RM, Pieterse JAZ, Ponec V (1997) Reactions of carboxylic acids on oxides: 1. Selective hydrogenation of acetic acid to acetaldehyde. J Catal 168(2):255–264. https://doi.org/10.1006/jcat.1997.1623

165. Pham LKH, Tran TTV, Kongparakul S, Reubroycharoen P, Karnjanakom S, Guan G, Samart C (2019) Formation and activity of activated carbon supported Ni2P catalysts for atmospheric deoxygenation of waste cooking oil. Fuel Process Technol 185:117–125. https://doi.org/10.1016/j.fuproc.2018.12.009

166. Phan D-P, Lee EY (2020) Phosphoric acid enhancement in a Pt-encapsulated metal-organic framework (MOF) bifunctional catalyst for efficient hydro-deoxygenation of oleic acid from biomass. J Catal 386:19–29. https://doi.org/10.1016/j.jcat.2020.03.024

167. Pimenta JLCW, Barreto RDT, dos Santos OAA, de Matos Jorge LM (2020) Effects of reaction parameters on the deoxygenation of soybean oil for the sustainable production of hydrocarbons. Environ Progress Sustain Energy 39(5):e13450. https://doi.org/10.1002/ep.13450

168. Platanitis P, Panagiotou G, Bourikas K, Kordulis C, Lycourghiotis A (2014) Hydrodeoxygenation of phenol over hydrotreatment catalysts in their reduced and sulfided states. The Open Catal J 7:18–25

169. Prasomsri T, Shetty M, Murugappan K, Román-Leshkov Y (2014) Insights into the catalytic activity and surface modification of MoO_3 during the hydrodeoxygenation of lignin-derived model compounds into aromatic hydrocarbons under low hydrogen pressures. Energy Environ Sci 7(8):2660–2669. https://doi.org/10.1039/C4EE00890A

170. Procházková D, Zámostný P, Bejblová M, Červený L, Čejka J (2007) Hydrodeoxygenation of aldehydes catalyzed by supported palladium catalysts. Appl Catal A: General 332(1):56–64. https://doi.org/10.1016/j.apcata.2007.08.009

171. Puértolas B, Keller TC, Mitchell S, Pérez-Ramírez J (2016) Deoxygenation of bio-oil over solid base catalysts: from model to realistic feeds. Appl Catal B: Environ 184:77–86. https://doi.org/10.1016/j.apcatb.2015.11.017

172. Qian EW, Chen N, Gong S (2014) Role of support in deoxygenation and isomerization of methyl stearate over nickel–molybdenum catalysts. J Mol Catal A: Chem 387:76–85. https://doi.org/10.1016/j.molcata.2014.02.031

173. Qin H, Guo C, Sun C, Zhang J (2015) Influence of the support composition on the hydrogenation of methyl acetate over Cu/MgO-SiO_2 catalysts. J Mol Catal A: Chem 409:79–84. https://doi.org/10.1016/j.molcata.2015.07.013

174. Rakmae S, Osakoo N, Pimsuta M, Deekamwong K, Keawkumay C, Butburee T, Faungnawakij K, Geantet C, Prayoonpokarach S, Wittayakun J, Khemthong P (2020) Defining nickel phosphides supported on sodium mordenite for hydrodeoxygenation of palm oil. Fuel Process Technol 198:106236. https://doi.org/10.1016/j.fuproc.2019.106236

175. Resasco DE, Crossley SP (2015) Implementation of concepts derived from model compound studies in the separation and conversion of bio-oil to fuel. Catal Today 257:185–199. https://doi.org/10.1016/j.cattod.2014.06.037

176. Romero A, Alonso E, Sastre Á, Nieto-Márquez A (2016) Conversion of biomass into sorbitol: Cellulose hydrolysis on MCM-48 and d-Glucose hydrogenation on Ru/MCM-48. Microporous Mesoporous Mater 224:1–8. https://doi.org/10.1016/j.micromeso.2015.11.013

177. Ruinart de Brimont M, Dupont C, Daudin A, Geantet C, Raybaud P (2012) Deoxygenation mechanisms on Ni-promoted MoS2 bulk catalysts: a combined experimental and theoretical study. J Catal 286:153–164. https://doi.org/10.1016/j.jcat.2011.10.022

178. Sad ME, Neurock M, Iglesia E (2011) Formation of C–C and C–O bonds and oxygen removal in reactions of alkanediols, alkanols, and alkanals on copper catalysts. J Am Chem Soc 133(50):20384–20398. https://doi.org/10.1021/ja207551f

179. Safa Gamal M, Asikin-Mijan N, Arumugam M, Rashid U, Taufiq-Yap YH (2019) Solvent-free catalytic deoxygenation of palm fatty acid distillate over cobalt and manganese supported on activated carbon originating from waste coconut shell. J Anal Appl Pyrol 144:104690. https://doi.org/10.1016/j.jaap.2019.104690

180. Santillan-Jimenez E, Morgan T, Lacny J, Mohapatra S, Crocker M (2013) Catalytic deoxygenation of triglycerides and fatty acids to hydrocarbons over carbon-supported nickel. Fuel 103:1010–1017. https://doi.org/10.1016/j.fuel.2012.08.035

181. Santillan-Jimenez E, Morgan T, Shoup J, Harman-Ware AE, Crocker M (2014) Catalytic deoxygenation of triglycerides and fatty acids to hydrocarbons over Ni–Al layered double hydroxide. Catal Today 237:136–144. https://doi.org/10.1016/j.cattod.2013.11.009

182. Sapunov VN, Stepacheva AA, Sulman EM, Wärnå J, Mäki-Arvela P, Sulman MG, Sidorov AI, Stein BD, Murzin DY, Matveeva VG (2017) Stearic acid hydrodeoxygenation over Pd nanoparticles embedded in mesoporous hypercrosslinked polystyrene. J Indus Eng Chem 46:426–435. https://doi.org/10.1016/j.jiec.2016.11.013

183. Scaldaferri CA, Pasa VMD (2019) Green diesel production from upgrading of cashew nut shell liquid. Renew Sustain Energy Rev 111:303–313. https://doi.org/10.1016/j.rser.2019.04.057

184. Scaldaferri CA, Pasa VMD (2019) Hydrogen-free process to convert lipids into bio-jet fuel and green diesel over niobium phosphate catalyst in one-step. Chem Eng J 370:98–109. https://doi.org/10.1016/j.cej.2019.03.063

185. Scaldaferri CA, Pasa VMD (2019) Production of jet fuel and green diesel range biohydrocarbons by hydroprocessing of soybean oil over niobium phosphate catalyst. Fuel 245 458–466. https://doi.org/10.1016/j.fuel.2019.01.179

186. Şenol Oİ, Viljava TR, Krause AOI (2005) Hydrodeoxygenation of methyl esters on sulphided NiMo/γ-Al$_2$O$_3$ and CoMo/γ-Al2O3 catalysts. Catal Today 100(3):331–335. https://doi.org/10.1016/j.cattod.2004.10.021

187. Seo J, Kwon JS, Choo H, Choi J-W, Jae J, Suh DJ, Kim S, Ha J-M (2019) Production of deoxygenated high carbon number hydrocarbons from furan condensates: hydrodeoxygenation of biomass-based oxygenates. Chem Eng J 377:119985. https://doi.org/10.1016/j.cej.2018.09.146

188. Sepúlveda C, Escalona N, García R, Laurenti D, Vrinat M (2012) Hydrodeoxygenation and hydrodesulfurization co-processing over ReS$_2$ supported catalysts. Catal Today 195(1):101–105. https://doi.org/10.1016/j.cattod.2012.05.047

189. Serrano-Ruiz JC, Pineda A, Balu AM, Luque R, Campelo JM, Romero AA, Ramos-Fernández JM (2012) Catalytic transformations of biomass-derived acids into advanced biofuels. Catal Today 195(1):162–168. https://doi.org/10.1016/j.cattod.2012.01.009

190. Shao Y, Xia Q, Dong L, Liu X, Han X, Parker SF, Cheng Y, Daemen LL, Ramirez-Cuesta AJ, Yang S, Wang Y (2017) Selective production of arenes via direct lignin upgrading over a niobium-based catalyst. Nat Commun 8(1):16104. https://doi.org/10.1038/ncomms16104

191. Shen W, Tompsett GA, Hammond KD, Xing R, Dogan F, Grey CP, Conner WC, Auerbach SM, Huber GW (2011) Liquid phase aldol condensation reactions with MgO–ZrO$_2$ and shape-selective nitrogen-substituted NaY. Appl Catal A: General 392(1):57–68. https://doi.org/10.1016/j.apcata.2010.10.023

192. Shi H, Chen J, Yang Y, Tian S (2014) Catalytic deoxygenation of methyl laurate as a model compound to hydrocarbons on nickel phosphide catalysts: remarkable support effect. Fuel Process Technol 118:161–170. https://doi.org/10.1016/j.fuproc.2013.08.010

193. Shin M, Kim J, Suh Y-W (2020) Etherification of biomass-derived furanyl alcohols with aliphatic alcohols over silica-supported nickel phosphide catalysts: effect of surplus P species on the acidity. Appl Catal A: General 603:117763. https://doi.org/10.1016/j.apcata.2020.117763

194. Silva LN, Fortes ICP, de Sousa FP, Pasa VMD (2016) Biokerosene and green diesel from macauba oils via catalytic deoxygenation over Pd/C. Fuel 164:329–338. https://doi.org/10.1016/j.fuel.2015.09.081

195. Simakova I, Simakova O, Mäki-Arvela P, Simakov A, Estrada M, Murzin DY (2009) Deoxygenation of palmitic and stearic acid over supported Pd catalysts: effect of metal dispersion. Appl Catal A: General 355(1):100–108. https://doi.org/10.1016/j.apcata.2008.12.001

196. Sitthisa S, Sooknoi T, Ma Y, Balbuena PB, Resasco DE (2011) Kinetics and mechanism of hydrogenation of furfural on Cu/SiO$_2$ catalysts. J Catal 277(1):1–13. https://doi.org/10.1016/j.jcat.2010.10.005

197. Smirnov AA, Khromova SA, Ermakov DY, Bulavchenko OA, Saraev AA, Aleksandrov PV, Kaichev VV, Yakovlev VA (2016) The composition of Ni-Mo phases obtained by NiMoOx-SiO2 reduction and their catalytic properties in anisole hydrogenation. Appl Catal A: General 514:224–234. https://doi.org/10.1016/j.apcata.2016.01.025

198. Smith GA (2019) Chapter 8—Fatty acid, methyl ester, and vegetable oil ethoxylates. In: Hayes DG, Solaiman DKY, Ashby RD (eds) Biobased surfactants, 2nd edn. AOCS Press, pp 287–301. https://doi.org/10.1016/B978-0-12-812705-6.00008-3

199. Soares AVH, Salazar JB, Falcone DD, Vasconcellos FA, Davis RJ, Passos FB (2016) A study of glycerol hydrogenolysis over Ru–Cu/Al$_2$O$_3$ and Ru–Cu/ZrO$_2$ catalysts. J Mol Catal A: Chem 415:27–36. https://doi.org/10.1016/j.molcata.2016.01.027

200. Soni VK, Dhara S, Krishnapriya R, Choudhary G, Sharma PR, Sharma RK (2020) Highly selective Co$_3$O$_4$/silica-alumina catalytic system for deoxygenation of triglyceride-based feedstock. Fuel 266:117065. https://doi.org/10.1016/j.fuel.2020.117065

201. Sooknoi T, Danuthai T, Lobban LL, Mallinson RG, Resasco DE (2008) Deoxygenation of methylesters over CsNaX. J Catal 258(1):199–209. https://doi.org/10.1016/j.jcat.2008.06.012

202. Srifa A, Faungnawakij K, Itthibenchapong V, Viriya-empikul N, Charinpanitkul T, Assabumrungrat S (2014) Production of bio-hydrogenated diesel by catalytic hydrotreating of palm oil over NiMoS2/γ-Al$_2$O$_3$ catalyst. Bioresour Technol 158:81–90. https://doi.org/10.1016/j.biortech.2014.01.100

203. Srifa A, Viriya-empikul N, Assabumrungrat S, Faungnawakij K (2015) Catalytic behaviors of Ni/γ-Al$_2$O$_3$ and Co/γ-Al$_2$O$_3$ during the hydrodeoxygenation of palm oil. Catal Sci Technol 5(7):3693–3705. https://doi.org/10.1039/C5CY00425J

204. Srihanun N, Dujjanutat P, Muanruksa P, Kaewkannetra P (2020) Biofuels of green diesel–kerosene–gasoline production from palm oil: effect of palladium cooperated with second metal on hydrocracking reaction. Catalysts 10(2):241. https://www.mdpi.com/2073-4344/10/2/241

205. Tago T, Konno H, Nakasaka Y, Masuda T (2012) Size-controlled synthesis of nano-zeolites and their application to light olefin synthesis. Catal Surv Asia 16(3):148–163. https://doi.org/10.1007/s10563-012-9141-4

206. Tauster SJ (1987) Strong metal-support interactions. Acc Chem Res 20(11):389–394. https://doi.org/10.1021/ar00143a001

207. Thilakaratne R, Brown T, Li Y, Hu G, Brown R (2014) Mild catalytic pyrolysis of biomass for production of transportation fuels: a techno-economic analysis. Green Chem 16(2):627–636. https://doi.org/10.1039/C3GC41314D

208. Thongkumkoon S, Kiatkittipong W, Hartley UW, Laosiripojana N, Daorattanachai P (2019) Catalytic activity of trimetallic sulfided Re–Ni–Mo/γ-Al$_2$O$_3$ toward deoxygenation of palm feedstocks. Renew Energy 140:111–123. https://doi.org/10.1016/j.renene.2019.03.039

209. To AT, Resasco DE (2014) Role of a phenolic pool in the conversion of m-cresol to aromatics over HY and HZSM-5 zeolites. Appl Catal A: General 487:62–71. https://doi.org/10.1016/j.apcata.2014.09.006

210. To AT, Resasco DE (2015) Hydride transfer between a phenolic surface pool and reactant paraffins in the catalytic cracking of m-cresol/hexanes mixtures over an HY zeolite. J Catal 329:57–68. https://doi.org/10.1016/j.jcat.2015.04.025

211. Tran NTT, Uemura Y, Chowdhury S, Ramli A (2016) Vapor-phase hydrodeoxygenation of guaiacol on Al-MCM-41 supported Ni and Co catalysts. Appl Catal A: General 512:93–100. https://doi.org/10.1016/j.apcata.2015.12.021

212. Tuli D, Kasture S (2022) Chapter 5—Biodiesel and green diesel. In: Tuli D, Kasture S, Kuila A (eds) Advanced biofuel technologies. Elsevier, pp 119–133. https://doi.org/10.1016/B978-0-323-88427-3.00010-6

213. Upare DP, Park S, Kim MS, Jeon YP, Kim J, Lee D, Lee J, Chang H, Choi S, Choi W, Park YK, Lee CW (2017) Selective hydrocracking of pyrolysis fuel oil into benzene, toluene and xylene over CoMo/beta zeolite catalyst. J Indus Eng Chem 46:356–363. https://doi.org/10.1016/j.jiec.2016.11.004

214. Upare DP, Park S, Kim MS, Kim J, Lee D, Lee J, Chang H, Choi W, Choi S, Jeon YP, Park YK, Lee CW (2016) Cobalt promoted Mo/beta zeolite for selective hydrocracking of tetralin and pyrolysis fuel oil into monocyclic aromatic hydrocarbons. J Indus Eng Chem 35:99–107. https://doi.org/10.1016/j.jiec.2015.12.020

215. Upare PP, Lee M, Lee S-K, Yoon JW, Bae J, Hwang DW, Lee UH, Chang J-S, Hwang YK (2016) Ru nanoparticles supported graphene oxide catalyst for hydrogenation of bio-based levulinic acid to cyclic ethers. Catal Today 265:174–183. https://doi.org/10.1016/j.cattod.2015.09.042

216. Venderbosch RH, Prins W (2011) Fast pyrolysis. In: Thermochemical processing of biomass, pp 124–156. https://doi.org/10.1002/9781119990840.ch5

217. Vlasova EN, Porsin AA, Aleksandrov PV, Nuzhdin AL, Bukhtiyarova GA (2021) Co-hydroprocessing of straight-run gasoil—rapeseed oil mixture over stacked bed Mo/Al$_2$O$_3$ + NiMo/Al$_2$O$_3$-SAPO-11 catalysts. Fuel 285:119504. https://doi.org/10.1016/j.fuel.2020.119504

218. Wagner JL, Jones E, Sartbaeva A, Davis SA, Torrente-Murciano L, Chuck CJ, Ting VP (2018) Zeolite Y supported nickel phosphide catalysts for the hydrodenitrogenation of quinoline as a proxy for crude bio-oils from hydrothermal liquefaction of microalgae. Dalton Trans 47(4):1189–1201. https://doi.org/10.1039/C7DT03318D

219. Wang H (2012) Biofuels production from hydrotreating of vegetable oil using supported noble metals, and transition metal carbide and nitride

220. Wang W-C, Hsieh C-H (2020) Hydro-processing of biomass-derived oil into straight-chain alkanes. Chem Eng Res Des 153:63–74. https://doi.org/10.1016/j.cherd.2019.10.030

221. Wang W, Zhang K, Li L, Wu K, Liu P, Yang Y (2014) Synthesis of highly active Co–Mo–S unsupported catalysts by a one-step hydrothermal method for p-Cresol hydrodeoxygenation. Ind Eng Chem Res 53(49):19001–19009. https://doi.org/10.1021/ie5032698

222. Wang Z, Pholjaroen B, Li M, Dong W, Li N, Wang A, Wang X, Cong Y, Zhang T (2014) Chemoselective hydrogenolysis of tetrahydrofurfuryl alcohol to 1,5-pentanediol over Ir-MoOx/SiO$_2$ catalyst. J Energy Chem 23(4):427–434. https://doi.org/10.1016/S2095-4956(14)60168-6

223. Widayatno WB, Guan G, Rizkiana J, Yang J, Hao X, Tsutsumi A, Abudula A (2016) Upgrading of bio-oil from biomass pyrolysis over Cu-modified β-zeolite catalyst with high selectivity and stability. Appl Catal B: Environ 186:166–172. https://doi.org/10.1016/j.apcatb.2016.01.006

224. Xia Q-N, Cuan Q, Liu X-H, Gong X-Q, Lu G-Z, Wang Y-Q (2014) Pd/NbOPO4 multifunctional catalyst for the direct production of liquid alkanes from aldol adducts of furans. Angewandte Chemie International Edition 53(37):9755–9760. https://doi.org/10.1002/anie.201403440

225. Xiong K, Yu W, Chen JG (2014) Selective deoxygenation of aldehydes and alcohols on molybdenum carbide (Mo2C) surfaces. Appl Surf Sci 323:88–95. https://doi.org/10.1016/j. apsusc.2014.06.100

226. Xiong W, Wang K-J, Liu X-W, Hao F, Xiao H-Y, Liu P-L, Luo H-A (2016) 1,5-Dinitronaphthalene hydrogenation to 1,5-diaminonaphthalene over carbon nanotube supported non-noble metal catalysts under mild conditions. Appl Catal A: General 514:126–134. https://doi.org/10.1016/j.apcata.2016.01.018

227. Yang Y, Gilbert A, Xu C (2009) Hydrodeoxygenation of bio-crude in supercritical hexane with sulfided CoMo and CoMoP catalysts supported on MgO: a model compound study using phenol. Appl Catal A: General 360(2):242–249. https://doi.org/10.1016/j.apcata.2009.03.027

228. Yang Y, Ochoa-Hernández C, de la Peña O'Shea VA, Pizarro P, Coronado JM, Serrano DP (2014) Effect of metal–support interaction on the selective hydrodeoxygenation of anisole to aromatics over Ni-based catalysts. Appl Catal B: Environ 145:91–100. https://doi.org/10. 1016/j.apcatb.2013.03.038

229. Yang Y, Wang Q, Zhang X, Wang L, Li G (2013) Hydrotreating of C18 fatty acids to hydrocarbons on sulphided NiW/SiO$_2$–Al$_2$O$_3$. Fuel Process Technol 116:165–174. https://doi.org/ 10.1016/j.fuproc.2013.05.008

230. Ye C-L, Guo C-L, Zhang J-L (2016) Highly active and stable CeO2–SiO2 supported Cu catalysts for the hydrogenation of methyl acetate to ethanol. Fuel Process Technol 143:219–224. https://doi.org/10.1016/j.fuproc.2015.12.003

231. Yoosuk B, Tumnantong D, Prasassarakich P (2012) Amorphous unsupported Ni–Mo sulfide prepared by one step hydrothermal method for phenol hydrodeoxygenation. Fuel 91(1):246–252. https://doi.org/10.1016/j.fuel.2011.08.001

232. Yuan P, Liu Z, Zhang W, Sun H, Liu S (2010) Cu–Zn/Al$_2$O$_3$ catalyst for the hydrogenation of esters to alcohols. Chin J Catal 31(7):769–775. https://doi.org/10.1016/S1872-2067(09)600 87-5

233. Zaccheria F, Ravasio N, Ercoli M, Allegrini P (2005) Heterogeneous Cu-catalysts for the reductive deoxygenation of aromatic ketones without additives. Tetrahedron Lett 46(45):7743–7745. https://doi.org/10.1016/j.tetlet.2005.09.041

234. Zacher AH, Elliott DC, Olarte MV, Santosa DM, Preto F, Iisa K (2014) Pyrolysis of woody residue feedstocks: upgrading of bio-oils from mountain-pine-beetle-killed trees and hog fuel. Energy Fuels 28(12):7510–7516. https://doi.org/10.1021/ef5017945

235. Zanuttini MS, Lago CD, Querini CA, Peralta MA (2013) Deoxygenation of m-cresol on Pt/γ-Al$_2$O$_3$ catalysts. Catal Today 213:9–17. https://doi.org/10.1016/j.cattod.2013.04.011

236. Zanuttini MS, Peralta MA, Querini CA (2015) Deoxygenation of m-cresol: deactivation and regeneration of Pt/γ-Al$_2$O$_3$ catalysts. Ind Eng Chem Res 54(18):4929–4939. https://doi.org/ 10.1021/acs.iecr.5b00305

237. Zarchin R, Rabaev M, Vidruk-Nehemya R, Landau MV, Herskowitz M (2015) Hydroprocessing of soybean oil on nickel-phosphide supported catalysts. Fuel 139:684–691. https:// doi.org/10.1016/j.fuel.2014.09.053

238. Zhang D, Yin H, Ge C, Xue J, Jiang T, Yu L, Shen Y (2009) Selective hydrogenation of maleic anhydride to γ-butyrolactone and tetrahydrofuran by Cu–Zn–Zr catalyst in the presence of ethanol. J Indus Eng Chem 15(4):537–543. https://doi.org/10.1016/j.jiec.2009.01.010

239. Zhang H, Lin H, Zheng Y (2014) The role of cobalt and nickel in deoxygenation of vegetable oils. Appl Catal B: Environ 160–161:415–422. https://doi.org/10.1016/j.apcatb.2014.05.043

240. Zhang R, Yin H, Zhang D, Qi L, Lu H, Shen Y, Jiang T (2008) Gas phase hydrogenation of maleic anhydride to tetrahydrofuran by Cu/ZnO/TiO$_2$ catalysts in the presence of n-butanol. Chem Eng J 140(1):488–496. https://doi.org/10.1016/j.cej.2007.11.031

241. Zhao C, He J, Lemonidou AA, Li X, Lercher JA (2011) Aqueous-phase hydrodeoxygenation of bio-derived phenols to cycloalkanes. J Catal 280(1):8–16. https://doi.org/10.1016/j.jcat. 2011.02.001

242. Zheng H-Y, Zhu Y-L, Bai Z-Q, Huang L, Xiang H-W, Li Y-W (2006) An environmentally benign process for the efficient synthesis of cyclohexanone and 2-methylfuran. Green Chem 8(1):107–109. https://doi.org/10.1039/B513584B

243. Zheng Y, Wang J, Liu C, Lu Y, Lin X, Li W, Zheng Z (2020) Efficient and stable Ni-Cu catalysts for ex situ catalytic pyrolysis vapor upgrading of oleic acid into hydrocarbon: effect of catalyst support, process parameters and Ni-to-Cu mixed ratio. Renew Energy 154:797–812. https://doi.org/10.1016/j.renene.2020.03.058

244. Zhu X, Lobban LL, Mallinson RG, Resasco DE (2011) Bifunctional transalkylation and hydrodeoxygenation of anisole over a Pt/HBeta catalyst. J Catal 281(1):21–29. https://doi.org/10.1016/j.jcat.2011.03.030

245. Ziolek M, Sobczak I (2017) The role of niobium component in heterogeneous catalysts. Catal Today 285:211–225. https://doi.org/10.1016/j.cattod.2016.12.013

246. Zulkepli S, Juan JC, Lee HV, Rahman NSA, Show PL, Ng EP (2018) Modified mesoporous HMS supported Ni for deoxygenation of triolein into hydrocarbon-biofuel production. Energy Convers Manage 165:495–508. https://doi.org/10.1016/j.enconman.2018.03.087

Chapter 4
Green Diesel Production by Hydroprocessing Technology

S. A. Farooqui, R. Kumar, A. K. Sinha, and A. Ray

Abstract Non-edible vegetable oils can be converted into diesel fuel by esterification (biodiesel) or hydroprocessing. Biodiesel contains oxygen atoms, whereas hydroprocessed vegetable oil, also known as green diesel or paraffinic diesel, is a pure hydrocarbon (drop-in fuel) with no oxygen atoms. Non-edible vegetable oil such as Jatropha curcus oil can be converted into a hydrocarbon mixture by hydroprocessing. The technology involves pre-treatment of plant oil followed by hydrodeoxygenation, hydrocracking, hydroisomerization, and aromatization. Hydrogenation is employed to saturate the double bonds and remove the oxygen from the triglyceride's fatty acid chains, either water or CO_2. Subsequently, the other reactions occurring during the process are hydrodesulfurization, hydrodenitrogenation, decarboxylation, and decarbonylation. The feed is processed over the catalyst under hydroprocessing conditions to convert the renewable source into n-paraffins and iso-paraffins. The selectivity of the produced n-paraffins and iso-paraffins range may be shifted to ATF and diesel by suitably selecting the active metals, support, and process conditions. Non-precious metals are used as active metals in the catalyst. The metal oxide or mixture of metal oxide, zeolite, and/or a combination thereof is used as support. Non-precious metals including nickel (Ni), cobalt (Co), molybdenum (Mo), tungsten (W), or a combination thereof, e.g., nickel-molybdenum (NiMo), cobalt-molybdenum (CoMo) and nickel-tungsten (NiW), are used as active metals. These active metals are supported in mesoporous Υ-alumina (Υ-Al_2O_3), silica-alumina, and/or zeolite, or a combination of thereof. The active metal(s) are generally in sulfided form. The renewable source is the oil originating from vegetable and animal fats. It includes, but is not limited to, waste restaurant oil, soyabean oil, jatropha oil, algae oil, etc. These oils mainly contain free fatty acids and triglycerides. Different commercial plants for the production of green diesel have been installed worldwide. Renewable Energy Group Inc., Neste, Eni, Total, BP, Cetane Energy, Eni/Honeywell-UOP, and Haldor Topsoe are the key players who developed the green diesel process and produce green diesel commercially. Indian Institute of Petroleum in India had made green diesel from non-edible waste oil via hydroprocessing technology and tested it on

S. A. Farooqui (✉) · R. Kumar · A. K. Sinha · A. Ray
CSIR-Indian Institute of Petroleum, Dehradun 248005, India
e-mail: farooqui@iip.res.in

© The Author(s), under exclusive license to Springer Nature Singapore Pte Ltd. 2022   109
M. Aslam et al. (eds.), *Green Diesel: An Alternative to Biodiesel and Petrodiesel*,
Advances in Sustainability Science and Technology,
https://doi.org/10.1007/978-981-19-2235-0_4

diesel engines and diesel generators. The COx, NOx, and particulates emission was reduced significantly while using green diesel.

Keywords Paraffinic diesel · Green diesel · Hydroprocessing · Hydrotreating · Hydrodeoxygenation

1 Introduction

Climate change, air quality issues, rising demand for energy, and diminishing petroleum reserves are significant factors driving the global search for sustainable, renewable, and environmentally friendly energy sources. Biofuels, particularly "drop-in" liquid hydrocarbons derived from renewable feed sources, are similar to conventional fuels. It is an attractive and promising option that can significantly reduce carbon footprints and greenhouse gas (GHG) emissions without any significant modification to existing fuel storage, pipelines, and internal combustion engines [1].

Among different emission components, CO_2 is the major contributor to global warming. We know that human-caused greenhouse gas emissions (from energy to transportation, industry to farming) are hastening global warming at a rate that threatens our planet's natural equilibrium as we know it. But we do not know when we would cross this line. Researchers developed an ad hoc indicator known as the carbon budget [2].

The carbon budget is the total quantity of carbon dioxide (CO_2) emissions allowed over time to stay under a specific temperature threshold [3]. The Carbon Tracker Initiative has been a critical contributor to mainstreaming the notion of a carbon budget, utilizing science established by leading academics and climate specialists. As of 2011, about 52% of the world's carbon budget has been used, and in another 30 years at the same consumption levels, the earth's carbon budget would be exhausted [4].

Utilizing renewable carbonaceous sources in an organized and planned manner can aid sustainable development by decreasing global CO_2 emissions from fossil energy sources, promoting rural areas' development as decentralized carbon sources, and reducing the world economy's dependence on diminishing fossil energy sources [5–10]. Industry and transport sectors, primarily based on declining fossil fuels that are progressively perceived as unsustainable, need to adopt biofuels or other renewable non-carbon alternatives such as green electricity or green hydrogen for their energy needs.

Among different technologies for renewable diesel production, the following pathways are globally in use at various stages of research, development, and deployment: (1) Hydrotreating/hydroprocessing technology—hydrotreating is a process used in petroleum refineries that includes reacting the feedstock (in this case, lipids) with hydrogen. This approach is the leading commercial route to green diesel globally. The corresponding green diesel is also referred to as HVO (hydroprocessed vegetable oil) in Europe and BHD (bio-hydrogenated diesel) in Thailand. (2) Biological sugar upgrading—This route employs a biochemical deconstruction process similar to cellulosic ethanol, but with the inclusion of microbes that convert carbohydrates to hydrocarbons. (3) Catalytic conversion of sugars—A catalytic process used in this route to transform a carbohydrate stream into hydrocarbon fuels. (4) Gasification followed by Fischer–Tropsch catalysis—This sequential route involves the conversion of biomass to syngas and then subsequently of syngas to long-chain hydrocarbons. Hydrotreating is required as a finishing step to improve the cold flow properties and stability of the fuel. (5) Pyrolysis—This process is the breakdown of organic molecules at high temperatures in the absence of oxygen. The product formed is liquid pyrolysis oil, which may be improved to hydrocarbon fuels either as a stand-alone process or as a feedstock for feeding crude oil into a conventional petroleum refinery. (6) Hydrothermal processing—High pressure and moderate temperature process to convert biomass or wet waste materials to oxygenated produce may be upgraded to hydrocarbon fuels.

Conversion of lipids through catalytic chemical processing is the simplest way to transform biomass into hydrocarbon fuels. The management of the feed supply chain, however, is a crucial challenge. If the feedstock for fuel is grown on agricultural land, it poses potentially unacceptable risks to food supply chains [11]. Used cooking oil, waste oils, oil-bearing trees that are grown in the wild (non-agricultural lands) without displacing existing flora, and short gestation crops that utilize idle months of already cultivated farmlands would all be relatively.

Biomass-derived oils and fats (classified more broadly as lipids), derived from plants, livestock, fisheries, or microbial sources, are relatively clean and consistent feedstocks compared to the complex mixtures of molecules that constitute crude oil. These lipids are usually glycerides molecules such as triglycerides, diglycerides, monoglycerides, or the corresponding free fatty acids. Free fatty acids are long-chain organic molecules primarily composed of carbon, hydrogen, and oxygen atoms [12]. Natural lipids may also contain other lipid derivatives such as phospholipids, sphingolipids, and lipoproteins—most of these are undesirable contaminants from a hydrocarbon production standpoint and need to be removed from the feedstock pre-treatment.

Some of the methods for converting oils/lipids to biofuels include transesterification [5, 13–15], hydrothermal conversion [15, 16], and hydroprocessing [5, 13–15]. All these approaches have their own set of limits and benefits. As technology advances, the most cost-effective and long-term method of producing drop-in biofuels would prevail commercially, with sustainability-associated costs—such as carbon taxes—the currently unknown factor in determining the eventual winners.

1.1 Green Diesel and Biodiesel

Green diesel has several advantages over conventional fatty acid methyl ester (FAME) biodiesel, such as higher energy density, superior oxidation stability, lower specific gravity, higher cetane number, and generally better cold flow properties whether as a neat fuel or in blends with petroleum diesel. Not being oxygenated, unlike FAME, green diesel is fully compatible with petroleum diesel and can be used with existing diesel engines without any changes or modification, thus being a drop-in fuel. Green diesel is also environment-friendly fuel as its use may generate lesser greenhouse gases than petroleum diesel, biodiesel, and fossil-derived syn-diesel (without carbon sequestration) [17]. However, it may be noted that biodiesel has higher lubricity than green diesel, but this can be compensated in green diesel systems with the incorporation of suitable lubricity-enhancing additives.

Several technical challenges have been identified with biodiesel when used as an alternative to diesel in marine applications. The key factors to be kept in mind while using biofuels in ships, mainly to avoid damage on board, include (a) Microbial growth: Bacteria and mold can form in FAME biodiesel fuel if condensed water accumulates. Microbial development causes excessive sludge accumulation, blocked filters, and pipework. It may be reduced or mitigated by frequent tank draining and the use of biocide. (b) Oxidative degradation: FAME biodiesel is prone to deteriorate over time, generating polymers and other insoluble materials. These may form deposits in pipelines, and engines may accumulate on the internal metal surfaces, reducing operating performance. Increased fuel acidity from hydrolysis of FAME could lead to corrosion in the fuel system and deposit formation in pumps and injectors. This is why FAME biodiesel is generally not stored for long periods before use but instead used as a new commodity and utilized within a few months of production. Adding antioxidants to the fuel can, however, increase the fuel storage shelf life. (c) Low temperature: FAME biodiesels typically have a higher cloud point than petroleum diesel (primarily depending on the feedstock used), resulting in poor flow characteristics and filter clogging at lower temperatures. As a result, it is critical to understand the product's cold flow qualities and keep storage and transfer temperatures above the cloud point threshold. (d) Corrosion: This is especially important for FAME biodiesel at higher concentrations (>80%). Some hoses and gaskets may decay, resulting in loss of integrity and contact with metallic materials such as copper, brass, lead, tin, zinc, and so on. As a result, it is critical to ensure that these fuel system components are robust and may be utilized with biofuel. (e) Conversion: Because biodiesel has solvent characteristics, it is likely that when switching over from diesel to biofuel, deposits accumulated in the fuel storage and flow systems will be flushed, clogging fuel filters. During this time, it is advised to flush the system and monitor the filters.

1.2   Hydroprocessing and Transesterification Process

Hydroprocessing—which produces HVO green diesel—has some significant advantages over the transesterification process: produces FAME biodiesel—for converting vegetable oils into transportation fuel. The hydroprocessing process is compatible with the current refinery infrastructure. There is a substantially reduced requirement for new refining assets compared to the transesterification process, which requires significant capital investment for new biodiesel plants. The hydroprocessing process is feedstock flexible and produces essentially the same range of specifications regardless of the feed composition, which is another advantage compared to the transesterification process since FAME properties are quite feedstock-dependent.

Poor catalyst recovery, products separation, and disposal of toxic waste (where base-catalysis is used) are significant drawbacks of the transesterification process. Also, transesterification generates large quantities of glycerol as a by-product that needs to find a suitable external market for further utilization. HVO green diesel, on the other hand, has products of naphtha, propane, and fuel gas, which can be utilized with ease within a petroleum refinery or petrochemicals complex.

For renewable diesel obtained by hydroprocessing technology, the cetane number is very high (>70 compared to 51 for biodiesel). Renewable diesel has higher energy density, high oxidation stability, better cold flow properties, and lower exhaust emissions (particulate emission, SOx, and possibly NOx) than biodiesel.

In addition to the hydroprocessing route, newer ways of making green diesel and aviation fuel from biomass are also emerging. Alcohol to hydrocarbon fuels is exemplified by the alcohol-to-jet (ATJ) pathway, where alcohols such as ethanol or butanol are deoxygenated and processed into hydrocarbon fuels [18, 19]. The ethanol or butanol for the process may be derived via the more conventional A-B-E fermentation of sugars and via novel gas fermentation that has reached demonstration deployment scales and attracted significant attention in recent years [20]. Global bioenergy has developed and scaled up a synthetic biology process for producing isobutene from feedstocks such as sugar, cereals, and agricultural or forestry waste. The second step involves catalytic oligomerization of isobutene molecules to produce iso-octane, iso-dodecane, iso-cetane, and other long-chain hydrocarbons [19]. A catalytic method for converting entire biomass into drop-in fuels has also been devised, including biomass pre-treatment, carbohydrate hydrolysis and dehydration, and catalytic upgrading of platform chemicals. Furfural and levulinic acid are produced from five- and six-carbon sugars found in hardwoods. These two platforms are then upgraded into a combination of branched, linear, and cyclic alkanes with a molecular weight range suitable for usage in the automobile and aviation industry [21]. A four-step catalytic method has been used to transform hemicellulose-derived C5 sugars into a high-quality petroleum refinery feedstock containing normal, branched, and cyclic alkanes [22]. Deployment to demonstration scales of drop-in liquid hydrocarbon fuel production for many non-lipid-based processes is in progress. Still, feedstock availability and variability challenges remain a key concern.

Global renewable diesel capacity is predicted to reach 14.63 million tonnes in 2024, increasing at a compound annual growth rate (CAGR) of 21.33% between 2020 and 2024. During the year 2020–2024, worldwide renewable diesel consumption is expected to reach 12.88 million tonnes, growing at a CAGR of 15.93%. Factors such as increased vehicle manufacturing increased biofuel demand, increased carbon dioxide emissions, availability of used cooking oils, and government emission control measures would boost market expansion. The market's growth would be hampered by a lack of feedstock and issues related to the quality control of renewable fuel [23].

This chapter discusses intellectual and technological advancements in the hydroprocessing of lipids from diverse sources to create biofuels. In the hydroprocessing of lipids, the influence of various catalytic systems, operational parameters, reaction routes, and kinetics are discussed.

1.3 Usage of Green Diesel for Marine Applications and Power Generators

Recent international efforts to reduce greenhouse gas (GHG) emissions have prompted international regulatory bodies such as the International Maritime Organization (IMO) and national environmental agencies to issue new rules and regulations that drastically reduce GHGs and emissions from marine sources. These new laws would have far-reaching consequences for the international maritime sector, the cruise industry, ship owners, and operators.

Regulations in Emissions Control Areas (ECAs) such as the North American ECA came into force in 2012, and SOx Emission Control Areas (SECAs) operate in the Baltic Sea, North Sea, and the English Channel since 2006 and 2007, respectively. Ships operating in ECAs and SECAs must use lower sulfur fuels or install sulfur oxide (SOx) exhaust scrubbers to reduce the emissions. The most significant challenges are short-term local (SECA/ECA) NOx and SOx reductions. Addressing GHG and PM emissions poses an additional environmental concern in the longer run. Alternative fuels that meet the criteria above and have particular characteristics might potentially replace the fossil fuels now in use. The following criteria must be present in these alternative fuels: (a) Fuels should not significantly influence the engine and shipboard fuel systems that must be considerably changed or replaced. (b) There should be no decrease in engine performance as a result of alternative fuel use. (c) Fuel should have low sulfur and also reduce NOx emission. (d) Due to the current heavy residual fuel oil pricing of around $700 (USD)/tonne, the alternative product must be reasonably priced. (e) Alternative fuels should be widely accessible, either globally or regionally. (f) The fuels could be blended with existing fossil fuels and pose no significant environmental problems.

However, low-sulfur liquid biofuels and fossil fuels can meet the ECAs' and MARPOL Annex VI's fuel sulfur criteria. In addition to utilizing low-sulfur fuels, scrubbers installed in the engine exhaust can be used to remove SOx. Renewable diesel with low sulfur (EN-159540), high energy density (refer to Sect. 3.2), good storage stability, low free fatty acid content (7 PPM), and excellent cold-flow qualities is a promising option to overcome the challenges discussed above (refer to Sect. 3.2). Similarly, in power generators, paraffinic diesel, being very low in sulfur and high energy density, used as an alternative option seems very promising.

2   Feedstocks and Catalysts for Hydroprocessing Technology

2.1   Feedstock

Lipids remain an industry staple for hydrocarbon biofuels, and technologies have begun to mature. Researchers from all over the world have hydroprocessed a variety of lipid sources, including soyabean oil [24–26], sunflower oil [27–30], palm oil [31, 32], rapeseed oil [33, 35], castor oil [36, 37], tall oil [38], jatropha oil [39–49], pomace oil [50], fresh and waste cooking oil [51–55], either as such or in combination with refinery feed such as gas oil [10, 39, 40, 51]. Model compounds like tristearin [56], tricaprylin, caprylic acid [57], stearic acid [58, 59], and oleic acid [60, 61], along with various lipids (jatropha oil, pongamia oil, palm stearin, palm fatty acid distillates, and used cooking oil) over different catalysts in hydrogen and inert atmospheres [43–45, 57–60, 62, 63] have been used for hydrocarbon production through hydroprocessing route. For these processes, fixed bed reactors, batch reactors, and semi-batch reactors have been investigated in great detail. Mono-functional [38, 42, 57] and bi-functional [24, 39, 41, 43, 44] catalysts have been employed depending on the intended reactions, such as hydrotreatment or hydrocracking. Hydrotreatment only eliminates hetero-elements like "O, N, and S" from hydrocarbons. It needs a hydrogenation catalyst, such as Pd, Pt-Re, or sulfided NiW, NiMo, CoMo, supported by non-acidic support, such as Al_2O_3 or activated carbon [30, 38, 40, 42, 56, 64]. On the other hand, hydrocracking reactions aim to produce a wide range of hydrocarbon distillates and require bi-functional catalysts that combine hydrogenation and acidic functionality, which are incorporated by supports such as zeolites silica-alumina, silicoaluminophosphate, titanosilicate, and others [24, 39, 41]. During lipid hydroprocessing, these bi-functional catalysts conduct hydrogenation/dehydrogenation, cracking, isomerization, cyclization, and aromatization reactions.

2.2   Catalyst Preparation for Lipid Hydroprocessing

Lipids are mostly made up of large triglyceride molecules and are relatively consistent in structure and carbon chain distribution for a given feedstock source. For hydroprocessing, it is necessary to have these molecules pass through the catalyst surface's porous structure to reach the active sites.

The generic catalyst production and selection parameters are catalytic activity, stability, longevity, and regenerability. Furthermore, lipids generate water upon hydrotreatment, emphasizing the importance of hydrothermal stability as an additional essential parameter for catalyst selection. Over the past two decades, with significant advances in molecular modeling, elucidation of catalytic mechanisms, and characterization of catalyst surfaces, hydroprocessing catalysts have evolved rapidly, with many traditional systems being replaced by novel non-conventional catalysts. Variations in catalyst support composition, active metal composition and loading, usage of different kinds of active metals, additives, and other factors have all been investigated to obtain the required characteristics [65–67]. Economics and sustainability being other vital factors, there has been a substantial global effort toward reducing precious metals content in catalyst compositions.

Many materials, including clays [68], carbon [30, 38, 40, 42, 56, 64, 69, 70], oxides [71–73] like SiO_2, MgO, ZrO_2, TiO_2, and mixed oxides derivatives of above-mentioned oxides such as TiO_2–ZrO_2, TiO_2–Al_2O_3, SiO_2–TiO_2 [69, 74–80], zeolites like Na–Y, USY, HY, SAPO [37, 46, 81–83], mesoporous materials like MCM-41 [84–88], HMS [89], SBA-15 [90, 91], nano-hydroxyapatite [49] have been identified as support materials for hydroprocessing catalysts. Traditional alumina and zeolite-based catalysts have lower hydrotreatment and hydrocracking activity compared to the mesoporous aluminosilicate MCM-41. MCM-41 provides a rather appealing mix of large surface area, homogenous pore size distribution, low acidity, and good stability [89]. Synergetic effects with enhanced activity and stability have been reported in TiO_2–ZrO_2, TiO_2–Al_2O_3, and SiO_2–TiO_2-supported catalysts [69, 74–80].

Kinetics and reaction pathway prediction studies for various reactions occurring during lipid hydroprocessing have also been studied, and their relative extent affects the yields of various biofuel components. Hydrocracking and hydroisomerization processes involve the following reactions simultaneously: depropanation (C_3H_8 elimination), hydrodeoxygenation (H_2O elimination), decarboxylation (CO_2 elimination), and decarbonylation (CO elimination) reactions during lipid hydroprocessing [24, 29, 39, 41, 43–45, 62, 63, 91]. Water–gas shift, methanation, oligomerization, cyclization, and aromatization are inevitable gas phase and side reactions [43–45, 61–63].

Different catalyst supports with varying porous structures and hydrogenating/dehydrogenating functionality on these supports in the form of active metals such as Pt, Pd, Ni, Co, Mo, and W are discussed in this section.

This section briefly outlines the synthesis procedure of different support catalytic materials.

Hierarchical mesoporous ZSM-5 is prepared by the modified method of Choi et al. [92]. Verma et al. [41] and Zhang et al. [93] synthesized mesoporous hierarchical SAPO-11 in an aqueous medium. Mesoporous titanosilicates (MTS) [94] were prepared using a modified sol–gel method, and MCM-41-supported catalyst and APTES–USY-modified MCM-41 catalysts were synthesized using Wang et al. [86] and Liu et al. [37] methods, respectively. After preparing the support material of the catalyst, active metals and promoters such as Pt, Pd, Ni, Co, Mo, and W are deposited over these different supports either simultaneously or consecutively using the incipient wetness impregnation method. Catalysts with metals in their reduced state (such as Pt, Pd, Ni) or sulfided metals (Ni/Mo, Co/Mo, Ni/W) have been evaluated and found to favor in commercial systems for hydroprocessing of various lipids.

3 Catalyst Effects on Lipid Hydroprocessing

3.1 Fatty Acid Profile of Lipids and Implications on Catalyst Selection

Triglycerides produced from fatty acids (Table 1) with carbon values ranging from C14 to C22 make up the lipids in almost all known vegetable oils, animal fats, and microbial oils. These oils are generally quite viscous, and the boiling point typically ranges from 250 to 350 °C. The density is in the range of 0.80–0.99 g/ml [95].

Hydroprocessing these fatty triglyceride molecules selectively to specific products can be challenging due to their high viscosities and large structure of the triglycerides. Table 1 shows the typical fatty acid composition of oil derived from different sources. Larger molecules of triglycerides are not easy to diffuse through catalyst pores to active sites. Optimum pore size, catalyst acidity, and hydrogenation functionality play an essential role in the reaction rate, catalyst life, and product yield distribution. Process parameters such as temperature (to control the cracking characteristics), pressure (to ensure the larger triglyceride molecule reaches the active metal sites), H_2/oil ratio (to ensure sufficient hydrogen availability at metal sites and reduce coke formation), and weight hourly space velocity (reaction time) are essential parameters to control the product yield and quality. Reactor type may sometimes play an important role in controlling the highly exothermic reaction occurring during the triglyceride conversion. Catalyst characteristics must retain the flexibility to be adjusted if multiple lipids (Refer to Table 1) are considered feedstock for a single process unit. Such adjustments enable the operator to maximize the desired components in the final fuel hydrocarbon composition.

The chemical composition of these lipids and the degree of unsaturation play an essential role in determining the reaction type in lipid hydroprocessing. To provide the appropriate cold-flow qualities, lipids from diverse feed sources such as soybean, sunflower, camelina, jatropha, castor, and spent cooking oil are hydrodeoxygenated, followed by isomerization. In some technologies, isomerization and deoxygenation steps are conducted simultaneously in a single reactor. From the lipids mentioned above, slight cracking, moderate isomerization, and hydrodeoxygenation processes are needed to meet the desired diesel specifications. Moderate to high cracking processes would enhance the production of C12–C15 range hydrocarbons in aviation kerosene under more severe circumstances. In the latter portions of this chapter, these correlations and response mechanisms are addressed.

3.2 Comparison of EN-15940 (HVO) and EN-590

Although a separate book chapter compares renewable diesel (EN-15940) to commercial crude-based diesel (EN-590), a summarized comparison is discussed here.

Table 1 Fatty acid composition of lipids from various sources

Fatty acid	Fatty acid (wt%) [16, 43, 110]								
	Myristc	Palmitic	Palmitoleie	Stearic	Oleic	Linoleic	Linolenic	Archidic	Behenic
Crop oil	14:00	16:00	16:01	18:00	18:01	18:02	18:03	20:00	22:00
Corn	0.1	12.2	0.1	2.2	27.5	57	0.9	0.1	
Peanut		11.6	0.2	3.1	46.5	31.4		1.5	3
Soyabean	0.1	11		4	23.4	53.2	7.8	0.3	0.1
Sunflower	0.2	6.8	0.1	4.7	18.6	68.2	0.5	0.4	
Camelina[a]		7.8		3	16.8	23.1	31.2		
Jatropha		19.5		7.9	45.4	27.3			
Palm	0.9–1.5	39.2–45.8	0.0–0.4	3.7–5.4	37.4–44.1	8.7–12.5	0.0–0.6	0.0–0.5	
Castor[b]		1		1	3	4.2	0.3	0.3	
Rape seed	4.3	1.7	63.7	15.4	14.3				
Algal		51		2	39	7			
Olive	0.1–1.2	7–16		1–3	65–80	4.0–10.0		0.1–0.3	
Rice bran	0.4–1	12–18	0.2–0.4	1–3	40–50	29–42	0.5–1		
Sesame		7–9		4–5	40–50	35–45		0.4–1	
Neem	0.2–2.6	13.6–16.2		14.4–24	49–62	2.3–15.8		0.8–3.4	4.2–5.3
Karanj		3.7–7.9		2.4–8.9	44.5–71.3	10.8–18.3		2.2–4.7	
Mahua		20–25		20–25	41–51	10–14		0–3.3	
Sal		4.5–8.6		34.2–44.2	41.4–42.2	2–3		6.3–12.2	

a 12% C20:1 and 2.8% C22:1, b 89.5% Ricinoleic acid (C18H34O3)

Renewable fuel shares many of the same characteristics as high-quality sulfur-free petroleum diesel. Previously, synthetic gas-to-liquids (GTL) diesel was one of the options for engines and emissions. Renewable diesel offers many of the same compositional advantages and characteristics as GTL diesel, and it is also totally renewable.

Like crude-based diesel (EN-590), synthetic or hydrotreated paraffinic diesel (EN-15940) fuel is a complex mixture of hydrocarbons that varies depending on the feedstock and manufacturing method. It is difficult to ascertain the precise composition of paraffinic diesel fuel using synthetic procedures. Paraffinic diesel fuel is derived from the synthesis of gas or hydrotreated oils, bio-oils, or fats such as cyclo-paraffins, aromatics, and iso-paraffins. It is suitable for diesel-powered vehicles, power generators, marine transport vessels, and other engines. Hydrotreated vegetable oil diesel, commonly known as green diesel or paraffinic diesel, can be generated using a catalyst in a single step or numerous steps. The fuel produced by hydrotreatment of oils, bio-oils, or fats is primarily paraffinic and may not meet the EN-590 density standard for traditional crude-based diesel. However, paraffinic diesel has better burning characteristics (higher cetane number) compared to commercial diesel.

The paraffinic diesel fuel generated may be divided into two major types based on the world's production procedures documented so far. (a) Paraffinic diesel fuel with better ignition quality (Class A) than automobile diesel fuel satisfies EN-590 criteria. (b) Paraffinic diesel with an average cetane number (Class B). The EN-15940, which is a specification on paraffinic diesel, differs from commercial diesel in terms of the following: (a) Compared to the EN-590, the cetane number for high cetane paraffinic diesel (Class A) has been increased to 70 from 51. However, the limit for Class B (low cetane paraffinic diesel) remains unaltered. (b) The cetane index is eliminated since there is insufficient data for the cetane index to restrict the same. (c) Pout point has been eliminated in EN-15940 because CFPP and cloud point specifications are the different fuel grades' cold flow parameters. (d) Distillation, the maximum volume recovered at 360 °C is 95% in the case of EN-15940 while it remains 65% @ 250 °C (max) and 85% @ 345 °C (min). (e) The flashpoint has been changed from 35 to 55 °C (min); (f) The density range has been changed from 765–800 (Class A) to 780–810 (Class B) kg/m^3 for EN-15940; it was 820–835 for EN-590. (g) For cloud point specification, EN-590 defines it as −5 °C (winter grade) and +3 °C (summer grade). The cold flow properties for EN-15940 are defined as the cold filter plugging point (CFPP). EN-defined CFPP in five classes (A–E) is based on the weather condition and temperature achieved in that geographical location (−44, −38, −32, −36, and −20 °C). h) The maximum manganese concentration (mg/L) is 2.0 for EN-15940, which is not part of the specification in EN-590.

Green diesel/paraffinic diesel is entirely compatible with standard petroleum-derived diesel fuel blends, offering refiners a significant advantage. The method for producing paraffinic diesel is environmentally friendly and compatible with existing petroleum refineries. It is almost indistinguishable from conventional diesel fuel and may be used as a straight replacement or as a beneficial blendstock to improve the quality of existing diesel pools. Because paraffinic diesel is chemically identical to ordinary diesel fuel, it may be used in today's tanks, pipelines, trucks, pumps, and cars without modification.

3.3   Co-processing of Lipids with Refinery Feed

Lipids have been co-processed with refinery streams, especially gas oil (both light and heavy gas oils), to make it similar to refinery feed so that the existing refinery infrastructure could be used for further processing without adding any extra capital costs. Non-acidic sulfided metal-based catalyst system $NiMo/Al_2O_3$ has been used to co-process jatropha [39], soybean [24, 96], sunflower [28, 29], waste cooking oil [97], and rapeseed oil [35] under hydrotreating conditions. Co-processing lipids with gas oil (heavy and vacuum) over hydrocracking catalysts has received less attention [28, 96]. Table 2 shows lipids co-processing (10–30% lipid in gas oil) in the temperature range of 350–370 °C, 50–130 bar H_2 pressure, and 1–5 h^{-1} space velocity over a hydrotreating catalyst. Lipid co-processing over a hydrocracking catalyst is described in Table 3. With respect to lipid feeds, the primary goal of hydrotreating is the removal of heteroatoms without significant cracking so that primarily diesel range components are formed. In contrast, over hydrocracking catalysts, due to the cracking of diesel range hydrocarbons into middle distillates, the yield of kerosene range hydrocarbons along with naphtha and gas range cracked by-products tend to increase.

The oxygen in lipids, which is present in significant 10–12% levels, can be analogous to a heteroatom found in fractionated crude oil products, such as sulfur and nitrogen. Before any further processing or use of products in downstream operations, these heteroatoms must be eliminated. Hydrodeoxygenation (–H_2O), decarboxylation (–CO_2), and decarbonylation (–CO) are all deoxygenation processes that compete for active sites with hydrodesulfurization (–H_2S). It was expected that these renewable feedstocks would have a detrimental impact on the catalyst and the deoxygenation process. Since the oxygen present in the lipids replaces sulfur present in the sulfided form of the catalyst (in cases of sulfided catalyst), there are significant chances of sulfur leaching from the catalyst surface. An additional agent may be required to avoid sulfur leaching so that the catalyst activity is retained.

Both Bezergianni et al. [97] and Huber et al. [29] have reported that increasing the lipid percentage (cooking and sunflower oil, respectively) in gas oil would increase the percent HDS achieved (Table 2). In comparison to other co-processed lipids, high HDS activity was reported for jatropha (96%) and soybean (89%) oil, while only 34% HDS activity was reported for sunflower oil. It may be attributed to less severe conditions of operation, i.e., at very high space velocities (5.2 h^{-1}) as compared to other co-processed lipids (Table 2).

Table 2 Yield of different products obtained by co-processing of lipids from various sources with refinery gas oil over **a** hydrotreating and **b** hydrocracking catalysts. [a]NiMo/Al$_2$O$_3$ catalyst at 50 bar pressure

Feed (space velocity)	350 °C (10–15% lipid)					370 (20–30% lipid)			
	Jatropha (2 h^{-1}) [39]	Soyabean* (4 h^{-1}) [24, 96]	Sunflower (5.2 h^{-1}) [29]	Rapeseed (2 h^{-1}) [35]	Cooking (1 h^{-1}) [97]	Soybean* (2 h^{-1}) [24, 96]	Sunflower (5.2 h^{-1}) [29]	Rapeseed* (2 h^{-1}) [35]	Cooking (1 h^{-1}) [97]
Naphtha (IBP-150)	0	0	0.09	0	0	0	2	0	1
Kerosene (150–250)	9	5	0.05	32.4	5	4.5	1	34.4	8
Diesel (250-FBP)	91	95	99.86	67.6	95	95.5	97	65.6	91
HDS %	96	89	34		92%		48		96%
C17/C18		1.6	0.49			0.75	0.9		
C15/C16		1.15	0.5			1	0.75		
CO			0.08				0.66		
CO$_2$			0.025				0.07		
Density, g/cc	0.839	0.846		0.835	0.8508			0.831	0.84

*380 °C

Table 3 Yield of different products obtained by co-processing of lipids from various sources with refinery gas oil over **a** hydrotreating and **b** hydrocracking catalysts. [b]Hydrocracking catalyst at 1000–1500 Nl/L H$_2$/feed ratio

Feed [Reference]	Soybean[*] [96]	Sunflower[+] [28]	Soybean[*] [96]
% Lipid	25	30	40
Naphtha (IBP-150)	0	10	0
Kerosene (150–250)	35	26	30
Diesel (250-FBP)	65	64	70
HDS%	84		
C17/C18	1.6		1.58
C15/C16	1.5		1.48
Density, g/cc		0.769	0.818

+ 350 °C, 1.5 h−1; *370 °C, 2 h−1

3.3.1 Effect of Space Velocity

Cracking reactions are maintained to a minimum, and only hydrotreatment reactions are targeted to maximize the diesel yield. Renewable diesel range (250-FBP) products are mainly generated during co-processing reactions (Table 2). For the co-processing of soya bean [24, 96] and sunflower [8] oil, high throughputs at 4 and 5 h^{-1} space velocity have been recorded. Cooking [97], rapeseed [35], and jatropha [5] oils, on the other hand, have all been co-processed at lower space velocities of 1–2 h^{-1}. Due to moderate cracking reactions, a decrease in diesel range hydrocarbons (250-FBP) was reported, which subsequently increased the kerosene range (150–250) products for the lipid mentioned above sources at a lower space velocity (1–2 h^{-1}). On the other hand, rapeseed oil yielded 32% kerosene range hydrocarbons and approx. 67% diesel range hydrocarbons, owing to co-processing with a light gas oil combination that included a 28% kerosene range in the feed (Table 2).

3.3.2 Effect of Temperature

At both 350 and 370 °C, products in the 10% kerosene range could be produced. When the percentage of lipid fraction in gas oil is increased, the acidity of the reaction medium increases due to increased production of acidic intermediates at high lipid concentrations and higher temperatures, resulting in a minor rise in cracked product yield (Table 2).

3.3.3   Feed Sources

In Table 2, when the lipid percentage is increased (for the same lipid), the diesel yield is observed to increase slightly in the case of soybean but decreases for rapeseed oil and waste cooking oil. Sunflower and used cooking oil show higher diesel yield compared to jatropha, soybean, and rapeseed oil.

Hydroprocessing of lipids from soybean and sunflower sources over hydrocracking catalyst provided higher yields of kerosene range hydrocarbons (25–35%) at 370 °C and 2 h^{-1} space velocity, in contrast to $NiMo/Al_2O_3$ (low acidity) catalysts (Table 3). As a result of breaking these diesel range hydrocarbons into lower range distillates, a reduced diesel yield was attained. Decarboxylation and decarbonylation are carbon rejection processes commonly assessed by odd-numbers carbon percentage, such as C17 or C15, whereas hydrodeoxygenation is a hydrogen addition reaction represented by the percentage of C18 or C16 [8, 24, 96]. C17/C18 and C15/C16 ratios for sunflower oil at 5 h^{-1} have been reported (Table 3), demonstrating an increase in carbon rejection processes at higher temperatures (370 °C) over lower temperatures (350 °C). The preceding result is supported by a significant rise in carbon monoxide content (8 times) at high temperatures (370 °C) owing to decarbonylation during co-processing. When comparing hydrotreating and hydrocracking catalysts, the latter showed more significant carbon rejection reactions (higher C17/C18 and C15/C16 ratios than the former) (Tables 2 and 3).

The density of products generated by co-processing sunflower oil over a hydrocracking catalytic system was lower at lower 350 °C temperatures than soybean feedstock due to improved cracking reactions (naphtha production). As expected, the density of products produced by hydrocracking catalytic systems was lower on average than that of products produced by hydrotreating catalytic systems (Table 2).

3.4   Direct Processing of Lipid Sources

Lipids have been treated directly in the presence of noble metal and sulfided base metal catalyst systems supported over acidic and non-acidic supports in both hydrotreating and hydrocracking conditions. Tables 4 and 5 show how different lipid sources are processed at various temperatures using hydrotreating and hydrocracking catalytic systems.

3.4.1   Hydrotreating and Hydrocracking of Lipids

Tables 2 and 3 shows the comparison of hydrotreating and hydrocracking catalyst effect on the product yield distribution, HDS percentage (in the case of the co-processed oil), and decarbonation-deoxygenation activity. The cracked product increased from 5 to 35% in the case of soybean oil for hydrocracking catalyst, and by 26% for sunflower oil in similar temperature ranges. Although the increased cracked

Table 4 Influence of temperature variation on product distribution during hydrotreatment of lipids from various sources

Lipid source	Castor [36]		Palm [31]			Sunflower [29]			Soyabean [24, 25]		Jatropha oil [43]			Cooking [47, 48]			Waste cooking oil [98]			
Catalyst system	5% Pd/C		NiMo/Al$_2$O$_3$			NiMo/Al$_2$O$_3$			NiMo/Al$_2$O$_3$		CoMo/Al$_2$O$_3$			Commercial HDT catalyst			NiMo/Al$_2$O$_3$			
Pressure, bar (space velocity, h^{-1})	25		50 (1)			50 (5.2)			50 (2)		80 (4)			81(1)			30			
Temperature	300	340	300	330	420	300	350	420	380	400	320	340	360	330	350	398	300	330	360	400
Naphtha	0	0				5.4	6.3	12	0	2.5	1	1.5	3.35	0	0	5	-	-	-	-
Kerosene	0	0	0.8	1	12.8	0.4	1	5	3	7.5	2.5	3	7.13	5	5	15	-	-	-	-
Diesel	40	100	95	93.7	41.6	93.2	91.8	76	97	90	85	88	85	95	95	80	87.83	91.82	94.34	98.72
C15/C16		0*	0.2	0.2	0.2	1	1	0.8		1.9				0.8	1.2	1.4	-	-	-	-
C17/C18	28*	9.7	0.2	0.2	0.2	0.4	0.7	0.8		2.5				1.1	1.3	1.8	-	-	-	-
i/n (kerosene)	0	0				0.2	0.05	0.3		0.5				0.1	0.1	0.6	-	-	-	-
Conversion	100	100	100	100	100	100	100	100	100	100	92.2	95.1	98.5	92.5	92	96.5				100

*Methyl stearate

product may also be attributed to the lower space velocity reported on the hydro-cracking catalyst (2 h^{-1} compared to 5 h^{-1}), the hydrocracking catalyst showed decarbonation activity twice that of the hydrotreating catalyst for soybean feedstock (Tables 2, 3).

From Table 4, for non-noble metal, the conversion of 100% lipid to hydrocarbon fuel increases with an increase in temperature from 300 to 340 °C. A complete conversion of castor oil to diesel range hydrocarbon has been reported by Meller et al. Over Pd/C catalyst at 25 bar pressure and 300 °C, only 40% of the diesel range yield was reported, compared to > 90% in the case of NiMo(S)-alumina catalyst. Meller et al. [36], Sirifa et al. [31], Tiwari et al. [24], Veriansyah et al. [25], Huber et al. [29], and a few others [36, 40, 41, 98] published the conversion of castor, palm, soybean, jatropha, used cooking, and sunflower oil at 300 °C (Table 4) to diesel and aviation range hydrocarbons.

Increasing the temperature resulted in a minor decrease in diesel production and increased kerosene and naphtha products yield for palm, sunflower, soybean, and jatropha oil. Used cooking oil also has a similar increase in diesel yield with increased temperature over NiMo/Al$_2$O$_3$ catalyst [24, 25, 29, 31, 36, 43, 47, 48, 98]. At higher temperatures, cracking and isomerization reactions are enhanced, resulting in higher kerosene and naphtha yields. Palm oil had a higher drop in diesel yield because of the lower space velocity (palm: 1 h^{-1} versus sunflower: 5.2 h^{-1}) and composition.

Increased cracking has been reported for soybean oil processed over a NiMo/Al$_2$O$_3$ catalyst in the range of 50–90 bar pressures. It was observed that the diesel range hydrocarbons are hydrocracked into smaller molecules (naphtha and kerosene) when the temperature was increased from 380 to 400 °C (Table 4).

At higher pressures of 80 bar, CoMo/Al$_2$O$_3$ and commercial hydrotreating catalytic systems were employed to process jatropha [43] and cooking [53, 54] oil. In the presence of CoMo/Al$_2$O$_3$ with jatropha oil, 92% conversion was observed with 85% diesel yield (Table 4), compared to commercial hydrotreating catalyst (95% diesel yield) using cooking oil as input at lower temperatures. For CoMo/Al$_2$O$_3$ systems and jatropha feed, low lighter cracked products such as naphtha and kerosene (yield = 3.5–10.5%) were reported in the temperature range of 320–360 °C, with the space velocity of 4 h^{-1} with conversion between 92 and 98%. The production of acidic intermediates on processing lipid feedstocks over a non-acidic Co-Mo/Al$_2$O$_3$ catalyst was observed by Anand et al. [43], which justified the rise in cracking events under less severe hydrotreating conditions. Increased temperature sensitivity (high cracking) was observed for palm and sunflower feed over NiMo/Al$_2$O$_3$ catalyst (Table 4).

The ratio of C15/C16 and C17/C18 indicates hydrodeoxygenation over decarbonation reaction. A higher decarbonation reaction is indicated by the higher C15/C16 or C17/C18 ratio. For the CoMo catalyst, 0.8 (C15/C16) and 0.7 (C17/C18) were reported. Because of the acidic SiO$_2$-Al$_2$O$_3$ support, which catalyzes C–C bond breakage and the generation of CO and CO$_2$, NiW/SiO$_2$-Al$_2$O$_3$ catalytic systems favored decarbonation processes over hydrodeoxygenation reactions (1.3 (C15/C16); 2.2 (C17/C18)) [39]. At higher temperatures, all feed showed an improvement in decarbonation selectivity (higher C15/C16 and C17/C18 ratio) over hydrodeoxygenation selectivity (H_2O elimination), except palm oil, which showed low decarbonation selectivity (low C15/C16 and C17/C18 ratio). It may be inferred that the heavier lipids like castor, jatropha, soybean, sunflower, and cooking oil favored the carbon rejection process at higher temperatures. Still, lighter lipid sources like palm oil showed hydrodeoxygenation for oxygen removal (Table 4). These data also suggest that hydrogen requirement in deoxygenation processes for palm oil may be higher than for other lipid sources.

3.4.2   Conventional Sulfided Catalysts

Catalytic systems based on NiMo, NiW, and CoMo have been used to convert lipids derived from soybean, algae, jatropha, and waste cooking oil to hydrocarbon fuel. Mo and W act as active metal sites, providing the catalyst with the necessary hydrogenation/dehydrogenation functionality, while Ni and Co act as promoters.

Table 5 compares the sulfided and non-sulfided catalyst systems. Except for used cooking oil (UCO), all lipid feedstocks were converted nearly completely. UCO conversion ranges from 72 to 82% as temperature increases (Table 5).

Bezergianni et al. [51] have defined unconverted feedstock as a boiling fraction over 360 °C in their conversion calculations. The temperature history of the feed may result in the dimerization of higher boiling lipid derivatives, resulting in the unconverted component in the final product.

Instead, Bezergianni et al. [51], Cheng et al. [26], Kumar et al. [39], Heriyanto et al. [98], Baladincz et al. [99], and Verma et al. [41, 46] investigated the conversion of lipids from various sources such as soybean, algae, jatropha, and used cooking oil. For jatropha and cooking oil lipids, high renewable diesel yields of about 60–80% were reported using NiMo and NiW-based catalysts at temperatures below 400 °C and pressures between 50 and 130 bar (Table 5) [26, 37, 39, 41, 42, 46, 51]. The highest diesel yield (98%) was observed for NiMo/Al_2O_3 (hydrotreating catalyst) for UCO as feedstock.

The yield of the diesel fraction decreased when the temperature was raised, and additional cracked products, such as kerosene and naphtha, were reported. By lowering the temperature from 450 to 400 °C, mesoporous silicoaluminophosphate SAPO-11 (MSP)-based catalytic systems supported by NiMo and NiW active metals demonstrated comparable yield patterns with higher green diesel yield and lower aviation and naphtha yield (Table 5).

From Table 5, maximum diesel yield was observed for NiW/MSP-2 catalyst at 400 °C, 70 bar, and 1–1.5 h^{-1}. However, when the temperature was decreased, the iso-paraffin to normal paraffin (i/n) ratios decreased significantly, showing a decrease in isomerization selectivity over SAPO-11-based catalytic systems at lower temperatures (Table 5). This decrease in isomerization yield with reducing temperature may not necessarily hamper the cold flow properties of the diesel significantly. Still, it would need to be optimized for each feedstock/catalyst/process conditions combination. It may be noted that cold flow properties and pour point are less critical in

Table 5 Product distribution for hydroprocessing lipids from various sources over different sulfided hydrocracking catalyst systems (WHSV1-1.5 h^{-1})

Lipid source [Reference]	Soyabean [26]		Algal [41]	Jatropha [39]	
Pressure, bar	40		50	50	
Catalyst system	NiMo/HY		NiMo/Meso-ZSM-5	NiW/SiO$_2$-Al$_2$O$_3$	CoMo/Al$_2$O$_3$
Temperature, °C	390	410	410	360	360
Naphtha					
Kerosene	49.1	55	78.5	19.7 (<C15)	40.2(<C15)
Diesel	18.2	5		80.8	49.2
C15/C16				1.3	0.8
C17/C18				2.2	0.7
i/n			2.5	1.1	0.3
Conversion	98	100	98	100	100

(continued)

Table 5 (continued)

Lipid source [Reference]	Jatropha [46]								Cooking [51]			10% waste lard + 90% Gas oil [99]	Waste Cooking Oil (Two stage reaction) [98]
Pressure, bar	70								130			50	30
Catalyst system	NiMo/MSP-1		NiMo/MSP-2		NiW/MSP-1		NiW/MSP-2		Commercial HC catalyst			NiMo/Al$_2$O$_3$	NiMo/Al$_2$O$_3$
Temperature, °C	400	450	400	450	400	450	400	450	350	370	390	340	First stage: 300 °C, Second stage: 400 °C
Naphtha	17.3	37	5.6	30.5	14	40	11	28.5	0	1.8	1.2		
Kerosene	8	28.5	8	34.5	3.5	24.5	4	37.5	5.3	8.2	19.9		
Diesel	74.7	34.5	86.4	35	82.5	35.5	85	34	66.7	65	61	95.1–95.5	98.93
C15/C16	0.5		0.3										
C17/C18	0.7		0.7										
i/n	0.8	2.3	2.5	3.1	1.5	3.5	0.8	3.8					
Conversion	100	100	100	100	100	100	100	100	72	75	82	~100	~100

+ cat: jatropha: H2O ratio (1:1:9; *1:10:9); CNT: Carbon nanotubes; HY: Y zeolite (H-form); Meso-ZSM-5: Hierarchical mesoporous H-ZSM-5 zeolite; MSP: Hierarchical mesoporous SAPO-11(MSP-1 (Si/Al: 0.4) & MSP-2 (Si/Al: 0.27) the case of green diesel than the freezing point in aviation fuel and that increased i/n ratio is generally accompanied by some increased cracking and loss of liquid yield. Thus, green diesel producers usually have to balance cold flow property targets and process economics.

On the other hand, poor diesel yields were reported over the CoMo/Al$_2$O$_3$ catalyst (Table 5), combined with a 40% yield of cracked (C15) products. Anand et al. [43, 45] reported a similar improvement in fractured product yield when using a CoMo/Al$_2$O$_3$ catalytic system. Increased cracking over a CoMo catalyst supported on low-acidity Al$_2$O$_3$ support implies the formation of acidic intermediates in the reaction medium that promoted the cracking processes. A higher decarbonation reaction is indicated by the higher C15/C16 or C17/C18 ratio; 0.8 (C15/C16) and 0.7 (C17/C18) were reported for CoMo catalyst. Because of the acidic SiO$_2$-Al$_2$O$_3$ support, which catalyzes C–C bond breakage and the generation of CO and CO$_2$, NiW/SiO$_2$-Al$_2$O$_3$ catalytic systems favored decarbonation processes over hydrodeoxygenation reactions (1.3 (C15/C16); 2.2 (C17/C18)) [39].

At 400 °C and 70 bar pressure, green diesel yield of 86.4% was observed for jatropha oil over NiMo/MSP catalyst, whereas at higher temperatures of 400 °C and higher pressures (130 bar), 61–66% yield of aviation kerosene range hydrocarbon was observed for cooking oil over commercial hydrocracking catalyst. Maximum aviation kerosene production of 78% (at 410 °C) was observed for algal oil lipids over a hierarchical mesoporous zeolite-supported (ZSM-5) NiMo catalyst [41], indicating that these zeolites have better lipid breaking ability than other catalytic systems.

In conclusion, lower reaction temperature and lower pressures showed higher green diesel yield for hydroconversion reaction over jatropha and castor oils. Soybean and algal oil have higher kerosene yield compared to green diesel.

Figure 1 describes the product yield distribution for various vegetable oil hydrocracking across multiple catalysts. Maximum diesel range (30–90%) hydrocarbon was observed in the temperature range of 350–380 °C and 80 bar pressure for different sulfided catalysts. The NiMo/H-ZSM-5 (HSAC) sample, which contains hierarchical porous, high surface area crystalline H-ZSM-5 zeolite with framework restricted mesoporosity as support, yielded the maximum production of aviation kerosene (60%) from jatropha oil, with minimal diesel. When algal oil was utilized as feed, the same catalyst produced the highest known output of aviation kerosene (80%) [10].

The highest naphtha yield was observed for jatropha oil supported on mesoporous silica-alumina, hierarchical mesoporous H-ZSM-5, and titanosilicate (MTS) support, loaded with sulfided NiMo, NiW, and CoMo active metals, in the temperature range of 380–420 °C. Even at a lower temperature (300°) and pressure (30 bar), high naphtha yield (60–90%) was reported on Ni catalyst supported on H-ZSM-5. The increased naphtha yield may be attributed to the Si/Al ratio (38–100).

However, cracking of jatropha oil was low over CoMo/MTS systems at decreased temperatures of 360 °C, with 80% renewable diesel as the main output, followed by aviation kerosene (20%) and insignificant naphtha production (Fig. 1). Even over low

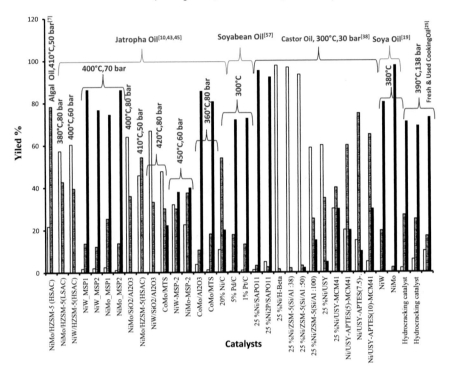

Fig. 1 Product yield distribution for hydroprocessed vegetable oils over different catalysts (Naphtha; Kerosene; Diesel)

acidic supports like MTS, an increase in acidic intermediates was observed [39, 43, 45], in addition to increased reaction severity at a higher temperature, which led to cracking and isomerization reactions, as reported for CoMo-alumina (Table 4) [43].

3.4.3 Non-conventional Non-sulfided Catalysts

Table 6 shows the comparison of non-sulfided catalysts for the conversion of lipids to hydrocarbons. Lipids from jatropha and castor oils were treated on various substrates using reduced Pt and Ni catalysts. The temperature has been observed to be the most significant parameter affecting the lipid conversion; low (15–30%) conversions were reported at lower temperatures (270 °C), while higher conversions (>95%) were observed beyond 300 °C temperature [42]. Nearly complete (99%) conversions for castor oil were achieved at temperatures (300 °C) and low pressures of 30 bar with Ni catalyst supported on different zeolites such as SAPO-11, H-Beta, USY, and APTES-MCM-41 [37]. Murata et al. [42] reported that using the Pt/H-ZSM-5 catalytic system for jatropha feeds enhanced the formation of cracked naphtha range

Table 6 Product distribution for hydroprocessing of lipids from various sources over different non-sulfided hydrocracking catalysts (WHSV1-1.5 h^{-1})

Lipid source	Jatropha [42]				Castor [37]				
Pressure, bar	65 (H$_2$/N$_2$: 85/15) $^+$				30				
Catalyst system	Pt/H-ZSM-5	Pt/USY	Pt/CNT	Pt/H-ZSM-5*	25% Ni/SAPO-11	25% Ni/H-Beta	25% Ni/ZSM-5	25% Ni/USY	25%Ni/USY-APTES-MCM-41
Si/Al	23	6.3		23			38		
Temperature	270				300				
Naphtha	8	0	0	18.2	1.2	97.7	96.8	79.4	13.8
Kerosene	3.8	1.4	13.3	6	3.2	1.5	2	19.2	80.3
Diesel	88.4	98.6	86.7	75.8	95.4	0	0	0.8	5.4
C15/C16	0	0.8	40.9	1.9					
C17/C18	0.4	0.5	0.7	0.1					
i/n					0.1	0	0	5.3	4.4
Conversion	31.2		13.6	14.2	99	99	99	99	99

(continued)

Table 6 (continued)

Lipid source	Palm Oil [100]			Palm Oil [111]			
Pressure, bar	50			60			
Catalyst system	Ni_pP_y	Co_2P	Cu_3P	0.50% Pd/γ-Al$_2$O$_3$	0.38% Pd, 0.12% Fe/γ-Al$_2$O$_3$	0.25% Pd, 0.25% Fe/γ-Al$_2$O$_3$	0.12% Pd, 0.38% Fe/γ-Al$_2$O$_3$
Si/Al							
Temperature	350			400			
Naphtha	–	–	–				
Kerosene	–	–	–	43.02	40.54	36.36	31.71
Diesel	62.5	30	n.d	50	51.35	56.06	62.2
C15/C16	–	–	–				
C17/C18	–	–	–				
i/n							
Conversion	100	95	75				

+cat: jatropha: H2O ratio (1:1:9; *1:10:9); CNT: Carbon nanotubes; HY: Y zeolite (H-form); Meso-ZSM-5: Hierarchical mesoporous H-ZSM-5 zeolite; MSP: Hierarchical mesoporous SAPO-11(MSP-1 (Si/Al: 0.4) & MSP-2 (Si/Al: 0.27) products, whereas using the Pt/carbon nanotubes catalytic system resulted in the higher forms of kerosene range molecules at equal conversion levels (Table 5). This might be due to Pt's carbon nanotube support that reduced acidity than H-ZSM-5's. When compared to other Pt-loaded catalytic systems (Table 6), ultra-stable Y (USY) zeolites gave two times the conversions (31%), suggesting that Pt/USY is a superior catalyst at lower temperatures (270 °C) for enhanced diesel generation (98.6%). When the temperature was raised to 300 °C, the 25% Ni/USY demonstrated complete conversions of castor oil, as well as a 79.4% naphtha yield and a 19.2% kerosene output, with insignificant diesel yield (Table 6). With a 25% Ni loading, a similar yield pattern was found for the H-Beta and USY catalytic systems, with enhanced naphtha yield reported. These results show that highly acidic catalytic systems with a 25% Ni loading improved cracking and isomerization processes even at 300 °C and 30 bar pressure. Conventional sulfided NiMo, CoMo, and NiW-based catalyst systems, on the other hand, enhanced these reactions at higher temperatures > 360 °C and pressures of 30 bar (Table 5). However, the Ni/zeolite-based catalysts deactivate faster than traditional sulfided catalysts supported on mildly acidic silica-alumina supports, which is a crucial constraint for commercial success.

Lower acidity and moderate isomerization activity are required for diesel range hydrocarbon production. Liu et al. [37] showed for castor oil lipids that a moderate acidity catalytic system could raise the aviation range hydrocarbon production to as high as 80% (Table 6). Combination of USY, APTES ((3-aminopropyl)-tri-ethoxysilane), and MCM-41, the kerosene range hydrocarbon yield was approximately six times higher (Fig. 1). The strong acid sites were inhibited when MCM-41 was added to USY zeolite, resulting in increased yields of aviation fuel range compounds and reduced yields of naphtha range compounds (Table 6) [37]. APTES enhanced the system's mesoporosity, which facilitated the diffusion (mass transfer) of bulkier triglyceride lipid molecules into the pores and to the active sites. The cracking of the feed was increased with a moderate rise in APTES concentration (5–7.5). However, when the APTES content grew further (7.5–10), the yield of cracked products, such as naphtha and kerosene range molecules, began to decline, resulting in higher green diesel yields. The decrease in cracked product output was ascribed to the blocking of active sites/zeolitic pores produced by excessive usage of APTES [37] and minimal mild acidity demonstrated by the high APTES content system. For a 25% Ni/ZSM-5 catalyst, Liu et al. [37] found that when the Si/Al ratio increases, the Lewis acidity lowers, lowering the yield of fractured lighter distillate products (Fig. 1), with a progressive rise in medium and heavy distillates (Fig. 2). Increased Si/Al ratio increases aviation kerosene production due to reduced secondary cracking reactions due to lower strong acidity.

Figure 1 shows that soybean oil was hydroprocessed at 300 °C using Pt, Pd, and Ni-based catalysts supporting carbon [56]. There were no secondary cracking reactions over Pt and Pd-based catalysts. Under comparable reaction conditions, lipid hydroprocessing over 20% Ni supported over carbon showed mainly kerosene range products (50%). Morgan et al. [56] and Liu et al. [37] have reported that Ni-supported catalytic systems are suitable for synthesizing aviation kerosene range hydrocarbons using lipids from soybean and castor oils. Carbon-based catalyst systems are naturally

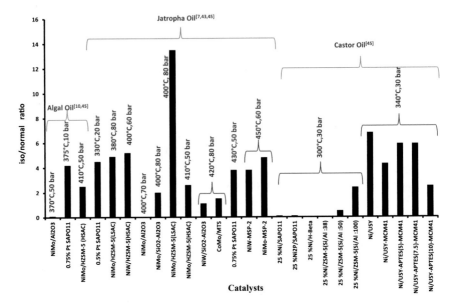

Fig. 2 Isomer/normal ratio for hydroprocessed vegetable oils over different catalysts

non-acidic and do not accelerate cracking processes. The increased output of middle distillates kerosene range products found over Ni-based catalyst systems (Fig. 1) indicates the development of acidic intermediates that facilitated lipid cracking at lower temperatures (300 °C).

Higher diesel range hydrocarbon yield has been observed on the neutral support (Pt, Pd, and Ni supported on carbon, refer to Fig. 1) and at lower temperature conditions (<380 °C) when CoMo is supported on the MTS catalyst system.

Saturated (Tristearin) and unsaturated (Triolein) model triglyceride compounds were likewise hydroprocessed in carbon-supported Ni, Pt, and Pd-based systems [56]. Similar findings were found over Ni-based catalysts, with higher cracking (kerosene 46–47%) as opposed to Pt and Pd-based systems (diesel range 45–75%, kerosene 27–34%), as well as increased conversion (81–85% for Ni/C) as compared to (22–47%) for Pt and Pd systems (Fig. 1). Product patterns were comparable across all catalytic systems for saturated and unsaturated lipid molecules, except for enhanced production of internal alkenes in triolein feedstocks, likely due to decreased hydrogenation capabilities at 300 °C temperature [49].

The catalytic performance of NiMo and NiW catalytic systems supported by overacidic mesoporous substrates such as SiO_2-Al_2O_3, ZSM-5 with kerosene, and diesel range products have also been demonstrated [10, 19, 25, 38, 43, 45, 57].

3.4.4   Isomerization Selectivity

Only if the products of vegetable oil hydroprocessing meet the necessary specifications, such as those established by ASTM, can they be used as liquid transportation fuels. The isomer-to-normal hydrocarbon ratio (i/n) is essential for all liquid transportation fuels to meet their specifications. To lower the pour point of diesel range hydrocarbons, or the freezing point of aviation range hydrocarbon, a high i/n ratio is required in the hydroprocessed product.

During lipid hydroprocessing, isomerization events can occur at the acidic sites on the catalyst surface. Acidic supports, such as silica-alumina and zeolites, offer the necessary acidity for these reactions, but alumina support-based catalysts have low acidity and do not favor isomerization processes. Figure 2 depicts the i/n hydrocarbon produced by hydroprocessing various lipids using various acidic and non-acidic catalytic systems. Over alumina-based catalytic systems such as $NiMo/Al_2O_3$ demonstrated a nearly insignificant i/n ratio for algal and jatropha-based hydroprocessed products at 370 °C and 400 °C, suggesting the minimal influence of temperature on the i/n ratio.

H-ZSM-5, SAPO-11, and SiO_2-Al_2O_3 supported with both sulfided and noble metal-based hydrogenation/dehydrogenation functionalities were used by Verma et al. [41] to hydroprocessing algal and jatropha oil over various hierarchical structured and porous catalytic systems, H-ZSM-5, SAPO-11, and SiO_2-Al_2O_3. In comparison to semi-crystalline H-ZSM-5, they found that crystalline H-ZSM-5 (low surface area crystalline, LSAC) stimulated more isomerization processes. Strong acidic sites (Lewis acidity) produced undesirable cracking reactions, whereas medium-strength acidic sites (Brönsted acidity) promoted good isomerization. At 300 °C and 30 bars, Liu et al. [37] processed castor oil lipids over 25% Ni/ZSM-5 catalytic systems with varied Si/Al ratios yielded comparable findings, i.e., increased isomerization reactions with increasing Si/Al ratio, ascribed to reduced Lewis acidity (with rising Si/Al ratio) owing to reduction in extra-framework Al content. Both Verma et al. [41] and Liu et al. [37] found that raising the Si/Al ratio decreases Lewis acidity (strong acid sites). The cracking was preferred over Lewis acidic sites, and isomerization processes over Brönsted acid sites were confirmed by a lower yield of lighter cracked products as the Si/Al ratio increased [37]. Liu et al. [37] investigated innovative mixed catalytic systems (USY, MCM-41, APTES) for hydroprocessing of castor oil and obtained i/n hydrocarbon ratios of 5–6, which are equal to values obtained with most other catalysts such as NiMo/H-ZSM-5 (Fig. 2).

At 410 °C for H-ZSM-5 (HSASC) supports, comparable i/n (2.2) was reported for algal oil lipids and jatropha oil lipids over NiMo catalytic systems. The i/n hydrocarbon ratio increased threefold when the temperature was increased from 380 °C to 400 °C over H-ZSM-5 (LSAC) supports. In contrast, no other catalytic system demonstrated a significant increase, suggesting that NiMo/H-ZSM-5 (LSAC) systems are better suited for isomerization processes at higher temperatures. The isomerization selectivity of Ni-W/H-ZSM-5 (HSASC) is higher (i/n = 4.5) than that of Ni-Mo/H-ZSM-5 (HSASC) (i/n = 2.5), demonstrating the function of W in increased hydrogenation/dehydrogenation activity over "Mo" supported catalysts for isomerization processes. Verma et al. [46] did not make a similar observation in the case of hierarchical mesoporous SAPO-11 supported NiMo and NiW catalytic systems. NiW (i/n = 3.8) catalytic systems had a somewhat lower i/n ratio than NiMo (i/n = 4.8) catalytic systems (450 °C, 60 bar). These discrepancies might be owing to the acidities of the two catalysts being different. For the noble metal-assisted catalysts (0.75% Pt/SAPO-11) identical i/n ratios (i/n = 4) could be achieved at somewhat lower temperatures and pressures (430 °C, 50 bar). Higher isomerization selectivity (i/n = 4) was observed when the temperature was reduced to 330 °C, confirming the isomerization process even at lower temperatures.

4   Kinetics, Reaction Mechanisms, and Pathways

The type of catalyst system used for lipid conversion into different hydrocarbons significantly impacts the reaction mechanism and pathway followed. Lipids, which are triglycerides, diglycerides, monoglycerides, and free fatty acids, are the main reacting component and hydrogen. The triglyceride molecules present in the oil are saturated and unsaturated, as depicted in Table 1. Unsaturated lipid molecules are first saturated and then deoxygenated over a hydrotreating catalytic system. The glycerol bond of the triglyceride molecule is broken during deoxygenation, yielding a propane molecule and three carboxylic acid molecules [31, 33, 62] (Scheme 1a). Cracking and isomerization reactions are observed simultaneously in the case of the multifunctional catalyst. At the active sites, cyclization and aromatization processes also occur if conjugated double bonds are present. Propane accounts for 30–60% of the gaseous products formed by cleavage and hydrogenation of the triglyceride molecule, according to the gas-phase composition of deoxygenation reactions of palm and jatropha oil products obtained using hydrotreating and hydrocracking catalysts (Table 7) (C–O bond). The intermediate acid molecule generated during the deprotonation event undergoes a deoxygenation reaction to yield hydrocarbon compounds. Depropanation and deoxygenation processes can happen simultaneously or in sequence, for example, depropanation to generate acid followed by deoxygenation of the acid molecule [33]. Deoxygenation processes, such as hydrodeoxygenation (H_2O), or decarbonation reactions (CO and CO_2), remove the oxygen present in the glyceride molecule. The deoxygenation routes, which determine how the lipid or acid-intermediate molecules are deoxygenated, are also influenced by the processing conditions and catalyst functionality (acidity/hydrogenation ability).

Hydrocracking catalyst systems, including acidic silica-alumina, silicoaluminophosphate, zeolites, and other materials as supports for active metals (sulfided NiW and NiMo), promote C–C bond breaking, resulting in more CO_2 and CO production than H_2O. The preceding result is supported by increased molar yields of CO and CO_2 over hydrocracking catalysts compared to hydrotreating catalysts (Table 7).

Anand et al. [43] and Sharma et al. [44] investigated lumped kinetic models for the hydroconversion of jatropha oil lipids over CoMo catalyst supported on Al_2O_3 and MTS (mesoporous titanosilicates). During the hydroconversion of lipids,

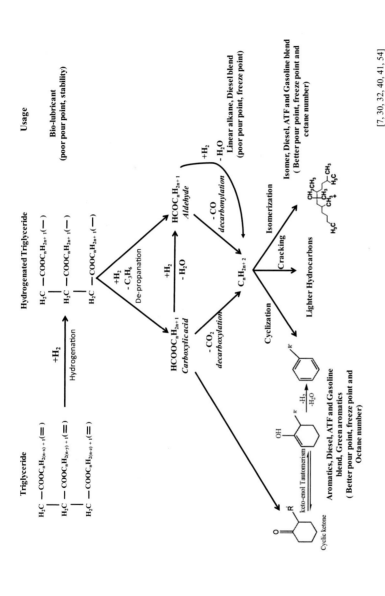

Schematic 1 Reaction schemes for the hydroconversion of lipids into transportation fuels

Table 7 Molar composition of gaseous products from the hydroprocessing of lipids at 50 bar and 1 h^{-1} over both hydrocracking and hydrotreating catalytic systems

Process	Hydrotreating				Hydrocracking		
Temperature, °C	**360**				**360**	**400**	
Lipid [reference]	Palm [31]	Jatropha [39]			Jatropha [39]	Jatropha [46] *	
Catalyst	NiMo/Al$_2$O$_3$	NiMo/Al$_2$O$_3$	CoMo/Al$_2$O$_3$		NiW/SiO$_2$-Al$_2$O$_3$	NiMo/MSP-1	NiW/MSP-2
Gas HC - C3	1.8	0.53	1.72		1.54	1.43	1.51
C3	1.7	0.69	0.74		0.83	1.09	0.66
CO$_2$	1.3	0.49	0.55		2.15	4.35	1.47
CO	0.1	2	1.8		5.98	2.25	0.93

*70 bar, HC: Hydrocarbons, MSP: Hierarchical mesoporous SAPO-11 (MSP-1 (Si/Al: 0.4) & MSP-2 (Si/Al: 0.27))

they examined all potential reactions and reaction products. The liquid hydrocarbon products produced after hydroprocessing were lumped according to their corresponding carbon ranges, i.e., gasoline (C5-C8), kerosene (C9-C14), gasoline, and kerosene (C15), diesel (C15-C18), and oligomerized product (>C18). Triglycerides in lipid feed were considered as a single lump. The experimental data constructed rate equations for the models, and reaction parameters for distinct reaction pathways were calculated (Table 8). Compared to alumina-supported catalyst systems, titanosilicates-based catalyst systems demonstrated a lower rate of triglyceride conversion at 320 °C. In contrast, in the CoMo/Al$_2$O$_3$ system at 320 °C temperatures, the triglycerides were first deoxygenated to yield diesel range products, which were then cracked into lower range naphtha and kerosene range hydrocarbons (schematic 1b). In contrast, in the CoMo/Al$_2$O$_3$ system at 320 °C temperatures, the triglycerides are initially deoxygenated to yield diesel range products (schematic 1c). Because of the various metal support interactions and the diverse physicochemical characteristics of these catalyst systems, the conversion routes for lipids are distinct. The MTS system catalyzed and offered an alternate route for synthesizing lower-range hydrocarbons directly from triglycerides at lower temperatures. Compared to the Al$_2$O$_3$ system, the production rate of principal cracked products, such as gasoline and kerosene range compounds, are lower in percentage over MTS catalyst (Table 8). Compared to MTS systems, Al$_2$O$_3$ catalyst systems had a higher conversion

Table 8 Rate constants and activation energies for hydroprocessing of lipids and co-processing of lipids with gas oil at various temperatures

Lipid (jatropha oil)—CoMo

Catalyst system	Titanosilicates [44]	Alumina [43]		
Temperature, °C →	320	320	360	Activation energy, kJmol^{-1}
$k_{Triglyceride}$, h^{-1}	14.4	17	24	26
$k_{Naphtha}$, C5-C8, h^{-1}	0.04	0.3	1.1	83
$k_{Kerosene}$, C9-C15, h^{-1}	0.1	0.3	2	127
k_{Diesel}, > C15, h^{-1}	1.2	17	20.9	47

Co-processing lipids—NiMo/Al$_2$O$_3$

Temperature, °C	325				350
% Lipid →	25% Soya [96]	10% Soya [96]	10% Jatropha [39]	5% Jatropha [39]	20% Jatropha[a] [47]
k_{HDS}, sec^{-1}	4.8	4.7	4.9	4.4	–
k_{HDO}, sec^{-1}	4.3	4.1	4.3	4.2	–
E_{HDS}, kJmol^{-1}	15	17.9	12.7	12.9	–
E_{HDO} kJmol^{-1}	136.6	140.2	129.3	132.3	148.4

[a]Co−processing with N−hexadecane / 4,6−dimethyldibenzothiophene

and corresponding diesel rate of production, showing that $CoMo/Al_2O_3$ systems are better suited for producing diesel range compounds at lower temperatures (320 °C).

When the temperature of the $CoMo/Al_2O_3$ catalyst was increased to 360 °C, the reaction pathway shifted from solely deoxygenation reactions to direct cracking into naphtha and kerosene range products, as well as the internal conversion of diesel range products into cracked kerosene range products (Schematic 1d). On increasing temperature (360 °C), the rate of production of naphtha, kerosene, and diesel range hydrocarbons increased (4–6 times) as compared to those produced at a lower temperature (320 °C) (Table 8). Acidic intermediates generated over the $CoMo/Al_2O_3$ catalyst system, which enhanced cracking reactions at 360 °C temperature, were ascribed to the significant cracking and shift in reaction pathway seen as temperature increased (320–360 °C). The acidic intermediates generated during the hydroconversion of lipids over various catalysts influenced reaction paths. These intermediates aided cracking processes while also altering the sort of reactions that took place. At lower temperatures of 300 and 320 °C, they induced moderate cracking and oligomerization events across MTS.

When lipids from soya and jatropha oil are co-processed, a minor increase in the rate of hydrodesulfurization (HDS) reaction is reported when the lipid concentration in gas oil is increased (Table 8). The preceding result is supported by reducing the activation energy needed for HDS reactions compared to $NiMo/Al_2O_3$ catalytic systems. Different feed/catalyst interactions occur in the presence of lipids, which might explain why activation energy decreases as lipid concentration rises (which produces acidic intermediates).

MTS systems [44] have activation energies of the order of 40 kJ/mole for triglyceride conversion, compared to only 26 kJ/mole for Al_2O_3 systems [43], indicating that different conversion routes pursued across these catalytic systems at lower temperatures (320 °C). For co-processing lipids over $NiMo/Al_2O_3$ catalyst complexes, the exceptionally high activation energy for hydrodeoxygenation reactions was reported (130–150 kJ/mole), showing that direct lipids processing is preferable to the absence of competing HDS reactions. The conversion of jatropha oil lipids into aviation fuels using a $CoMo/Al_2O_3$ catalyst system requires 127 kJ/mole high activation energy.

5   Globally Available Technologies for Renewable Diesel Production

Renewable diesel is a mixture of hydrocarbon compounds produced via various methods, including hydrotreating, isomerization, gasification, pyrolysis, many other chemicals, biochemical, and thermochemical processes. Commercial diesel complies with the ASTM D975 standard for petroleum diesel. Transesterification produces biodiesel, which is a mono-alkyl ester. Biodiesel complies with ASTM D6751 and is suitable for blending with petroleum diesel.

Several small businesses across the globe use the pyrolysis process to produce gasoline from plastics and other MSW (end-of-life tires, organic wastes). On average, plastic pyrolysis produces 45–50% oil, 35–40% gases, and 10–20% tar. In a few cases, pyrolysis of plastics can produce a significant liquid yield (more than 80 wt. percent), which is higher than pyrolysis of wood-based biomass in general [101]. Separating the pyrolysis oil into various fractions with different boiling point ranges, such as light (0–160 °C), mid-distillate (160–370 °C), and heavy (>370 °C), improves its valorization. Pyrolysis oil or fractions can be further enhanced by conversion procedures, such as catalytic hydrotreatment (Ratnasari et al. [102]). Another process for diesel production is a two-stage pyrolysis/catalysis process, which uses high-density polyethylene as feedstock. In the first stage, the plastic pyrolysis is carried out, followed by the catalysis of the derived hydrocarbon pyrolysis gases in the second stage. The results revealed a high yield of hydrocarbon products (>80%) obtained from high-density polyethylene [103].

Several petroleum-related firms have been commercially producing sustainable green liquid fuels from the hydroprocessing of diverse lipid feedstocks during the last 10 years. The technology has just lately begun to be commercialized (Table 9). UOP Honeywell Co. offers one of the hydroprocessing technology for producing green fuels from various feedstocks. The UOP/ENI Ecofining™ method [104] converts non-edible second-generation natural oils into green diesel, which may be utilized in engines in any proportion in existing fuel tanks.

Similarly, Haldor Topsoe, a Danish catalyst firm, enables the manufacture of green diesel and jet fuel from raw tall. Tall is a non-edible substance, unlike other feedstocks used in renewable fuel generation, so the method does not contribute to world food shortages [105].

Neste Oil Co. produces high-quality diesel fuel from vegetable and animal fat in the same line from similar hydroprocessing technologies. By hydrotreating vegetable oils or waste fats, the NExBTL method creates sustainable diesel with a CO_2 reduction of 40–80% depending on the feedstock. Neste Oil has established a factory in Singapore that would use NExBTL technology to create over 800,000 tonnes of renewable diesel per year using feedstocks, including palm oil and waste animal fat [106].

Tyson and Syntroleum Corporation, a synthetic fuels technology firm located in Texas, has created a synthetic fuels technology aimed at the sustainable diesel, jet, and military fuel sectors. Dynamic Fuels, a joint venture between Syntroleum and Tyson Foods, has established a factory in Geismar, Louisiana, to produce sustainable fuel from non-food grade animal fats by hydrotreating. Beef tallow, pork lard, and chicken fat are among the animal fats [107]. Valero Energy Corporation plans to build a commercial unit at its St. Charles Refinery in Norco, Louisiana, using hydrogenation and isomerization to convert spent cooking oil and animal fat into long-lasting diesel fuel [108]. Darling International, Valero's joint venture partner, would supply the fat and oil feedstocks [109].

ConocoPhillips' Whitegate Refinery in Cork, Ireland, also developed the technology by hydrogenating vegetable oils to provide a green diesel fuel component that complies with European Union requirements. The refinery produces 1,000 barrels

Table 9 Green diesel production technologies available around the globe via hydroprocessing route [100, 111]

Technology name	Company	Feedstocks	Plant capacity (million tons/yr) and location
Existing commercial units			
NExBTL	Neste	Palm oil, vegetable oil, and waste animal fat	1 (Netherlands) 1 (Singapore) 0.38 (Finland)
Ecofining™	Diamond green diesel	Palm oil, non-edible vegetable oils, and animal fats	0.9 (USA)
	UOP/Eni	Vegetable oils, animal fats, and used cooking oils	0.78 (Italy)
	Altair fuels	Non-edible natural oils and agricultural waste	0.13 (USA)
Dynamic Fuels LLC (Syntroleum and Tyson Foods joint venture)	Energy Group (REG) Inc	Animal fats Soybean oil	0.25 (USA)
UPM Bio Verno	UPM Biofuels	Crude tall oil	0.1 (Finland)
Planned commercial units			
Valero and joint venture		Used cooking oil, animal fats Soybean oil	0.5 (Norco, Louisiana)
UPM biofuels		Tall oil (pine)	0.1 (Lappeenranta, Finland)
Toyota Motor Corporation, Hino Motors, Nippon Oil Corporation		Vegetable oil like soybean oil, grape seed oil, and other vegetable oils and animal fats	
ConocoPhillips		Vegetable oils	

of sustainable diesel fuel per day for sale throughout Ireland. TMC, Hino Motors, the Tokyo Metropolitan Government, and Nippon Oil Corporation (NOC) have started commercializing bio hydroformed diesel (BHD). This second-generation green diesel fuel is produced by hydrogenating a vegetable oil feedstock.

6   Path Forward and Challenges

Because of the above (both fuel demand and environmental concerns), green fuel derived from renewable sources has much promise. It may be a viable alternative to fossil fuels. The following are the critical problems in hydroprocessing technology for the commercial synthesis of hydrocarbons from vegetable oil (lipid): (1) a high hydrogen demand, (2) a high exothermic reaction, (3) a constant feed supply for large scale manufacturing, and (4) a feed pre-treatment issue.

6.1   High Hydrogen Requirement

A hydrocracking unit is generally the refinery's leading hydrogen consumer, and hydroprocessing of vegetable oil requires significantly more hydrogen (nearly 2–3 times more) than a conventional hydrocracker. Other processes, such as hydrodeoxygenation and saturation of large percentages of unsaturated components in some vegetable oils, such as jatropha, are the primary reason for such high hydrogen consumption. The hydrogen demand may rise when unsaturated hydrocarbons produced due to undesired cracking processes become saturated, necessitating more hydrogen, raising the cost. Purification and recycling of wasted hydrogen are required to make the process commercially viable. Although simple water washing or amine washing are commonly employed to recover leftover hydrogen, they cannot remove contaminants, particularly CO, which might impair catalyst performance. In comparison to traditional methods, new technologies such as membrane filtration provide higher hydrogen purity. Impurities like CO, CO_2, and methane may be eliminated from the recycling stream using such membrane systems, resulting in the necessary (99.9%) hydrogen purity.

6.2   High Exothermicity of the Reaction

Renewable feedstocks such as plant, animal, and algal oils, triglycerides, and free fatty acids are highly exothermic when converted into aviation fuel and other hydrocarbons. The need for quench hydrogen, reactor bed temperature control, and a high catalyst deactivation rate are significant problems for the commercialization of the processes due to the highly exothermic reactions involved. Excess cracking and coke formation reactions in the catalyst pores are caused by the high exothermicity of these reactions, resulting in a significant pressure drop, a short catalyst life, and, therefore, a less cost-effective process.

6.3   Continuous Feed Supply for Large Production

To make the process viable, it must have a constant supply of feedstock that can be turned into fuel while avoiding concerns like food versus fuel and deforestation. Aside from discarded cooking oils, planting oil-seed bearing trees (afforestation) might be one of the environmentally beneficial strategies to boost feed supplies. For renewable aviation fuel technology to advance, government assistance in terms of policymaking and support is required.

6.4   Feed Pre-treatment Issue

Most vegetable oils include substantial contaminants, such as metals, Na, K, Ca, and phosphorus, which must be eliminated before conversion into aviation fuel to increase the catalyst life, which may cause catalyst poisoning. Metals, as well as the phosphorus molecule known as phospholipids, are commonly found in vegetable oil. Water degumming or acid degumming are generally known pre-treatment processes for metal removal. Guard bed catalysts also provide an alternative option to remove these metals and phospholipids.

By overcoming the obstacles mentioned above, renewable diesel fuel made from lipid hydroprocessing can compete with fossil-based fuels. Renewable fuels have a higher calorific value, lower sulfur, better combustion characteristics than fossil fuels, and lower aromatics content, resulting in fewer emissions.

7   Conclusion

In view of the preceding considerations, it is evident that an alternate source of automobile fuel, especially diesel, is desperately needed due to increasing particulate emission with conventional compression ignition (CI) engines. Alternative automobile fuels, which are drop-in in nature, are the best solutions to meet increasing fuel and environmental demands due to stringent specifications (sulfur limits for Euro VI < 10PPM, paraffinic diesel < 5 PPM), higher investment cost, and engine/aircraft modifications.

The hydroprocessing method for creating alternative fuels is a well-established technology. However, it is still not cost-competitive due to the higher cost of animal and plant-derived triglycerides/lipids. The sulfided mesoporous catalysts with mild acidity and greater surface area are the best among the many catalysts researched and reviewed in this chapter. These catalysts are simpler to integrate into existing refinery infrastructure for large-scale production, even under co-processing conditions, because they are highly comparable to already employed hydrocracking catalysts. Suppose the challenges such as consistent feedstock availability, exothermicity of the reaction using novel strategies (liquid quench, heat integration), pre-treatment of feedstock to extend catalyst life, recycle gas purification, and other issues are addressed. In that case, green diesel could be produced from animal fats, waste cooking oils, and plant oils (lipids) competitively.

References

1. Ray A, Anumakonda A (2011) Production of green liquid hydrocarbon fuels. Biofuels: alternative feedstocks and conversion processes. Elsevier, pp 587–608
2. https://www.lifegate.com/carbon-budget
3. https://carbontracker.org/carbon-budgets-explained
4. The Intergovernmental Panel on Climate Change's (IPCC) Fifth Assessment Report (AR5) (2014)
5. Kubickova I, Kubicka D (2010) Waste Biomass Valor 1:293–308
6. Klass DL (1998) Academic Press, San Diego
7. Lynd LR, Cushman JH, Nichols RJ, Wyman CE (1991) Science 251:1318
8. Wyman CE (1994) Appl Biochem Biotechnol 45–46:897
9. Huber GW, Iborra S, Corma A (2006) Chem Rev 106:4044
10. Sinha AK, Anand M, Rana BS, Kumar R, Farooqui SA, Sibi MG, Kumar R, Joshi RK (2013) Catal Surv Asia 17:1–13
11. Liu CLC, Kuchma O, Krutovsky KV (2018) Global Ecol Conserv 15:e00419
12. Sheehan PM, Yeh YY (1984) Lipids 19(2)
13. Stocker M (2008) Angew Chem Int Ed 47:9200–9211
14. Huber GW, Corma A (2007) Angew Chem Int Ed 46:7184–7201
15. Calero J, Luna D, Sancho ED, Luna C, Bautista FM, Romero AA, Posadillo A, Berbel J, Verdugo-Escamilla C (2015) Ren Sus Ener Rev 42:1437–1452
16. Li L, Coppola E, Rine J, Miller JL, Walker D (2010) Energy Fuels 24:1305–1315
17. Stephen S Doliente, Aravind Narayan, John Frederick D Tapia, Nouri J Samsatli, Yingru Zhao, Sheila Samsatli (2020) Front Energy Res
18. Prak DJL, Jones MH, Trulove P, McDaniel AM, Dickerson T, Cowart JS (2015) Energy Fuels 29:3760–3769
19. http://www.lanzatech.com/innovation/markets/fuels/. Accessed August 2015
20. Holladay J (2012) Workshop on Advanced Bio-based Jet Fuel Cost of Production, US Department of Energy, Energy Efficiency and Renewable Energy. http://www1.eere.energy.gov/bio energy/pdfs/holladay_caafi_workshop.pdf. Accessed August 2015
21. Bond JQ, Upadhye AA, Olcay H, Tompsett GA, Jae J, Xing R, Alonso DM, Wang D, Zhang T, Kumar R, Foster A, Sen SM, Maravelias CT, Malina R, Barrett SRH, Lobo R, Wyman CE, Dumesic JA, Huber GW (2014) Energy Environ Sci 7:1500–1523
22. Olcay H, Subrahmanyam AV, Xing R, Lajoie J, Dumesic JA, Huber GW (2013) Energy Environ Sci 6:205–216; b). http://www.virent.com/news/virent-delivers-plant-based-jet-fuel-to-u-s-air-force-research-laboratory-for-testing/. Accessed August 2015
23. https://www.globenewswire.com/news-release/2021/04/09/2207391/28124/en/Global-Ren ewable-Diesel-Capacity-and-Demand-Market-Insights-Forecast-Report-2020-2024-Featur ing-Total-S-A-Valero-Energy-Eni-Neste-HollyFrontier-and-Renewable-Energy-Group.html
24. Tiwari R, Rana BS, Kumar R, Verma D, Kumar R, Joshi RK, Garg MO, Sinha AK (2011) Catal Commun 12:559–562
25. Veriansyah B, Han JY, Kim SK, Hong SA, Kim YJ, Lim JS, Shu YW, Oh SG, Kim J (2012) Fuel 94:578–585
26. Cheng J, Li T, Huang R, Zhou J, Cen K (2014) Bioresour Technol 158:378–382
27. Simanek P, Kubica D, Kulikov I, Homolka F, Pospisil M (2011) Fuel 90:2473–2479

28. Biriyani S, Kalogiannis A, Vassals IA (2009) Bioresour Technol 100:3036–3042
29. Huber GW, Connor PO, Coram A (2007) Appl Catal A: Gen 329:120–129
30. Karri M, Kivas S, Kale D, Hanscom J (2010) Bioresour Technol 101:9287–9293
31. Sirifa A, Faungnawakiji K, Itthibenchapong V, Viriya-empikul N, Charinpanitkul T, Assabum-rungrat S (2014) Bioresour Technol 158:81–90
32. Vonortas A, Kubika D, Papayannakos N (2014) Fuel 116:49–55
33. Kubicka D, Simacek P, Zilkova N (2009) Top Catal 52:161–168
34. Šimácek P, Kubicka D, Šebor G, Pospíšil M (2009) Fuel 88(456):460
35. Walendziewski J, Stolarski M, Łużny R, Klimek B (2009) Fuel Proc Technol 90:686–691
36. Meller E, Green U, Aizenshtat Z, Sasson Y (2014) Fuel 133:89–95
37. Liu S, Zhu Q, Guan Q, He L, Li W (2015) Bioresour Technol 183:93–100
38. Rozmyslowicz B, Maki-Arvela P, Lestari S, Simakova OA, Eranen K, Simakova IL, Murzin DY, Salmi TO (2010) Top Catal 53:1274–1277
39. Kumar R, Rana BS, Tiwari R, Verma D, Kumar R, Joshi RK, Garg MO, Sinha AK (2010) Green Chem 12:2232–2239
40. Liu Y, Sotelo-Boyas R, Murata K, Minowa T, Sakanishil K (2009) Chem Lett 38:552–553
41. Verma D, Kumar R, Rana BS, Sinha AK (2011) Energy Environ Sci 4:1667–1671
42. Murata K, Liu Y, Inaba M, Takahara I (2010) Energy Fuels 24(2404):2409
43. Anand M, Sinha AK (2012) Bioresour Technol 126:148–155
44. Sharma RK, Anand M, Rana BS, Kumar R, Farooqui SA, Sibi MG, Sinha AK (2012) Catal Today 198:314–320
45. Anand M, Sibi MG, Verma D, Sinha AK (2014) J Chem Sci 126:473–480
46. Verma D, Rana BS, Kumar R, Sibi MG, Sinha AK, Apcata J (2015) 490:108–116
47. Rizo-Acosta P, Linares-Vallejo MT, Munoz-Arroyo JA (2014) Catal Tod 234:192–199
48. Guo J, Xu G, Shen F, Fu Y, Zhang Y, Guo Q (2015) Green Chem 17:2888–2895
49. Fan K, Liu J, Yang X, Rong L (2014) Int J Hydrogen Energy 39:3690–3697
50. Pinto F, Varela FT, Goncalves M, Andre RN, Costa P, Mendes B (2014) Fuel 116:84–93
51. Bezergianni S, Kalogianni A (2009) Bioresour Technol 100:3927–3932
52. Bezergianni S, Voutetakis S, Kalogianni A (2009) Ind Eng Chem Res 48:8402–8406
53. Bezergianni S, Dimitriadis A, Kalogianni A, Pilavachi PA (2010) Bioresour Technol 101:6651–6656
54. Bezergianni S, Dimitriadis A, Sfetsas T, Kalogianni A (2010) Bioresour Technol 101:7658–7660
55. Bezergianni S, Dimitriadis A, Kalogianni A, Knudsen KG (2011) Ind Eng Chem Res 50:3874–3879
56. Morgan T, Grubb D, Jimenez ES, Crocker M (2010) Top Catal 53:820
57. Boda L, Onyestyak G, Solt H, Lonyi F, Valyon J, Thernesz A (2010) Appl Catal A: Gen 374:158–169
58. Lestari S, Maki-Arvela P, Bernas H, Simakova O, Sjoholm R, Beltramini J, Lu GQM, Myllyoja J, Simakova I, Murzin DY (2009) Energy Fuels 23:3842–3845
59. Snare M, Kubickova I, Maki-Arvela P, Eranen K, Murzin DY (2006) Ind Eng Chem Res 45:5708–5715
60. Ahmadi M, Nambo A, Jasinski JB, Ratnasamy P, Carreon MA (2015) Catal. Sci Technol 5:380–388
61. Immer JG, Lamb HH (2010) Energy Fuels 24:5291–5299
62. Melis S, Mayo S, Leliveld B (2009) Biofuels Technol 1:43–47
63. Donnis B, Egeberg RG, Blom P, Knudsen KG (2009) Top Catal 52:229–240
64. Lappas AA, Bezergianni S, Vasalos IA (2009) Catal Today 145:55–62
65. Okamoto Y, Breysse M, Dhar GM, Song C (2003) Catal Today 86:1
66. Dhar GM, Srinivas BN, Rana MS, Kumar M, Maity SK (2003) Catal Today 86:45
67. Breysse M, Afanasiev Geantet P, Vrinat M (2003) Catal Today 86:5
68. Maity SK, Srinivas BN, Prasad VVDN, Singh A, Dhar GM, Rao TSRP (1998) Stud Surf Sci Catal 113:579
69. Prins R, Beer VHJD, Somorjai GA (1989) Catal Rev Sci Eng 31:1

70. Sankaranarayanan TM, Banu M, Pandurangan A, Sivasanker S (2011) Biores Technol 102:10717–10723
71. Caero LC, Romero AR, Ramirez J (2003) Catal Today 78:513
72. Barrera MC, Viniegra M, Escobar J, Vrinat M, Reyes JADL, Murrieta F, García J (2004) Catal Today 98:131
73. Maity SK, Rana MS, Srinivas BN, Bej SK, Dhar GM, Rao TSRP (2000) J Mol Catal A-Chem 153:121
74. Damyanova S, Petrov L, Centeno MA, Grange P (2002) Appl Catal A-Gen 224:271
75. Rana MS, Capitaine EMR, Leyva C, Ancheyta J (2007) Fuel 86:1254
76. Rana MS, Maity SK, Ancheyta J, Dhar GM, Rao TSRP (2004) Appl Catal A-Gen 268:89
77. Massoth FE, Dhar GH, Shabtai J (1984) J Catal 85:53
78. Daly FP, Ando H, Schmitt IL, Sturm EA (1987) J Catal 108:401
79. Rana MS, Srinivas BN, Maity SK, Dhar GM, Rao TSRP (2000) J Catal 195:31
80. Zhaobin W, Qin X, Xiexian G, Sham EL, Grange P, Delmon B (1991) Appl Catal 75:179
81. Li D, Nishijima A, Morris DE (1999) J Catal 182:339
82. Welters WJJ, Vorbeck G, Zandbergen HW, Ven LJMVD, Oers EMV, Haan JWD, Beer VJHD, Santen RAV (1996) J Catal 161:819
83. Zarchin R, Rabaev M, Vidruk Nehemya R, Landau MV, Herskowitz M (2015) Fuel 139:684–691
84. Wang A, Wang Y, Kabe T, Chen Y, Ishihava A, Qian W, Yao P (2002) J Catal 210:319
85. Klimova T, Calderon M, Ramirez J (2003) Appl Catal A-Gen 240:29
86. Wang A, Wang Y, Kabe T, Chen Y, Ishihara A, Qian W (2001) J Catal 199:19
87. Turaga UT, Song C (2003) Catal Today 86:129
88. Corma A, Martinez A, Soria VM, Monton JB (1995) J Catal 153:25
89. Chiranjeevi T, Kumar P, Rana MS, Dhar GM, Rao TSRP (2002) J Mol Catal A- Chem 181:109
90. Dhar GM, Kumaran GM, Kumar M, Rawat KS, Sharma LD, Raju BD, Rao TSRP (2005) Catal Today 99:309
91. Alwan BA, Salley SO, Simon KYN (2014) Appl Catal A-Gen 485:58–66
92. Choi M, Chao HS, Srivastava R, Venkatesan C, Choi DH, Ryoo R (2006) Nat Mater 5:718
93. Zhang S, Chen SL, Dong P (2010) Catal Lett 136:126–133
94. Sinha AK, Seelan S, Tsubota S, Haruta M (2004) Angew Chem Inter Ed 43:1546–1548
95. Diamante LM, Lan T (2014) J Food Proc (234583)
96. Rana BS, Kumar R, Tiwari R, Kumar R, Joshi RK, Garg MO, Sinha AK, Biombioe J (2013) 56:43–52
97. Bezergianni S, Dimitriadis A, Meletidis, G (2014) Fuel 125:129–136
98. Heriyanto H, Murti S, Is Heriyanti S, Sholehah I, Rahmawati A (2018) MATEC Web of Conferences 156, 03032
99. Baladincz P, Hancsók J (2015). Chem Eng J. https://doi.org/10.1016/j.cej.2015.04.003
100. Ruangudomsakul M, Osakoo N, Wittayakun J, Keawkumay C, Butburee T, Youngjan S, Faungnawakij K, Poo-arporn Y, Kidkhunthod P, Khemthong P. Molecular Catalysis. https://doi.org/10.1016/j.mcat.2021.111422
101. Wong SL, Ngadi N, Abdullah TAT, Inuwa IM (2015) Renew Sustain Energy Rev 50:1167–1180
102. Ratnasari DK, Nahil MA, Williams PT (2017) J Anal Appl Pyrol 124:631–637
103. Kaimal VK, Vijayabalan P (2016) Waste Manag 51:91–96
104. http://www.uop.com/processing-solutions/biofuels/green-diesel.Lastaccessed.Februaryth (2012)
105. http://www.topsoe.com/business_areas/refining.Renewable_fuels
106. http://www.upi.com/Science_News/Resource-Wars/2011/03/09/Singapore-opens-renewable-dieselplant/UPI-74561299678336/
107. http://www.environmentalleader.com/2010/11/09/tyson-foods-syntroleum-partner-to-turn-grease-into-fuel/
108. http://www.nola.com/politics/index.ssf/2011/01/new_biodiesel_plant_headed_to.html

109. http://www.biodieselmagazine.com/articles/7826/diamond-green-diesel-plant-secures-financing
110. Applewhite TH, Fats, Fatty Oils (1979) Kirk-Othmer, Encyclopaedia of Chemical Technology, Grayson M, Eckorth D (eds). John Wiley and Sons: New York, vol 9, pp 795–831
111. Srihanun N, Dujjanutat P, Muanruksa P, Kaewkannetra P (2020) Catalysts 10:241

Chapter 5
Commercial Green Diesel Production Under Hydroprocessing Technology Using Solid-Based Heterogeneous Catalysts

Nur Izyan Wan Azelee, Danilo Henrique da Silva Santos, Lucas Meili, and Hilman Ibnu Mahdi

Abstract The decimation of fossil fuel and rises in environmental issues are induced by the increasing global consumption of fossil fuel. This phenomenon is forecasted to increase gradually and has caught serious attention from researchers and governments around the world to discover and develop a cleaner and greener fuel-like biofuel. Green diesel has been crowned as the best biofuel to successor fossil-based diesel owing to abundant feedstocks, eco-friendly process, and lower technology price. Green diesel is produced using heterogeneous vegetable oils and catalysts under hydrogenation process, supercritical reaction, alkali catalyzed process, deoxygenation technology, and hydroprocessing (HP) technology. Nonetheless, the HP technology is enthroned as the most popular technology in producing green diesel in which its process can be maximally enhanced by optimizing operating parameters, harnessing the proper oil, improving and modifying the catalysts and utilizing micro- and nano-sized catalysts. This chapter focuses on the improvement of the catalyst capability and increases the HP technology performance. The improvement of the catalysts has positively contributed to the strengthening of Bronsted-Lewis acidity

N. I. W. Azelee
School of Chemical and Energy Engineering, Faculty of Engineering, Universiti Teknologi Malaysia, UTM Skudai, 81310 Skudai, Johor, Malaysia
e-mail: nur.izyan@utm.my

Institute of Bioproduct Development (IBD), Universiti Teknologi Malaysia, UTM Skudai, 81310 Skudai, Johor, Malaysia

D. H. da Silva Santos · L. Meili
Laboratório de Processos, Centro de Tecnologia, Universidade Federal de Alagoas, Av. Lourival de Melo Mota, Tabuleiro Dos Martins, Maceió, Alagoas, Brazil
e-mail: danilo.santos@ctec.ufal.br

L. Meili
e-mail: lucas.meili@ctec.ufal.br

H. I. Mahdi (✉)
Chemical and Materials Engineering, College of Engineering, National Yunlin University of Science and Technology, 123 University Road, Section 3, Douliou, Yunlin 64002, Taiwan
e-mail: dave087.ronel@gmail.com

and enlarged catalyst active sites leading to escalate the HP efficiency, conversion (~100%), selectivity (95%), yield (97%), and reusability (~5 cycles), and further producing less oxygen-contained green diesel.

Keywords Green diesel · Heterogeneous catalysts · Vegetable oils · Optimizing process · Hydroprocessing technology

1 Introduction

The increasing demand for fuel supply and the depletion of the petroleum raw material have urged researchers across the globe to develop and improvise current technologies for edible and non-edible oils processing including animal fats into green diesel. The increasing demand for fuel supply is parallel with the increased in the human population, modernization, industrialization, urbanization, and socioeconomic activities [1]. Nowadays, as people are having more awareness on the depletion of raw materials and the massive greenhouse gasses production such as CO_2, CH_4, NOx, and ozone the search for alternative fuels that is more sustainable and pollution-free is highly on the go.

Generally, among the largest contribution to global warming comes from the transportation and industrialization sectors [2]. Table 1 shows the increasing number of carbon dioxide emissions in billion metric tons from 2010 to 2021. Worldwide governments and policymakers have now began to ban certain sectors which produce or release toxic compounds to the environment with the aim of achieving zero carbon footprint in the near future. Due to these facts, hydrogenation-derived renewable energy or simply known as green diesel has become one of the recent most feasible and sustainable alternative fuel which can suit in compression ignition engine without the need of any major modifications [3]. Compression ignition engine is the main component in power generation, transportation, industrial and agricultural sectors where they are used in metro vehicles, agricultural machinery, and factory engines [4].

Catalytic hydroprocessing of oils, especially the non-edible ones has attracted more attention recently as it can avoid the competition with the food supply [6]. Hydrodeoxygenation (HDO) is the main process for biofuel production especially green diesel as it produces diesel with superior selectivity, high cetane number, tuned

Table 1 Increasing number of carbon dioxide emissions in billion metric tons (BMTs) from 2010 to 2021 [5]

Year	BMTs	Year	BMTs
2010	33.34	2016	35.45
2011	34.47	2017	35.93
2012	34.97	2018	36.65
2013	35.28	2019	36.7
2014	35.53	2020	34.81
2015	35.5	2021	36.4

cold flow properties, and high calorific value [7]. The catalyst used was primarily from noble metals (i.e., Pd/C, Co, Pd, Pt, Ni over Al_2O_3) on SiO_2, γ-Al_2O_3, and activated carbon supports [8]. The selection of good catalyst gives impact toward excellent product with enhanced properties. The catalysts are chosen based on their excellent activity and stability and thus is the most challenging task which requires extensive study with trial and error.

The main goal of this catalyst in the hydroprocessing is to obtain the highest conversion of triglycerides and fatty acids (consisting of homologous carbon chain) to produce the highest yield and quality biofuel product by successfully lowering the activation energy [9]. Notwithstanding, the use of these catalysts in the large-scale application has been constricted by their extremely high cost. The acidity of the catalyst support positively contributes to the HDO reactions. One of the challenges in producing new generation biofuels is the design and development of multifunctional catalyst, integrating redox and acidic moieties for a single-step HDO [10]. Hence, metal-zeolites appear as a highly potential catalyst to achieve a single-step HDO due to the existence of redox centers and acidic sites in the zeolites matrix even under critical diffusion limitations. Second level porosity in zeolite-based catalyst has proven to enhance the ability and selectivity of the catalyst in the reaction medium [11]. Nevertheless, the exact localization of the acidic and hydrogenation sites and catalytic behavior contributed to the most important factor in the catalytic efficacy.

Characterization of catalyst used will usually be performed by nitrogen adsorption–desorption isotherm analysis, X-ray fluorescence and diffraction methods, high-resolution transmission electron microscopy with elemental mapping, and IR spectroscopy. Larger particle size catalyst is prone to associate with carbon formation and coke formation which may block certain pores in the catalyst and deactivate the active site and eventually reducing the catalytic activity [12]. The study on the type of catalyst used for green diesel is currently evolving and optimization process is still on-going which include to maximize green diesel yield, having excellent tolerance against coke formation, faster production rate, and high dispersion inside the reaction [13]. This chapter generally gives an overview on the various technologies implementing heterogenous catalyst for green diesel production, the application of solid-based heterogeneous catalyst (metal and zeolites), and the role of different processing parameters (type of oils, metal supports, phosphides contents, and catalyst size) utilizing edible and non-edible oils for green diesel.

2 Green Diesel Produced Over Various Technologies

Commercially, green diesel is produced over numerous processes such as hydrodeoxygenation (HDO), deoxygenation (DO), decarboxylation (DCX), decarbonylation (DCN), catalytic cracking (CC), isomerization, oligomerization, hydroprocessing, hydrogenation, hydro-treatment and more. HDO technology is carried out in the green diesel production to remove chemically bound oxygen (in water formation) [14] by adding extreme H_2 compound into the reactor [15, 16]. This technology is implemented and developed by Neste Oil company called NExBTL process and ENI-UOP company known as Ecofining process [17] habitually using sulfided NiMo, CoMo, and other noble metal catalysts [18–21]. Additionally, the process results in high conversion, selectivity, yield, and HDO efficiency [22, 23]. Yet, it needs a high hydrogen consumption leading to massive H_2 off-gas products [16] and is conducted over highly extreme pressure and temperature causing elevated explosive risk [24].

DO technology is undertaken to remove oxygen content in the green diesel product by adding less H_2 molecules and further the resulted heavier hydrocarbon chains ($>C_{20}$) are cracked into the lighter hydrocarbon chains like green diesel (C_{12}–C_{18}). On the other hand, DCX process enforced to remove oxygen content yields CO_2 molecules while DCN process generates CO molecules [25, 26] causing the main problems of the environmental pollution and severe human health. These technologies employ cost-effective Ni-derived non-sulfided metal catalysts [26, 27]. To jack up the catalyst capability and activity, the catalyst is supported with metal oxides leading to the increase of the DO/DCN/DCX efficiency caused by synergistic interaction between catalysts and metal oxides [27]. These technologies have been widely implemented at large-scale industrial application with the optimum results, low-technology budget, high DO/DCN/DCX efficiency, and low H_2 consumption as well as mild operating parameters. Notwithstanding, it plays a crucial role in global warming and drops public health by dint of enormous CO_2 and CO production [28].

Apart from HDO, DO, DCN, and DCX processes, the green diesel product can be produced over hydro-treatment process by catalytic hydrogenation reaction of the whole unsaturated C=C bonds and by disconnection of the oxygen content in the generated product [29]. This technology harnesses cost-effective solid acid catalysts and is performed at mild operating reaction to produce diesel range hydrocarbon chains [30, 31]. Suitable modification of the solid catalysts plays an important factor to achieve the maximum results [32] and the less energy supply donates a cleaner process [29, 33]. To acquire the optimal results and the best process efficiency, several improvements have been recently implemented by many authors and researchers [34–41] as following discussed.

2.1 Commercial Green Diesel Production Under Heterogeneous Processes

In the previous section, a mini overview regarding heterogeneous technologies implemented in the commercial green diesel process using various edible and non-edible oils to substitute petroleum-originated diesel fuel is discussed. In this part, authors are going to describe some of the current improvements applied in the process. The common oils used in the green diesel process are sunflower [42], rapeseed [43], algae and lipases [44], soybean [45], palm-based oils [46], Jatropha [47, 48], anisole [49], WCO [50], and other vegetable oils [51].

The vegetable oils-based green diesel technologies are known well as cost-effective and eco-friendly process and receive a huge concern these days thus emerge an attractive to develop and to modify the process as listed in Table 2. For instance, Mahdi et al. [52] used nano-sized catalysts particles, catalyst activation, and different operating conditions in the green diesel production via the DO technology. These modifications and developments dramatically contributed to strengthen and to activate Bronsted-Lewis acidity as well as to enlarge crystallite pore areas, which enable to enhance the DO efficiency (>80%), conversion (>80%), selectivity (>70%), yield (>70%), and catalyst reusability (~10 runs). Furthermore, it resulted in high-quality green diesel containing less oxygen molecules.

Modification in the DO technology was not only carried out by Mahdi et al. [52], but also it was conducted by Papadopoulos et al. [53] by adding 8wt.% of tungsten (W) and 35 wt.% of Ni metal into Al_2O_3 catalyst using sunflower oil and waste cooking oil as the feedstocks. The W/Ni-assisted Al_2O_3 catalyst demonstrated more spacious active sites, stronger dispersion of nickel content, higher catalyst acid–base sites, and lower inactive $Ni-Al_2O_4$ catalyst causing the increase of the product yield and deoxygenation efficiency [53]. This argument was strengthened by Hongloi et al. [54] who claimed that Ni-modified catalyst played a critical role in increasing the conversion (100%), yield (77.8%), selectivity (82%), and the higher DO technology efficiency [73, 81].

Meanwhile, Lycourghiotis et al. [55] employed Ni metal in nanoparticles. The nano-sized Ni catalyst and metal oxide disclosed more extensive accessibility of the feedstocks and extraordinary catalyst activity [55]. Hafeez et al. [56] reported that the increase of the conversion was obtained over higher reaction temperature. Nonetheless, the highly extreme reaction temperature (>300 °C) supplied the lower conversion because the $Ni-ZrO_2$ catalyst was deactivated over too high reaction temperature [56]. Additionally, Aliana-Nasharuddin et al. [68] exploited binary metal components like Ni-Mg, Ni-Mn, Ni-Cu, and Ni-Ce doped into multi-walled carbon nanotubes catalyst. These binary metal oxides afforded greater catalyst ability and higher DO efficiency causing the increase of the yield (>84%) and selectivity (>85%) [68].

Apart from the DO technology, hydro-treatment technology applied in the green diesel production has been considerably modified in the recent experiments. Fernández-Villamil et al. [57] firstly used the modified Aspen Custom Modeler® system in Aspen Plus to improve more proper models. The hydro-treatment technology produced cleaner and high-grade quality green diesel product as well as low oxygen and sulfur contents. The Aspen Plus-simulated hydro-treatment technology produced 468,000 tons per year of green diesel, 15,900 tons per year of bio-jet fuel, 9,100 tons per year of bio-naphtha, and 39,500 tons per year of biogas [57]. Zikri and Aznury [67] exploited various pressures (40–90 bar) and temperatures (280–380 °C) in the hydrogenation technology using crude palm oil (CPO) as the feedstock and NiMo-assisted Al_2O_3 catalyst. The optimum yield achieved at 315 °C was 68.2% [67]. Furthermore, green diesel characteristics produced possessed similar characteristic as commercial green diesel European standard thus it can be applied in well in diesel engine [62]. A modification undertaken by Vázquez-Garrido et al. [63] for

Table 2 Green diesel produced from heterogeneous technologies

Technologies	Oils	Catalysts	Process parameters	Result	Remark	References
Deoxygenation (DO)	Vegetable oils	Metal-supported zeolites	200–400 °C, 1–50 bar, 1–10 h, >2% of catalyst	X = >80; S = >70; Y = >70	Capability of the deoxygenation technology was improved over the best process conditions, catalyst treatment, metal oxides, and nanoparticle catalysts leading to stronger Bronsted-Lewis acidity and enlarged active sites	Mahdi et al. [52]
Deoxygenation (DO)	Sunflower oil and waste cooking oil	W-doped Ni-Al$_2$O$_3$	310 °C, 40 bar, 9 h, 1 gr of catalyst	X = 70–100; Y = 63.6	Addition of W content contributed to larger active sites, stronger dispersion of nickel content, higher catalyst acidity, and lower inactive Ni-Al$_2$O$_4$ catalyst causing the increase of the green diesel yield and the deoxygenation efficiency	Papadopoulos et al. [53]
Deoxygenation (DO)	Rubber seed oil	15 wt.% Mo/γ-Al$_2$O$_3$	350 °C, 35 bar, 3 h	X = 98; S = 30; Y = 61.77	Mo metal donated to jack up the conversion, yield, and selectivity as well as the better deoxygenation process performance	Hongloi et al. [54]

Table 2 (continued)

Technologies	Oils	Catalysts	Process parameters	Result	Remark	References
Deoxygenation (DO)	WCO, FADO, SCGO, and CFO	MP-modified nickel	310 °C, 40 bar, 4 h, 0.33 gr of catalyst	X = 90–100; Y = 10–98	Nanoparticle and metal oxide affected larger ease of access of the oils and the better catalyst ability and the catalyst deactivation was caused over some large molecules of oils	Lycourghiotis et al. [55]
Deoxygenation (DO)	Palm oil	Ni-ZrO$_2$	250–400 °C, 30 bar, ~2 h, 1,000 of H$_2$:oil molar ratio	X = >95;	Deactivation of the catalyst was induced by higher reaction temperature	Hafeez et al. [56]
Hydro-treatment (HT)	Palm oil	NiMo-Al$_2$O$_3$	300–390 °C, 30–60 bar	X = 99.9	Hydro-treatment technology produced cleaner and high-grade quality green diesel product as well as low oxygen and sulfur contents	Fernández-Villamil and de Mendoza Paniagua [57]
Hydrodeoxygenation (HDO)	Palm oil	Phosphides-nickel (Ni$_2$P), cobalt (Co$_2$P), and copper (Cu$_3$P)	300–400 °C, 50 bar, 3 h	X = 100; Y = 83.2	The nickel catalyst supported by phosphides revealed the best performance and activity followed by cobalt and copper catalysts due to the highest Bronsted-Lewis acid–base areas	Ruangudomsakul et al. [58]

(continued)

Table 2 (continued)

Technologies	Oils	Catalysts	Process parameters	Result	Remark	References
Decarboxylation (DCX)	Castor oil	Ca(OH)$_2$	475 °C, 1.5 h	X = 65	The product yield was enhanced over the increasing reaction temperature	Chumaidi et al. [59, 60]
Decarboxylation (DCX)	Waste palm cooking oil	Ca(OH)$_2$	460 °C, 1.5 h	X = 32	The product yield was enhanced over the increasing reaction temperature	Chumaidi et al. [59]
Decarboxylation (DCX)	Soybean oil	NiMo	360 °C, 1.5 h	Y = 72	The product yield was enhanced over the increasing reaction temperature	Chumaidi et al. [59, 61]
Decarboxylation (DCX)	Palm stearin oil	Mg-Zn	350 °C, 1 bar, 5 h	Y = >50	Molar ratio of Mg-Zn catalyst did not affect the catalytic green diesel production and the decrease of the selectivity was influenced over higher reaction temperature	Chumaidi et al. [59]
Hydrotreating	Crude palm oil	Natural zeolites	350–400 °C, 0.7–2.1 bar, 3 h, 1–4 wt.% of catalyst	Y = 37.30	Green diesel properties produced from CPO possessed similar characteristic as green diesel European standard thus it can be applied in well in diesel engine	Zikri et al. [62]

(continued)

Table 2 (continued)

Technologies	Oils	Catalysts	Process parameters	Result	Remark	References
Hydroconversion	Waste soybean oil	Mn-Al$_2$O$_3$/NiMo	400 °C, 40 bar, 4 h, 10% of H$_2$S:H$_2$	X = 48.8; Y = 0.82	Mn supporting had a crucial impact on the NiMo catalyst performance due to good stabilization of Mn component leading to re-disperse the Ni-doped MoS$_2$ active sites	Vázquez-Garrido et al. [63]
Hydroprocessing	Waste cooking oil	SiO$_2$, Al$_2$O$_3$, and CaO-doped biowaste fly ash	–	Y = 93.85	The active sites of the catalyst were significantly improved from 0.8611 m^2/g to 41.2571 m^2/g due to SiO2 doping into the catalyst	Pelemo et al. [3]
Hydrocracking	Palm, sunflower, soybean, and rice bran oils	Pd/TiO$_2$-Al$_2$O$_3$	450 °C, 50 bar, LHSV of 1/h	Y = 82 & 94	Green diesel produced from hydrocracking technology possessed more high-grade quality if compared to biodiesel produced from transesterification process	Attaphaiboon et al. [64]

(continued)

Table 2 (continued)

Technologies	Oils	Catalysts	Process parameters	Result	Remark	References
Hydrocracking	Waste palm cooking oil	Mo-modified SBA-15	500 °C, catalyst:oil molar ratio of 1:100	X = 50.35; Y = 53.86	Mo metal contributed to escalate the catalyst performance and the increasing conversion and yield due to higher porosity and active sites	Alisha et al. [65]
Hydroprocessing	Methyl palmitate oil	Ni/beta zeolite	240 °C, 25 bar, 4 h, and 20 mg of catalyst	X = 38–77; S = 11.3–61.9	The largest accessibility of the reactants and the highest acid sites were achieved over addition of Ni nanoparticle into the dealuminated beta catalyst	Papanikolaou et al. [7]
Oligomerization	Waste cooking oil and microalgae oil	Pt/γ-Al$_2$O$_3$, Pt/TiO$_2$, and Pt-MoOx/TiO$_2$	320 °C, 50 bar, 4 h, 0.24 gr of catalyst	Y = 67.0	Oligomerization process resulted in green diesel product with low sulfur content and high viscosity	Jin et al. [66]
Hydrogenation	Crude palm oil (CPO)	NiMo/γ-Al$_2$O$_3$	280–380 °C, 40–90 bar, 1 h, 0.15 gr of catalyst	Y = 68.2	The green diesel produced has similar characteristic to commercial diesel fuel and the highest yield was reached at 315 °C	Zikri and Aznury [67]

(continued)

Table 2 (continued)

Technologies	Oils	Catalysts	Process parameters	Result	Remark	References
Hydrodeoxygenation (HDO)	Karanja oil	Ni/γ-Al$_2$O$_3$	360 °C, 2 h, 20 wt.% of catalyst	X = 100; S = 70	Conversion was augmented by more elevated pressure and higher Ni concentration and further selectivity was obtained over higher reaction temperature	Yenumala et al. [38]
Deoxygenation (DO)	Chicken fat oil (CFO)	Ni–Mg, Ni–Mn, and Ni–Ce–doped MWCNTs	350 °C, 2 h, 3 wt.% of catalyst	S = 81–85; Y = 75–84	Mg-Mn component supported by Ni afforded highly remarkable DO capability if compared to other metal oxides	Aliana-Nasharuddin et al. [68]
Hydrodeoxygenation (HDO)	Rubber seed oil (RSO)	NiMo/γ-Al$_2$O$_3$	300–400 °C, 30–80 bar, WHSV of 1–3/h	X = 100; Y = 84.94	The catalyst was very stable within 18 h with high yield and conversion as well as without occurring to sulfidization process	Ameen et al. [69]
Hydrodeoxygenation (HDO)	Rubber seed oil (RSO)	NiMo/γ-Al$_2$O$_3$	300–400 °C, 30–80 bar, 5 h	X = 100; S = 19.1	Hydrodeoxygenation technology was positively affected over various reaction temperatures and WHSV parameters	Ameen et al. [70]

(continued)

Table 2 (continued)

Technologies	Oils	Catalysts	Process parameters	Result	Remark	References
Decarboxylation (DCX)	Rubber seed oil (RSO)	NiMo/γ-Al$_2$O$_3$	300–400 °C, 30–80 bar, 5 h	X = 100; S = 81.7	Decarboxylation technology was negatively influenced over extreme reaction temperature and pressure leading to bring down the selectivity	Ameen et al. [70]
Pyrolytic-deoxygenation	Jatropha curcas oil	MWCNTs assisted by NiO–Fe$_2$O$_3$ and NiO–ZnO	350 °C, 1 bar, 1 h, 3 wt.% of catalyst	Y = 75	NiO–Fe$_2$O$_3$-modified MWCNTs generated the green diesel with the lowest oxygen content and the highest heating value	Asikin-Mijan N et al. [71]
Hydrogenative decarboxylation in microwave	Fatty acids	Nanoparticle Rh doped by SiNA-Rh	200 °C, 10 bar, 24 h, 500 mol ppm of catalyst	Y = 99	The microwave irradiation provided a crucial key in the increase of the yield and in improvement of the green diesel process	Baek et al. [72]
Deoxygenation (DO)	Palm fatty acid distillate (PFAD)	NiO–ZnO	350 °C, 1 bar, 2 h, 5 wt.% of catalyst	S = 86.0; Y = 83.4	The green diesel was alternatively appropriate for the engine machine to replace the commercial diesel fuel	Baharudin et al. [73]

(continued)

Table 2 (continued)

Technologies	Oils	Catalysts	Process parameters	Result	Remark	References
Deoxygenation (DO)	Palm fatty acid distillate (PFAD)	Activated carbon-assisted CoMo	350 °C, 1 h, 5–20 wt.% of catalyst	S = 89; Y = 92	AC-modified CoMo catalyst demonstrated higher deoxygenation activity due to higher acidity and had a good thermal stability	Gamal et al. [74]
Deoxygenation (DO)	Palm fatty acid distillate (PFAD)	Ni, Co, and Ni–Co-doped SBA-15	350 °C, 3 h, 10 wt.% of catalyst	S = >85; Y = 85.8 & 88.1	Co metal caused formation of carbon and coke because of larger particle diameter thus blocking the catalyst porosity and deactivating catalyst leading to mitigate the catalyst ability	Kamaruzaman et al. [12]
Hydrotreating	Methyl esters	Co–Ni-assisted Zr-SBA-15	300 °C , 30 bar, 6 h, 5 wt.% of catalyst	X = >90; Y = >60	Synergistic interaction between Co and Zr revealed a fabulous interactivity and resulted in high conversion and yield	Ochoa-Hernández et al. [75]
Deoxygenation (DO)	Oleic acid	Ni$_2$P-supported USY, H-ZSM-5 and Al-SBA-15	260–300 °C, 50 bar, 6 h, 1.5 gr of catalyst	Y = 24, 29, & 42	Ni$_2$P-supported Al-SBA-15 was crowned as the best catalyst followed by Ni$_2$P-ZSM-5 and Ni$_2$P-USY owing to the largest crystalline size	de Oliveira Camargo et al. [76]

(continued)

Table 2 (continued)

Technologies	Oils	Catalysts	Process parameters	Result	Remark	References
Deoxygenation (DO)	Macauba acid oil	Co-Carbon (activated biochar)	350 °C, 30 bar, 4 h, 10 wt.% of catalyst	Y = 98.4	Co-Carbon as the catalyst and macauba acid oil as the feedstock were implemented to lower the process cost and to develop a sustainable process	de Barros Dias Moreira et al. [77]
Hydrocracking (HC)	Palm oil	Pd/Al$_2$O$_3$ and Pd-Fe/Al$_2$O$_3$	400 °C, 60 bar, 2 h, 0.9% of catalyst	Y = 50–62.20	The biofuel yield attained by Pd-Fe/Al$_2$O$_3$ was higher than the one obtained by Pd/Al$_2$O$_3$ catalyst	Srihanun et al. [78]

(continued)

Table 2 (continued)

Technologies	Oils	Catalysts	Process parameters	Result	Remark	References
Hydrodeoxygenation (HDO)	Rubber seed oil (RSO)	H-Y zeolite	350 °C, 35 bar, 3, 12, & 15 wt.% of catalyst	X = >99; S = 10.4	H-Y zeolite was highly suitable for hydrocracking process in green diesel production using vegetable oils rather than the fossil-derived diesel fuel production by dint of highly elevated acid sites	Ameen et al. [79]
Hydroprocessing	Soybean oil	NbOPO$_4$	300–350 °C, 10 bar, 3–5 h, 0–25% of catalyst	Y = 40	Hydroprocessing technology produced around 62% of bio-jet fuel, green diesel (40%), and biogasoline (18%) in a single process step	Scaldaferri and Pasa [80]

Note X = conversion (%); S = selectivity (%); Y = yield (%); W = tungsten; WCO = waste cooking oils; FADO = fatty acid distillate oil; SCGO = spent coffee grounds oil; CFO = chicken fat oil; MP = mineral palygorskite; MWCNTs = multi-walled carbon nanotubes; LHSV = liquid hourly space velocity; WHSV = weight hourly space velocity; SiNA-Rh = silicon nanowire array; NbOPO$_4$ = niobium phosphate green diesel produced over hydroconversion technology was to employ Mn metal. The Mn supporting had an important influence on the NiMo catalyst ability in consequence of good stabilization of Mn component leading to re-disperse the Ni-doped MoS_2 active sites [63].

In the hydrocracking technology, Alisha et al. [65] employed Mo metal doped into SBA-15 catalyst in the reaction at 500 °C with oil:catalyst molar ratio of 100:1. The Mo metal contributed to jack up the catalyst capability and to escalate the conversion and yield because of higher porosity and active sites [65]. Moreover, green diesel obtained from the modified hydrocracking technology possessed more high-grade quality if compared to biodiesel produced from the transesterification process as investigated by Attaphaiboon et al. [64]. Addition of Mo metal into the Pt catalyst was also observed by Chumachenko et al. [82] in the hydrocracking process which resulted in yield of 82% at 400 °C within 20 h whereas Trisunaryanti et al. [83] used nano-sized Ni metal for SBA-15 catalyst showing remarkable catalyst capability due to higher active sites.

In conclusion, developments and modifications in the green diesel production performed over these commercial technologies are catalysts nanoparticles, catalyst activation, optimizing process parameters, metal oxides application, and binary catalysts. These improvements supply the higher results, the better catalyst capability, and the greater technology efficiency.

The green diesel produced over the DO technology is portrayed in Fig. 1. As previously explained in which the DO technology is implemented in the green diesel process employing miscellaneous vegetable oils to expunge the oxygen content [76]. Vegetable oil at 30 °C is transferred into a reactor by a magnetic pump at 6 bar while hydrogen make up is supplied to bond the oxygen molecules to become water. Variety

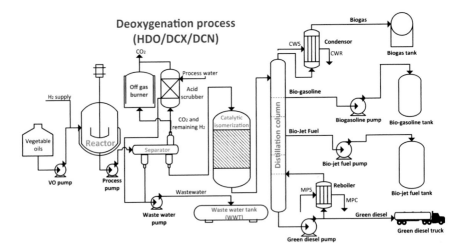

Fig. 1 Commercial green diesel production over deoxygenation technology adapted from Mahdi et al. [52] with the licence number of 5207580112774

of the solid catalysts is firstly activated using organic solvents, supported by nobel metals (i.e., Pt, Pd, Ru, etc.) [84], and implemented based on the proper oils [61].

The reaction can occur within 2–5 h in the reactor at certain temperatures and pressures. Jeon et al. [85] used non-edible fatty acids as the feedstock and Al_2O_3 catalyst assisted by Ni–MgO metal under the DO technology. Meanwhile, Makertihartha et al. [86] harnessed palm oil as the reactant and H-ZSM-5 catalyst reacted at 500 °C within 2 h. This reaction resulted in conversion of 100% with the catalyst reusability of 3 batches [86]. Furthermore, the produced residue as a side-product which was palm fatty acids distillate (PFAD) can be reused to reproduce green diesel using zeolite as the catalyst as investigated by Oliveira et al. [87]. The reaction was conducted at 250–350 °C within 30 min and 5 wt.% of the catalyst modified by Ni and Co metals. The process resulted in yield of 84.8% and reusability of the residue afforded the DO technology as the green and economic process [87, 88] and cost-effective catalyst is superior [89].

The green diesel crude produced from the reactor is transferred by a process pump and still comprises light hydrocarbons (3–9 wt.%) [90], H_2, CO, CO_2, and wastewater [91]. Consequently, a separation process is utilized to remove side-products from the primary products (biofuels-range hydrocarbons) [92]. The wastewater is accumulated in the wastewater tank (WWT tank) for the further treatment while CO, CO_2, and H_2 gases are treated in acid scrubber using process water to absorb the gases. The acid scrubber produces wastewater and either H_2SO_4 or NaOH solutions are injected to neutralize the wastewater. The treated wastewater is then accumulated in the WWT tank. On the other hand, these toxic gases can be utilized as raw material for off-gas burner to heat thermal oil heating (TOH) used to the reactor and glycerin is used to obtain intermediate products [93].

The biofuels-range hydrocarbons are reacted in a catalytic isomerization by adding hydrogen [94]. The product is then processed in a distillation column at certain temperature and pressure. In the top distillation column produces biogas whereas the bottom column generates green diesel (85 wt.%). Besides, it produces biogasoline that can be used for a substitution of gasoline and bio-jet fuel that can be harnessed for the aviation sector [95–99]. The DO technology is an easy process, less pollution, low energy consumption, green and clean process as studied by authors [52, 95, 100, 101].

2.2 Superiority and Drawback

Despite the incredible green diesel technologies, each of the common technologies used in the commercial green diesel process has its own advantageous and disadvantageous as listed in Table 3. The DO process is mostly applied in the green diesel application because it is cost-effective, eco-friendly, and easy technology [52]. Additionally, it needs less energy consumption, results in high conversion, selectivity, and yield, has high catalyst reusability, and is suitable for high FFAs-containing oil [101].

Table 3 Advantageous and disadvantageous aspects of technologies applied in the green diesel production

Technology	Predominance	Deficiency	References
Deoxygenation (DO)	• Low sulfur content • Minor oxygen content • Less production cost • Eco-friendly technology	• Complex process with less reusability catalyst • Required extra processes such as purification and separation process • Produces wastewater • Required H_2 make up • Pretreatment process was employed to remove impurities	[23, 52, 94, 96, 100, 102]
Hydroprocessing	• Less sulfur content • Modest oxygen content • Cost-effective process budget • Eco-friendly technology • Simple process • Recovery nutrient molecules • Less wastewater production • Higher cetane number-contained products	• High energy consumption • Operated at high reaction temperature	[3, 103–105]
Hydrodeoxygenation (HDO)	• Simple maintenance • Easy process • Moderate operating parameters	• Produces wastewater as by-product • Needs high H_2 consumption • Required high sulfur material consumption • Yields residue of sulfur	[25, 26, 52, 88, 106, 107]

(continued)

Table 3 (continued)

Technology	Predominance	Deficiency	References
Hydrotreating	• Decreases washing for active metal on catalyst surface sites • Creates a route for C–O bond breaking in the employed oil • Less oxygen, sulfur, and nitrogen content • High stable reaction	• Required extra H_2 supply • Rapid deactivation of the catalyst • Coke formation	[106, 108, 109]
Hydroconversion	• Good thermal stability • Mild operating conditions	• Rapid deactivation of the catalyst • Coke formation • Quick catalyst deactivation	[63, 108, 110]
Decarboxylation (DCX)	• No need hydrogen supply • High yield and selectivity	• Generates CO_2 as by-product	[106, 107]
Decarbonylation (DCN)	• No need hydrogen supply • High yield and selectivity	• Results in CO as by-product	[106, 107]
Hydrotreated vegetable oils (HVOs)	• Slight sulfur content and small oxygen content • Less aromatic products • Strong engine compatibility • Good storage stability and low water solubility • High-grade green diesel quality with high cetane number • High conversion, yield, and selectivity	• Required an extra process to remove the catalyst from the main products • Higher NOx pollution • Deposit and coke formation • Elevated operating conditions • High explosive risk	[57, 111–113]

(continued)

Table 3 (continued)

Technology	Predominance	Deficiency	References
Hydrocracking (HC)	• Good thermal stability • High oxygen removal • Less H_2S emissions • Higher oxidation stability and cetane number	• Extreme reaction temperature • Explosive risk • High energy consumption	[114–116]
Oligomerization	• By-product like light cracked naphtha (LCN) is utilized to produce other light hydrocarbons products • Stable reaction	• Extreme reaction temperature • Explosive risk	[117–119]
Isomerization	• Produces heavier hydrocarbon products • Suitable process for high FFAs content • Low freezing point and high cetane number	• Extreme reaction temperature and pressure • Required high hydrogen consumption • Coke formation	[108, 120]
Hydrogenation of WVO	• Eco-friendly process • Appropriate technology for high free fatty acid content	• More elevated energy consumption • Less stable process for high oxygen-contained oils	[121, 122]
Supercritical process, SCA	• Higher conversion • Appropriate technology for high free fatty acid content	• High process parameters • Waste gas and deposit formation donates to mitigate the yield • Less stable reaction for high oxygen-contained oils	[121, 123, 124]

(continued)

Table 3 (continued)

Technology	Predominance	Deficiency	References
Alkali catalyzed process, HACA	• Mild operating conditions • Cost-effective technology • Eco-friendly process • Suitable technology for high free fatty acid content	• Required a separation process to separate catalyst • Large-scale wastewater production • Refinement and segregation processes are undertaken to remove the by-products from the predominant products	[121, 122, 125]

Note WVO = waste vegetable oil; SCA = supercritical alcoholysis; HACA = homogeneous alkali catalyzed alcoholysis

Lastly, the DO technology is capable of producing some biofuels such as green diesel, biogas, biogasoline, and bio-jet fuel [56, 94].

Due to its eminence aspects, the DO technology has been implemented nowadays. As done by Zeng et al. [126] who studied the green diesel production via the DO technology using stearic acid oil and CuO-NiO nano-sized catalyst. This reaction resulted in 99.9% of conversion and 94.4% of yield [126]. Meanwhile, Hanafi et al. [127] explored the DO technology using waste chicken fat oil as the raw material and Ni–MoS$_2$/Al$_2$O$_3$–(15%) TiO$_2$ catalyst for the first time reacted in fixed bed reactor at 400–450 °C, LHSV of 1.0–4.0/h, 60 bar, and H$_2$:oil molar ratio of 600 v/v. The lower reaction temperature positively afforded the green diesel yield while negatively affected the biogasoline and bio-jet fuel yields [127].

Yet, the DO technology requires a purification and separation process, needs hydrogen supply, produces wastewater, and involves a pretreatment process for the oils [23, 52, 102]. These problems are also demonstrated by HDO technology. Even, it necessitates high sulfur consumption and generates sulfur residue [25, 26, 52, 88, 106, 107]. However, the HDO technology possesses easy maintenance and is undertaken at low temperature and pressure thus showing a low explosive risk. Meantime, the DCX and DCN technologies have no hydrogen consumption and result in high yield and selectivity [106, 107]. But, those technologies generate poisonous gases such as CO and CO$_2$ contributing to environmental pollution and global warming.

By-product like light cracked naphtha (LCN) produced by oligomerization technology can be exploited to produce other light hydrocarbons products and the reaction is highly stable [117–119]. Nonetheless, it is performed at an extreme reaction temperature and has a high explosive risk. The oligomerization technology was recently investigated by Gorimbo et al. [128] and Mirzaei et al. [129]. On the other hand, isomerization technology yields heavier hydrocarbon products, is proper process for high FFAs-containing oils and low freezing point, and further the green diesel produced has elevated cetane number [108, 120]. A high explosive probability caused by extreme reaction temperature and pressure, high hydrogen consumption, and coke formation become the most important concern that has not been addressed yet to now [108, 120]. Exploration of the isomerization technology was, in this year, reviewed by Maghrebi et al. [130].

The technologies that are suitable for high FFAs-containing oils are hydrogenation of waste vegetable oils (WVO) [121, 122, 131], supercritical process, SCA [121, 123, 124], and alkali catalyzed process, HACA [121, 122, 125]. Either WVO technology or HACA technology is eco-friendly process for the green diesel production whereas SCA technology results in higher conversion. On the other hand, the WVO and HACA technologies entail a high energy consumption and are not appropriate process for the high oxygen-containing oils [121, 122, 131]. Even, the HACA technology produces large-scale wastewater and necessities extra processes such as purification and separation processes leading to the increase of the process budget [121, 122, 125]. Meanwhile, the SCA technology is operated at high pressure and temperature causing high explosive risk and forms waste gas and coke leading to bring down the yield and the catalyst activity [121, 123, 124]. Hence, these technologies provide many challenges for more exploration and studies before applied in the commercial green diesel production.

Among those green diesel technologies, the hydroprocessing technology is the most effective, efficient, promising, and potential process because green diesel produced comprises low sulfur content, less oxygen content, and higher cetane number content. This technology is also known as cost-effective, eco-friendly, and simple process as well as easy maintenance. Additionally, the hydroprocessing technology produces less wastewater and is capable of recovering nutrient molecules as reported by authors [3, 103–105]. In the meanwhile, it necessitates high energy consumption and is operated at high temperature.

These shortages become a gold chance to be more explored by researchers. Pelemo et al. [3] studied the hydroprocessing technology using used cooking oil as the feedstock and SiO$_2$, Al$_2$O$_3$, and CaO catalysts without using hydrogen supply. Furthermore, Gieleciak et al. [132] employed low-grade canola oil to produce the green diesel at modest temperatures and pressures under the hydroprocessing technology. The green diesel produced had the same physical and chemical property as the commercial diesel fuel characteristic [132]. Comparison of those whole green diesel technologies demonstrates that the advantages of hydroprocessing technology are highly better than its disadvantages. Authors highlight that the hydroprocessing technology has become the most effective, efficient, potential, and promising green diesel process in the future.

3 Application of the Use of Solid-Based Heterogeneous Catalysts in Green Diesel Produced

The implementation of the heterogeneous solid base catalysts in the green diesel reaction has received a lot of attention due to the high activity and selectivity, better reuse, reduction in processing steps, and waste [26]. Conventionally in industrial processes, metal sulfide catalysts (as Mo or W metal sulfides), doped with transition metals Ni or Co, deposited on oxide or mixed oxide supports, such as silica (SiO$_2$), titanium oxide (TiO$_2$), and alumina (Al$_2$O$_3$) are the most frequently used [133, 134]. However, the use of these catalysts can lead to sulfur contamination of products due to catalyst leaching, which is environmentally undesirable, since the presence of sulfur in fuels results in the emission of sulfur oxides. Sulfur leaching also leads to catalyst deactivation, which decreases efficiency [133].

Precious metals like Pd, Pt, and Rh components proved to be the most active metal promoters of DO ability and had a higher energy affinity for C-O bond fission [135]. However, high-cost restrictions made them unattractive. It is necessary to develop cheap and highly active non-sulfur-based catalysts for deoxygenation. In this sense, a wide range of alternative catalysts, such as oxides, bimetallic materials, acid catalysts, have been widely explored [68, 136].

The composition of the catalyst influences the catalytic activity and performance of the catalysts, which are crucial factors in deoxygenation processes and directly influence the composition and properties of the products [137]. Metal-based catalysts have usually become the most efficient for the DO process and the acid catalysts act mainly in cracking reaction, isomerization process, cyclization reaction, and aromatization process [138, 139]. Generally, bifunctional or multifunctional catalysts are designed to combine the catalytic effect of metals with the action of acid supports, such as zeolites, activated carbon, silica, among others [138].

There are two catalyst types commonly harnessed in the green diesel production, which are solid synthetic catalysts and zeolite catalysts. Each of these catalysts negatively and positively contributes to conversion, selectivity, and yield of green diesel achieved. Thereunto, these different catalysts disclose good stability, high reusability, and elevated efficiency of the catalysts which play a highly critical key in high-grade green diesel quality and the greener and cleaner process. In this section, authors will discuss in detail regarding influence of these catalyst types on process performance and the catalyst capability as further discussed below.

3.1 Green Diesel Produced Over Metal Catalysts

Alternatively, metallic catalysts reveal attractive concern to the DO process by dint of the elevated reactivity at modest temperatures, no sulfur content in liquid biofuels, and less need for H_2 supply [140]. A number of researchers have maximally exploited metal catalysts used for the DO technology as listed in Table 4. Metallic catalysts can be monometallic [133], bimetallic [59], or even supported on various materials, such as silica [12], activated charcoal [74], and others.

According to the data obtained in Table 4, it is possible to observe that either the type of metal or the chosen support significantly affects the catalytic capability and consequently the selectivity, conversion, and other aspects of the reaction [136]. Jeon and collaborators $_{0.6}$], for example, compared the properties of different catalysts $Ni-Ce_{0.4}Zr_{0.6}O_{0.4}$ $Pd–Ni/Ce_{0.6}Zr_{0.4}O_{0.6}$, $Ru-Ni/Ce_{0.4}Zr_{0.6}O_{0.4}$ and $Pt-Ni/Ce_2Zr_2O_2$ as well as its distinct performances in the DO of Oleic acid. The reactions were undertaken at 20 vol. % H_2 / N_2 (1 bar) during 3 h a 300° C and at a stirring speed of 300 rpm. Among these solid catalysts, the perfect conversion of oleic acid, high selectivity of diesel range biofuel, and less oxygen-contained product were achieved by $Pt/Ni-Ce_2Zr_2O28$, and this result is mainly attributed to its higher fraction of oxygen defect and acidity [28].

Ameen et al. [70] evaluated the performance of $NiMo/S\gamma-Al_2O_3$ catalyst in the conversion of Rubber Seed Oil through different technologies including Hydrodeoxygenation (HDO) and Decarboxylation (DCX) pathways. Experimental runs were performed under four different process conditions such as reaction temperature of 300–400 °C, pressure of 30–80 bar, WHSV of 1–3/h, and H_2:oil molar ratio of 400–1000 Ncm^3/cm^3. This recent investigation has demonstrated that the oil is entirely converted

Table 4 Green diesel produced over heterogeneous acid solid catalysts

Acid solid catalysts	Raw materials	Technologies applied in	Results (%)	References
Cobalt–alumina	Sunflower oil and WCO	Deoxygenation (DO)	Y = >70	Nikolopoulos et al. [141]
Acid solid catalysts	Palm oil	Deoxygenation (DO)	X = 100; S = >80, and Y = >80	Mahdi et al. [52]
Mg- Zn	Kapok oil	Deoxygenation (DO)	X = 65	Chumaidi et al. [59]
Acid solid catalysts	Triglycerides	Deoxygenation (DO)	X = 96 and Y = >78	Silva and de Andrade [142]
Ni/γ-Al$_2$O$_3$ and γ-Al$_2$O$_3$/La$_2$O$_3$	Palm oil	Deoxygenation (DO)	X = 100 and Y = 60	Papageridis et al. [101]
Pt/Al$_2$O$_3$, TiO$_2$, and TiO$_2$	Vegetable oils	Deoxygenation (DO)	Y = 67.0	Jin et al. [66]
NiMo/γ-Al$_2$O$_3$	Crude palm oil (CPO)	Hydrogenation	Y = 68.2	Zikri and Aznury [67]
Ni/γ-Al$_2$O$_3$	Karanja oil	Hydrogenation	X = 100 and S = 70	Yenumala et al. [38]
NiMg, NiMn, NiCu, and NiCe-supported MWCNTs	Chicken fat oil (CFO)	Deoxygenation (DO)	S = 81–85 and Y = 75–84	Aliana-Nasharuddin et al. [68]
NiMo/γ-Al$_2$O$_3$	Rubber seed oil (RSO)	Hydrodeoxygenation (HDO)	X = 100 and Y = 84.94	Ameen et al. [69]
NiMo/γ-Al$_2$O$_3$	Rubber seed oil (RSO)	Hydrodeoxygenation (HDO)	X = 100 and S = 19.1	Ameen et al. [70]
NiMo/γ-Al$_2$O$_3$	Rubber seed oil (RSO)	Decarboxylation (DCX)	X = 100 and S = 81.7	Ameen et al. [70]
MWCNTs modified by NiO–Fe$_2$O$_3$ and NiO–ZnO	Jatropha curcas oil	Pyrolytic-deoxygenation	Y = 75	Asikin-Mijan et al. [71]
Nanoparticle Rh assisted by silicon nanowire array (SiNA-Rh)	Fatty acids	Hydrogenative decarboxylation (HDCX) in microwave	Y = 99	Baek et al. [72]
NiO-ZnO	Palm fatty acid distillate (PFAD)	Deoxygenation (DO)	S = 86.0 and Y = 83.4	Baharudin et al. [73]

(continued)

Table 4 (continued)

Acid solid catalysts	Raw materials	Technologies applied in	Results (%)	References
Activated carbon-supported by CoMo	Palm fatty acid distillate (PFAD)	Deoxygenation (DO)	$S = 89$ and $Y = 92$	Gamal et al. [74]
Ni, Co, and Ni-Co-assisted SBA-15	Palm fatty acid distillate (PFAD)	Deoxygenation (DO)	$S = >85$ and $Y = 85.8$ and 88.1	Kamaruzaman et al. [12]
Co/Ni-modified Zr-SBA-15	Methyl esters	Hydrotreating	$X = >90$ and $Y = >60$	Ochoa-Hernández et al. [75]
USY, H-ZSM-5 and Al-SBA-15-doped Ni_2P	Oleic acid	Deoxygenation (DO)	$Y = 24$, 29, and 42	de Oliveira Camargo et al. [76]
Co-Carbon (activated biochar)	Macauba acid oil	Deoxygenation (DO)	$Y = 98.4$	de Barros Dias Moreira et al. [77]
Pd/Al_2O_3 and $Pd-Fe/Al_2O_3$	Palm oil	Hydrocracking (HC)	$Y = 50-62.20$	Srihanun et al. [78]
$Mn-Al_2O_3/NiMo$	Waste soybean oil	Hydroconversion	$X = 2.9-21.5$ and $Y = 39-82$	Vázquez-Garrido et al. [63]
HY zeolite	Rubber seed oil (RSO)	Hydrodeoxygenation (HDO)	$X = >99$ and $S = 10.4$	Ameen et al. [79]
Niobium phosphate ($NbOPO_4$)	Soybean oil	Hydroprocessing	$Y = 40$	Scaldaferri and Pasa [80]
Niobium phosphate ($NbOPO_4$)	Soybean oil	Deoxygenation (DO)	$S = 37$ and $Y = 15$	Scaldaferri and Pasa [39]
Pd-Carbon	Soybean oil	Deoxygenation (DO)	$Y = 38$	Scaldaferri and Pasa [39]
Fluid catalytic cracking	Soybean oil	Deoxygenation (DO)	$Y = 27$	Scaldaferri and Pasa [39]
CoP-USY and Ni_2P-USY	Oleic acid	Deoxygenation (DO)	$X = 100$ and $Y = 52$ and 36	Kochaputi et al. [35]
$Pt-Ni/Ce_{0.6}Zr_{0.4}O_2$	Oleic acid	Deoxygenation (DO)	$X = 98.7$ and $S = 54.2$	Jeon et al. [28]
$Ni-Ce_{0.6}Zr_{0.4}O_2$	Oleic acid	Deoxygenation (DO)	$X = 98.3$ and $S = 33.9$	Jeon et al. [28]

(continued)

Table 4 (continued)

Acid solid catalysts	Raw materials	Technologies applied in	Results (%)	References
Pd–Ni/Ce$_{0.6}$Zr$_{0.4}$O$_2$	Oleic acid	Deoxygenation (DO)	X = 90.8 and S = 19	Jeon et al. [28]
Ru-Ni/Ce$_{0.6}$Zr$_{0.4}$O$_2$	Oleic acid	Deoxygenation (DO)	X = 88.3 and S = 19.9	Jeon et al. [28]
NiMo/CoMo	Canola oil	Deoxygenation (DO)	X = 100	Taromi and Kaliaguine [29]
CoMo-SG	Oleic acid	Deoxygenation (DO)	X = 88.9 and S = 48.1	Shim et al. [40]
CoMo-IWI	Oleic acid	Deoxygenation (DO)	X = 62 and S = 19.7	Shim et al. [40]
CoMo-HT	Oleic acid	Deoxygenation (DO)	X = 84.9 and S = 22.8	Shim et al. [40]
CoMo-CP	Oleic acid	Deoxygenation (DO)	X = 87.2 and S = 31	Shim et al. [40]
NiMo/γ-Al$_2$O$_3$ and Pd–C	Coffee ground oil	Hydrotreating	X = >90 and Y = >20	Phimsen et al. [37]
Pd-Carbon	Palm kernel fat	Hydrodeoxygenation (HDO)	X = 96 and Y = 98	de Sousa et al. [24]
Pd-Carbon	Macauba oils	Deoxygenation (DO)	S = 97–100 and Y = 85	Silva et al. [41]
Ni$_2$P/H-ZSM-22	Palmitic acid	Deoxygenation (DO)	X = 99.6 and Y = 42.9	Liu et al. [36]
Pd-Carbon	Crude palm oil (CPO)	Hydroprocessing	X = >60 and Y = 51	Kiatkittipong et al. [34]

Note MWCNTs = multi-walled carbon nanotubes; C = carbon; Ni$_2$P = nickel phosphide; WCO = waste cooking oil into diesel range biofuels with HDO selectivity of 19.1 wt.% and DCX selectivity of 81.7 wt.% achieved at 400 °C, 80 bar, 1 h^{-1}, and ratio H$_2$:oil 400 cm^3/cm^3 [70].

As for works that use metals supported on silica mesoporous, γ-Al$_2$O$_3$, and activated charcoal, some of which are discussed below. Srifa et al. [133] compared the catalytic activity of cobalt, nickel, palladium, and platinum metals modified on γ-Al$_2$O$_3$ for DO reaction of palm oil and oleic acid under conditions of 330 °C, the pressure of 50 bar of H$_2$, in a continuous flow reactor with H$_2$/oil ratio of 1000 Nm^3/m^3, and LHSV (spatial velocity in terms of liquid per unit of hour) of 2 and 8 h^{-1}. The authors reported that the ascending order of catalytic activity comparing the metals was Co > Pd > Pt > Ni, being the largest yields obtained in hydrocarbons of 94.3% for oleic acid and 88.5% for the palm oil using Co/γ-Al$_2$O$_3$ catalyst. It was proposed that the mechanism predominant for Pd, Pt, and Ni catalysts was decarbonylation, while for the Co catalyst a greater contribution of hydrodeoxygenation was observed [133].

Wang et al. [143] studied the conversion of oleic acid methyl ester to green diesel, using different molybdenum and/or nickel-based catalysts. You catalysts used were molybdenum carbide supported on activated carbon (Mo$_2$C/AC), molybdenum oxide supported on activated carbon (MoO/AC), carbon oxide alumina-supported molybdenum (MoO/γ-Al$_2$O$_3$), sulfide-supported molybdenum alumina (MoS$_2$/γ-Al$_2$O$_3$), and nickel supported on alumina (Ni/γ-Al$_2$O$_3$). The reactions were conducted in a batch reactor for 3 h at 370 °C, 20 bar H2, with one ratio catalyst/oil 1:10. A greater degree of deoxygenation of the ester was obtained from the Molybdenum-based catalysts, following the order Mo$_2$C/AC (100%) > MoO/γ-Al$_2$O$_3$ (85%) > MoS2/γ-Al$_2$O$_3$ (83%) > MoO/AC (56%) > Ni/γ-Al$_2$O$_3$ (22%). Alongside from the catalytic performance of molybdenum catalysts, they showed considerable selectivity for short-chain alkanes arising from cracking reactions [143].

Kamaruzaman et al. [12] evaluated cobalt and nickel catalysts supported on silica (Co-doped SBA-15, Ni-modified SBA-15, and Ni/Co-assisted SBA-15) to obtain green diesel from oil palm. The OD process was conducted in a semi-batch reactor with a catalyst charge of 10% by weight at 350 °C for 3 h. The utilization of Ni/SBA-15 and Ni-Co/SBA-15 provided products with high contents of liquid hydrocarbons (C$_8$–C$_{17}$) with yields of 85.8% and 88.1%, respectively, and selectivity for hydrocarbons in the range of diesel (C$_{13}$–C$_{17}$) above 85% was achieved. Cobalt appears to possess a more extensive size, so it correlates to carbon conformation and familiarizes deposit formation. It blocks some pores and disables the catalyst active sites [12].

In addition to these, numerous works using different catalysts are found out in the literature. The challenges in this area include the improvement of catalysts with extraordinary catalytic capability for deoxygenation and isomerization, as well as studies of optimization of process parameters aiming to use mild reaction conditions.

3.2 Green Diesel Produced Over Zeolite Catalysts

The production of green diesel obtained through zeolite-supported catalysts is extensively investigated [34]. Zeolites are crystalline aluminosilicates possessing 3D tetrahedral dimension with conclusive porosity which is adjustable acid centers, essential characteristics to improve the selectivity of the diesel range [144–146]. These materials are usually used in the form of bifunctional catalysts, which are characterized by the presence of hydrogenating and an acidic function. The acid function is responsible for the cracking itself and is associated with a support. The hydrogenating function is associated with noble metals, transition metals, and transition metal sulfides [138].

The combination of these metals with acid supports, in general, results in catalysts with greater catalytic activity and selectivity [134, 147–149]. Several modified zeolites including ZSM-5, ZSM-22, SAPO-11, beta, and H–Y have been considerably exploited in the production of green diesel are summarized in Table 5 [150].

The effect of the zeolite type on the catalyst activity was studied by Camargo et al. [76]. For this, the authors synthesized nickel phosphide catalysts modified with various types of zeolites (i.e., USY, H-ZSM-5, and Al-SBA-15) and evaluated their activity in the DO of the oleic acid. The catalytic tests were performed at 50 bar hydrogen pressure in an autoclave reactor. The highest biofuels (C_{10}–C_{18}) yield obtained at 300 °C was resulted by Ni_2P/Al-SBA-15 catalyst followed by Ni_2P/H-ZSM-5 catalyst and Ni_2P/USY catalyst. The Ni_2P/Al-SBA-15 catalyst demonstrated unbelievable DO performance if compared to noble metals catalysts. All catalysts have achieved more C_{17} hydrocarbons than C_{18}, therefore decarbonylation/decarboxylation reactions prevail [76].

Another important factor in catalysts produced with zeolite is the effect of zeolite crystal size as per Choo et al. [147]. The authors evaluated the influences of Y zeolite crystallite (Y20: 20 nm, Y65: 65 nm, Y380: 380 nm, and Y2750: 2.75 μm) on the triolein DO process and observed that as the size of the zeolite crystal decreased, the oil conversion and the product yield increased. In addition, high-quality product is attained: greater formation of hydrocarbons, greater formation of diesel, less formation of heavy hydrocarbons, and a greater proportion of diesel to gasoline. The elevated catalyst capability is related to the high concentration of acidic sites, with accessible average acid strength composed of the Brønsted and Lewis acidic sites located on the outer surface of the crystals [147].

Freitas et al. [152] studied the bio hydrocarbon products obtained from macaúba oils as the raw material. These fatty molecules were fed into DO reactions employing beta zeolite and ZSM-5 catalysts with different acidity (Si/Al ratios) of 100% in 3 h. The products have been classified into different compounds, which include linear hydrocarbons, cyclic hydrocarbons, branched hydrocarbons, and aromatic hydrocarbons [152]. In addition to these, numerous works aimed at obtaining green diesel that use zeolite as a catalyst are listed in Table 5.

Table 5 Green diesel produced over zeolite catalysts

Zeolite types	Raw materials	Technologies applied in	Results (%)	References
Natural	Crude palm oil (CPO)	Hydrogenation	Y = 37.30	[Zikri et al. 62]
Beta	Oleic acid	Hydrodeoxygenation	X = 99 and S = 83	Azreena et al. [150]
Beta	Palm olein	Hydroconversion	Y = 28.7	Chintakanan et al. [151]
Beta and ZSM-5	Macauba oils	Deoxygenation	X = 100	Freitas et al. [152]
Y	Oil sludge	Hydroconversion	Y = 90.6	Kang et al. [153]
ZSM-5	Oleic acid	Deoxygenation	Y = 65	Arumugam et al. [154]
Al-MCM-41	Oleic acid	Deoxygenation	Y = 82.30	Wang et al. [155]
USY and ZSM-5	Oleic acid	Deoxygenation	Y = 24 and 29	de Oliveira Camargo et al. [76]
Y	Triolein	Deoxygenation	X = 80 and S = 93.7	Choo et al. [148]
HY	Triolein	Deoxygenation	X = 76.21 and S = 92.61	Choo et al. [149]
ZSM-5	Palmitic acid	Hydrodeoxygenation	X = 99	Lee et al. [156]
SAPO-11	Stearic acid	Deoxygenation	X = >95	Liu et al. [157]
Beta	Methyl palmitate	Hydroprocessing	X = 77 and S = 61.9	Papanikolaou et al. [7]
ZSM-5	HDPE	Cracking	Y = 61.05	Ghaffar et al. [158]
HY	Triolein	Deoxygenation	X = 71.16 and S = 72.15	Choo et al. [147]
HY	Rubber seed oil	Hydrodeoxygenation	X = >99, Y = 30 and S = 6.8–14.5	Ameen et al. [79]
Beta	Palm kernel oil and palm oil	Deoxygenation, cracking and isomerization	X = 96	Sousa et al. [159]
INZ	Palm oil	Hydrodeoxygenation	X = 58 and 89	Putra et al. [91]
ZSM-5	Soybean oil	Hydrocracking	X = 99.5 and S = 42.11	Zandonai et al. [160]
Beta	n-hexadecane	Hydroisomerization	X = 36.5 and Y = 35.5	Gomes et al. [161]
HY	Stearic acid	Hydrodeoxygenation	X = 100 and S = >20	Hachemi et al. [162]

Note X = Conversion; S = Selectivity; Y = Yield; HDPE = high density polyethylene; INZ = Indonesian natural zeolites

4 Catalytic Hydroprocessing Technology of Edible and Non-Edible Oils

During the last decades, the hydroprocessing technology is one of the processes that is mostly used in the green diesel reaction and further exhibits the most potential and effective process among others [57]. The hydroprocessing is widely applied in the oil refinery to crack the larger hydrocarbon chains into the smaller hydrocarbon chains such as green diesel, biodiesel, bio-jet fuel, and biogasoline [163] as delineated in Fig. 2. Furthermore, it is used to remove sulfur (S) content, nitrogen (N) content, and metal contents contained in the biofuels [164, 165] and equipped with a distillation column in the final process to easily facilitate a separation process [166, 167]. Diversiform vegetable oils such as fresh oils, used oils, edible oils, non-edible oils, and waste oils have been employed in order to discover the most cost-effective and eco-friend process as well as less waste production [168–170]. These diversified oils were inspected by Ramírez et al. [171].

The hydroprocessing technology consists of two process steps which are hydro-treatment and syngas treatment. For the first step, the vegetable oils are reacted in a batch reactor using solid acid catalysts and water content is removed in the hydro-treatment process [172]. Meanwhile, gasification syngas process is carried out using fresh oxygen under Fischer Tropsch process at the second one [172]. The main reaction is catalytically conducted over the hydroprocessing in which the mostly used catalyst is Al_2O_3 doped with Ni-Mo metal [164, 171]. Afterward, the crude product is fed into a distillation column to obtain the biofuels (i.e., biogas, bio-jet fuel, biogasoline, and green diesel). A cooler is utilized to bring down the product temperature because the high product temperature contributes to the poor quality and degradation of the product [3, 173, 174].

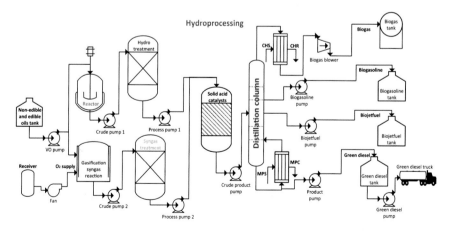

Fig. 2 Commercial green diesel production over hydroprocessing technology [172]

Apart from the distinct vegetable oils, variegated catalysts and reactor types are also investigated by many authors as reported by Furimsky [175] who observed the preference for appropriate catalysts and proper reactors for the hydroprocessing technology. In the report, fixed-bed reactor is suitably used for light oils whereas moving-bed and ebullated bed reactors are successfully employed for high asphaltene and metal-contained oils. Additionally, multi-reactors are sensibly implemented for heavy oils [175]. Bezergianni et al. [168] used two dissimilar solid catalysts like $CoMo/Al_2O_3$ and $NiMo/Al_2O_3$ catalysts in the co-hydroprocessing technology using heavy oils like waste cooking oil. The NiMo-modified Al_2O_3 catalyst was more appropriate for the co-hydroprocessing technology rather than $CoMo/Al_2O_3$ catalyst due to low deactivation at extreme reaction temperatures. Hereinafter, Marafi et al. [176] studied that spent catalyst consisting of vanadium content disclosed a remarkable capability either for the fixed bed reactors or for the ebullated bed reactors [176].

In the hydroprocessing method, this process produces high-grade green diesel with the similar characteristics to the diesel fuel properties and less oxygen content and further involves hydrogen and triglycerides as the feedstocks [144]. Thereunto, it results in three different products namely long hydrocarbon chains, short hydrocarbon chains, and branched-chain ones [177]. Normally, the hydroprocessing technology is operated at reaction temperature of over 280 °C and pressure up to 50 bar [178]. As reported by Liu et al. [30] who hydroprocessed the oils using $NiMo/SiO_2$-Al_2O_3 catalyst at 350 °C and 40 bar and this reaction generated the yield of 82.1% [30]. Yulia and Zulys [179] also employed $NiMo/Al_2O_3$ catalyst at 400 °C and 35 bar within 5 h. Approximately, 57.79% of kemiri Sunan oil was catalytically converted into the green diesel product [179].

To jack up performance and efficiency of the hydroprocessing technology, some modifications and improvements have been exploited during the last decade as listed in Table 6. Kaewchada et al. [180] applied distinct reaction temperatures (400–500 °C), pressures (7–34 bar), WHSVs (10–43.7/h), and H_2:oil ratios (36.6:1–220.9:1) to escalate the green diesel production. The process used palm oil as the reactant and Rh-doped ZSM-5 catalyst under the hydroprocessing technology. The most maximum green diesel yield (51.13%) was reached at 500 °C, 34 bar, 0.05 g of the Rh-ZSM-5, WHSV of 43.7 h^{-1}, and H_2:oil molar ratio of 36.6:1. On the contrary, green diesel produced possessed physical and chemical characteristics as the properties of the commercial US and European standards diesel fuel [180]. Either the process parameters or the raw material diversity obviously contributes to the increase of the yield and the green diesel purity obtained in the final process as observed by Velden et al. [181] and Carpenter et al. [182].

McIntosh et al. [184] employed a multistage process which was a pyrolysis process and the hydroprocessing. A milk processing waste was maximally harnessed as the raw material to reduce the environmental issue. The pyrolysis process was operated at 450 °C and resulted in 57.7% of the biofuels consisting of 19–22% of aliphatic hydrocarbons, 11–14% of alkenes, and 57–63% of alkyl nitriles with <2% of aromatic compounds. Meanwhile, the hydroprocessing technology used sulfided Ni-Mo/Al_2O_3 catalyst and produced 100% of green diesel in a distillation process [184].

Table 6 Hydroprocessing technology of vegetable oils using multiform solid catalysts

Vegetable oils	Catalysts type	Metal oxide	Result	Recent improvements	References
Waste cooking oil	Eskom-derived biowaste fly ash	SiO_2, Al_2O_3, and CaO	SA = 41.2571	Waste fly ash was harnessed as the catalyst due to cost-effective catalyst price	Pelemo et al. [3]
Canola oil and Canadian oil	Solid catalysts	–	Y = 40.39	The green diesel process was carried out under different oil molar ratio, temperatures, pressures, and LHSVs	Gieleciak et al. [132]
Pennycress oil seed	Al_2O_3	Pt	Y = 90	Hydroprocessing technology using pennycress oil seed as the raw material indicated industrial feasibility and low process costs	Mousavi-Avval and Shah [183]
Milk industrial waste	Al_2O_3	NiMo	X = 100	The catalyst was supported with sulfided NiMo metal and the process was combined with pyrolysis reaction	McIntosh et al. [184]

(continued)

Table 6 (continued)

Vegetable oils	Catalysts type	Metal oxide	Result	Recent improvements	References
Palm oil	H-ZSM-5	Rh	$Y = 51.13$	Rh-assisted ZSM-5 was employed as the catalyst to facilitate hydroprocessing reactions such as cracking, hydrogenation, and isomerization processes	Kaewchada et al. [180]
Palm oil	Al_2O_3	$NiMoS_2$	$R = 46.3$	Hydrogen was required to achieve the maximum conversion	Phichitsurathaworn et al. [185]
Methyl palmitate	Beta	Ni nanoparticles	$X = 77; S = 61.9$	Ni nano-sized was added into the beta catalyst under sol immobilization method	Papanikolaou et al. [7]
Methyl palmitate	Al-SBA-15	Ni	$X = 47; S = 73.7$	Ni nano-sized was synthesized into the SBA-15 catalyst with Si:Al ratio of 40 using Pluronic P123 as a carrier	Papanikolaou et al. [7]
Kemiri Sunan oil	Al_2O_3	NiMoCe	$X = 57.79$	Catalyst was synthesized over wet impregnation method using Ce contents of 1, 5, and 15%	Yulia and Zulys [179]

(continued)

Table 6 (continued)

Vegetable oils	Catalysts type	Metal oxide	Result	Recent improvements	References
Jatropha oil	ZSM-5	Pt	Y = 67	Pt metal was added into the catalyst to enhance the process performance	Pelemo et al. [174]
Palm oil	NiMo	NiMo	Y = 97	Various reaction temperatures, pressures, and H_2:oil molar ratios were investigated using Aspen plus 7.3 under reactor equilibrium	Plazas-González and Guerrero-Fajardo [186]
Palm oil	Carbon and Al_2O_3	Pd and NiMo	X = 100; C = >200	Optimum conversion and yield were achieved by maximizing reaction temperatures, pressures, LHSVs, and H_2:oil molar ratios	Wang and Hsieh [187]
Polypropylene	Mordenite	Ni-Au	Y = 75	Polypropylene plastic waste was harnessed to bring down the environmental emission caused by the massive plastic production	Mangesh et al. [188]
WCO, WLO and VGO	Al_2O_3	NiMo	X = 99.786; Y = 48.224	Mixture of the oils was implemented in order to enhance the capability of the hydroprocessing reaction	El-Sawy et al. [163]

(continued)

Table 6 (continued)

Vegetable oils	Catalysts type	Metal oxide	Result	Recent improvements	References
Soybean oil	$NbOPO_4$	–	Y = 40	Green diesel process was carried out under mild operating parameters using lipid catalyst	Scaldaferri and Pasa [80]
Palm oil	SiO_2 and Al_2O_3	Pt and Ni_2P	Y = 70	Catalysts supported with Pt and Ni_2P metals donated the best solution for a problem regarding the rapid deactivation of the catalyst and the poor thermal stability	Jeong et al. [189]
Waste sunflower oil	MoS_2 and Al_2O_3	CoNi and CoMo	S = 95	Co and Ni-supported MoS_2 catalyst was synthesized over incipient wetness impregnation method using 12-molybdophosphoric heteropolyacid as a carrier agent	Varakin et al. [190]

Note SA = surface active area (m^2/g); X = conversion (%); Y = yield (%); S = selectivity (%); R = recovery of methanol (%); C = concentration (g/L); WCO = waste cooking oil; WLO = waste lubricating oil; VGO = vacuum gas oil

The Ni-Mo/Al$_2$O$_3$ catalyst was also exploited by Plazas-González and Guerrero-Fajardo [186] who implemented Aspen Plus 7.3 and studied the effect of multiform operating parameters. Approximately, 97% of product purity was perfectly accomplished at 60 °C and 1 bar in the second distillation column [186]. Implementation of the Aspen Plus 7.3 completed with Aspen Custom Modeler® in the green diesel production was perfectly designed by Fernández-Villamil et al. [57].

The hydroprocessing technology not only produces the green diesel product, but also it results in bio-jet fuel as reported by Mousavi-Avval and Ajay Shah [183] with the production capacity of 18.9 million liters/year.

As studied above in which catalysts supply a highly crucial factor in the increase of the green diesel yield and the technology performance. Application of the appropriate catalysts applied in the hydroprocessing technology provides a challenging and gold opportunity for more discussions. Hence, some modifications of the catalyst application and synthesis have been conducted [174]. One of those modifications by using a catalyst derived from waste like eggshell, mud shell, and soda-lime known as the most active solid catalysts [191] acquires a very attractive to exploit these days [192]. The modified catalysts donate to lower the activation energy causing the higher green diesel yield and high-grade products [193]. Pelemo et al. [3] used the biowaste fly ash obtained from Eskom modified with SiO$_2$, Al$_2$O$_3$, and CaO components. The metal-doped catalyst disclosed larger active sites, higher porosity, and enlarged volume giving rise to the best performance [3].

Phichitsurathaworn et al. [185] attached a heterogeneous catalyst in the reactor by feeding high hydrogen consumption and methanol to maximally convert the palm oil into the diesel range biofuel. The hydrogen recovery system was implemented to mitigate the lost hydrogen component. This process was capable of hydrogen recovery (46.3%) and reduction of CO$_2$ and CO emissions up to 14%. Furthermore, this methanol synthesis-supported process could abbreviate a payback period within 2.81 years. On the other hand, Mangesh et al. [188] harnessed polypropylene plastic aiming to bring down the petroleum-derived plastic pollution using Nickel/Gold-assisted mordenite catalyst at 70 bar and 350 °C. This process resulted in 90% of the green diesel [188] and subscribed the decrease of the plastic waste pollution [194, 195]. Utilization of the plastic waste as the raw material to produce the green diesel under the hydroprocessing technology has been previously explored by Mangesh et al. [196] and Mangesh et al. [197].

Diversity of the vegetable oils, varied operating conditions, the plastic waste feedstock, methanol synthesis, multistage process, and the utilization of the modified catalysts are precarious factors in the green diesel production. These developments dramatically contribute to jack up the capability of the hydroprocessing technology and the increase of the green diesel yield with high quality. Authors declare that the developed catalyst is the most important role among others due to the most superior contribution for the process. A deep discussion is obviously explained in detail in the next sections.

4.1 The Impact of Process Parameters

Stable oxidation, high thermal stability, and steady reaction can be perfectly attained under variegated operating conditions including temperature, hydrogen pressure, WHSV, LHSV, H_2:oil ratio, catalyst weight, etc. [198, 199]. The temperature becomes one of the highly crucial roles that significantly influences performance and capability of the hydroprocessing technology in the green diesel production of vegetable oils [180]. The product can be converted into a wax at atmosphere temperature due to free fatty acids (FFAs) content and lower conversion is reached at low temperature thus the process should be operated at high reaction temperature [43, 200]. For instance, the palm-based conversion (<5%) was lowly obtained at 375 °C via the hydroprocessing technology [34] whereas the impeccable palm-based conversion (~100%) was acquired at the higher reaction temperatures (400–500 °C) [180]. Hence, the hydroprocessing technology is commonly carried out at 380–525 °C to obtain the optimum conversion and yield [121, 201].

As studied by Mahdi et al. [52] who claimed that the higher reaction temperatures provided stronger Bronsted-Lewis acid–base sites, higher Van der Waals force, enlarged active areas, and larger accessibility of the reactant causing the increase of the catalyst ability [52]. Additionally, the high temperature plays a critical role in cracking of hydrocarbon molecules because of rupture of either C=C bond or C–O bond [202]. These factors promote to upgrade the conversion, yield, and selectivity. Kaewchada et al. [180] proved that the higher reaction temperature donated the higher conversion with the one of >95% at 450 °C and 100% at 500 °C at the same reaction time. Notwithstanding, the extreme reaction temperatures strongly affected the deactivation of the catalyst activity leading to reduction of the conversion and yield as investigated by Ramlee et al. [88]. The higher reaction temperature plays an important key for the higher results [62] while the over temperature offers the poor hydroprocessing performance and the catalyst deactivation.

Hydrogen pressure is another factor that can affect the hydroprocessing performance in the green diesel production [203] in consequence of hydrogen gas donor onto the catalyst active sites [204]. So, the more elevated the pressure supplied into the reaction, the more hydrogen adsorbed into the catalyst surface active sites thus the better catalyst performance is reached. The hydrogen pressure is commonly undertaken at 25–150 bar [121]. Kaewchada et al. [180] did an experiment by applying in the hydrogen pressure of 7–34 bar. The yield obtained at 7 bar was 33.7% while the yield achieved at 343 bar was 44.3%. The higher the applied reaction pressure, the higher the obtained yield. This is caused by shifting distribution of heavier hydrocarbon molecules which are from the fraction of >C_{18} into the diesel range hydrocarbons (C_8–C_{18}) [180]. The same statement was also disclosed by Srifa et al. [203] who reported that the higher hydrogen pressure favored the hydroprocessing technology to gain the maximum conversion. Nonetheless, the over hydrogen pressure negatively redounded the process performance caused by the deactivation of the catalysts and the ruished catalyst framework as investigated by Mahdi et al. [52] and Ramlee et al. [88]. The higher hydrogen pressure demonstrates the remarkable process and the highest results whereas the extreme pressure negatively furnishes the hydroprocessing capability and the catalyst deactivation.

Apart from the temperature and pressure, reactant flowrate distributes toward the capability of the extraordinary process in which the flowrate is varied with the constant catalyst mass. The catalyst fed into the reaction is constantly maintained as low as possible due to the expensive catalyst cost [202]. The molar ratio of the reactant and the catalyst is called LHSV or WHSV which is calculated according to Yang et al.'s report [205]. Either conversion or yield is enhanced at the higher WHSV as observed by Kaewchada et al. [180] who reported that the yield obtained at WHSV of 2 l/h was 55% whereas the yield achieved at WHSV of 43.7/h was 60.1%. The optimum yield was gained at the higher WHSV due to the lower space time. Meanwhile, the higher LHSV can significantly drop the obtained yield as investigated by Srifa et al. [203]. The yield acquired at LHSV of 5/h was 84.3% while the yield achieved at LHSV of 0.25/h was 95%. The highest yield was resulted in the lower LHSV because of the longer contact duration between the catalyst and the feedstock promoting the cracking and isomerization processes [206]. The higher WHSV and the lower LHSV provide a positive effect for the received yield, vice versa.

On the other hand, higher H_2:oil molar ratio gives a good contribution to the higher yield and high-quality product because excessive H_2 molecule can facilitate the process and reduction of the catalyst mass [30]. This parameter was proven by Patil and Vaidya [202] who reported that the conversion and yield were increased by applying in the H_2:oil ratio from 200 to 600 v/v. The same trend was also exhibited by Liu et al. [30] and Bezergianni et al. [207].

Authors highlight that the lower and the higher operating parameters offer positive or negative influence on the catalyst performance and the hydroprocessing technology capability. As long as the process is carried out at the proper condition, it will have contributed to the good enforcement. Otherwise, it is conducted at the improper parameters, it will have donated the bad activity and ability leading to the decrease of the conversion, yield, and selectivity.

4.2   The Influence of Different Vegetable Oils

The green diesel can be catalytically produced from a variety of the edible and non-edible oils. These different vegetable oils play a highly important role in the increase of the conversion, yield, and selectivity as well as the high biofuel quality particularly the green diesel produced from the hydroprocessing technology. In addition, the particular oil type dramatically contributes to the wastewater produced in the final step process and to the environmental pollution. Consequently, each of the industries employs the certain oils based on the massively available vegetable oils in its own countries aiming to maximally utilize its natural potency [208].

For instance, Nestle oil company settled in Singapore harnesses palm oil as the feedstock with the production capacity of 906 million liters per year (MLPY)

[209] whereas Nestle oil company located in Netherlands and Finland uses rapeseed oil, palm oil, and animal fats. The production capacity in Roterdam, Netherland, and Porvoo, Finland, was 906 MLPY and 215 MLPY, seriatim [209, 210]. On the other hand, Syntroleum and Tyson Foods joint venture company settled in Geismar, Louisana, has the capacity of 283 MLPY [211] and Valero and joint venture company established in Louisiana, Norco, produces the green diesel of 509 MLPY [212]. Eventually, UPM biofuels company built in Finland, Lappeenranta, has the production capacity of 117 MLPY [213].

For more information regarding the vegetable oils used in every country and continent, it can be noticed in Table 7 in which the total of vegetable oil production in Europe and the Americas is 55% of soybean oil, 55% of rapeseed oil, 10% of sunflower oil, and 10% of cottonseed oil [214]. Meanwhile, the worldwide vegetable oil production is notably dominated from palm oil (48.10 million tons) followed by soybean, rapeseed, sunflower, palm kernel oils, and more [52, 215].

In Asia countries such as Indonesia, Malaysia, Singapore, and Thailand mostly use palm oil as the feedstock to generate the green diesel under the hydroprocessing technology. Moreover, Malaysia has always developed the palm oil as the potential and sustainable oil for the renewable energy [221, 222] and Indonesia has optimally explored the palm oil to generate the biofuel products such as biogas, biodiesel, biojet fuel, biogasoline, and green diesel [223, 224]. Even, the crude palm oil (CPO) produced from Indonesia is partially exported to some other countries including Europe, Africa, Americas, and Asia as reported by Khatiwada et al. [224]. The other CPO is utilized to yield the biofuel as a mixture of 30% with the petroleum-based diesel fuel as studied by Yasinta and Karuniasa [225]. The palm-based green diesel production generated a residue called palm fatty acid distillate (PFAD) which can be re-produced to the diesel range biofuel as studied by Mahdi et al. [52]. Exploration of the palm oil was applied by Plazas-González [226] carried out at 60 °C and 1 bar using NiMo-Al$_2$O$_3$ catalyst in which the obtained green diesel purity was 97%. Superiorities of the palm oil feedstock are abundant raw materials, operated at the modest parameters, high yield, high-grade product, and the reused residue [52, 67].

Several countries in Europe utilize the sunflower oil and rapeseed oil [227] as the prominent feedstocks for the biofuel production. Approximately, 40%–50% of the vegetable oils is coming from the sunflower oil for the green diesel production in Ukraine, Turkey, Russian Federation, France, and Argentina [174]. Šimáček et al. [228] explored the sunflower oil in the process carried out at 420 °C and 180 bar. The green diesel produced possessed the very good modest-temperature characteristic [228]. Nonetheless, it was operated at the high operating condition which has a high explosive risk. Meanwhile, Finland, Netherland, Germany, Russia, Italy, and Sweden commonly harness the rapeseed oil as the raw material [217]. The use of the rapeseed oil was observed by Šimáček et al. [200] with an elevated yield. Yet, the rapeseed-based green diesel had the worse low-temperature characteristic and the process was conducted at the high temperature (360 °C) and the extreme pressure (150 bar) as investigated by Šimáček et al. [43]. These extreme operating conditions are not safe and frightening risk in the explosion.

Table 7 Raw materials commonly used in the worldwide green diesel production [174, 216–220]

Continent	Countries	Raw materials
Asia	Indonesia	Palm oil, coconut, and jatropha oils
	Philippines	Coconut and jatropha oils
	Singapore	Palm oil
	Thailand	Palm oil, coconut, and jatropha oils
	China	Rapeseed, waste cooking oil, and jatropha oils
	Japan	Waste cooking oil
	Malaysia	Palm oil
	India	Waste plastic oil, waste type oil, sunflower, peanut, jatropha, and kangra oils
Europe	Louisiana	Soybean oil and animal fats
	Finland	Rapeseed oil, palm oil, and animal fats
	Netherland	Rapeseed oil, palm oil, and animal fats
	Germany	Rapeseed oil
	Russia	Rapeseed oil
	Italy	Rapeseed, sunflower, and waste plastic oils
	France	Sunflower and rapeseed oils
	Sweden	Rapeseed oil
	Greece	Cottonseed oil
	Ireland	Animal fat and frying oils
	United Kingdom	Waste cooking oil and rapeseed oil
	New Zealand	Waste cooking oil and tallow oil
	Argentina	Soybean oil
Americas	Brazil	Soybean, Cottonseed, palm oil, and castol oils
	United State of America	Waste cooking oil, soybean, and peanut oils
	Colombia	Fuel cell
	Canada	Rapeseed, animal fat, yellow grease/tallow, and soybean/mustard oils
	Mexico	Animal fat and waste oil
	Venezuela	Soybean oil

On the other hand, the used cooking oil (UCO) or waste cooking oil (WCO) has potentially become the cheapest feedstocks for the green diesel production conducted via the hydroprocessing method [229]. Over 200,000 tons per year (TPY) of the UCO is produced in South Africa as claimed by Rocha Filho et al. [51, 230] whereabouts 205 MLPY of the green diesel are produced from these waste oils [3]. China produces around 4.5 million tons per year (MTPY) of the green diesel produced from the UCO [231]. The use of the UCO and WCO in the green diesel production donates a huge profit by reducing the process cost [232]. Pelemo et al. [174] reviewed a potency of the UCO as a feedstock in the green diesel process via the hydroprocessing technology.

The UCO-based process resulted in the high yield and a high-grade product with the lower operating process [174]. These potencies motivate the Neste oil company to develop and to modify a technology by optimizing the waste oils namely NExBTL technology which operated in Singapore with the capacity of 800,000 TPY [233].

In authors view, the diesel range hydrocarbons produced from the used/waste cooking oils (UCO/WCO) via the hydroprocessing technology is enthroned as the best oil if compared to the other oils because it has plenteous feedstock, low operating conditions, optimum result, and high-quality green diesel. Furthermore, the UCO/WCO-based green diesel is capable of reduction of the pollution caused by the massive UCO/WCO production resulted from restaurant, domestic, and industries. Reutilization of these used oils provides a good solution for our green and clean universe.

4.3 The Effect of Metal Supports

There are many methods to jack up a performance of the solid catalysts. One of them is by modifying the catalysts supported with noble metal components and transition metals as their carriers [234] as well as metal sulfides [235–239]. The noble metals and the transition metals commonly used in the catalyst synthesis are ruthenium, rhodium, palladium, silver, iridium, platinum, gold, zinc, copper, vanadium, iron, cobalt, molybdenum, nickel, and more [136, 240]. The metal-doped catalysts demonstrate a remarkable stability during the process while the sulfide-based catalysts show a good capability only during the initial reaction time [241].

Yet, the performance of the sulfide-contained catalysts dramatically bring down during over time reaction by dint of deposit formation on active sites and the decrease of the sulfides content leading to the catalyst deactivations [237, 239]. Hence, addition of sulfiding agents called co-hydroprocessing is highly useful to protect the MoS_2 active sites from poisoning leading to less catalyst deactivation [242–245]. As studied by Varakin et al. [190] who employed the unsupported MoS_2/Al_2O_3 catalyst and the sulfiding agent-supported MoS_2 catalyst (Et-MoS_2-Ind catalyst). The selectivity obtained by the MoS_2/Al_2O_3 catalyst was 0.85 while the one achieved by the Et-MoS_2-Ind catalyst was 0.95 during the longer reaction time [190]. The similar trend was also revealed by Varakin et al. [246] earlier. The Al_2O_3-assisted MoS_2 catalyst exhibited the higher selectivity than the one obtained by the MoS_2 catalyst. It obviously shows that the sulfiding agents contribute to upgrade the yield and to strengthen the stability.

The solid catalysts supported with the metal oxides have disclosed fabulous capability and stability and been capable of increasing the green diesel yield [247, 248]. For instance, Srihanun et al. [78] compared two distinct catalysts which were the unsupported catalysts (Pd/Al_2O_3) and the metal-supported catalyst (Fe-assisted Pd/Al_2O_3) in the palm-based diesel range hydrocarbon production. The Fe-supported Pd/Al_2O_3 catalyst possessed enlarged active sites and higher porosity causing the increment of the biofuel yield up to 94%. Meanwhile, the Pd/Al_2O_3 catalyst resulted in the yield of 86% [78]. Comparison of two solid catalyst types, NiMo-doped Al_2O_3 catalyst and Pd-modified carbon catalyst, was investigated by Wang [187] conducted under the hydroprocessing technology at the same process parameters. The NiMo-doped Al_2O_3 catalyst revealed the higher conversion (100%) if compared to the other (~97%) [187] as a result of stronger Bronsted-Lewis acidity and more spacious active sites [52]. Using the NiMo-Al_2O_3 catalyst was also studied by Kovacs et al. [249] who produced the green diesel using sunflower oil via the hydroprocessing technology at 350–370 °C, 20–40 bar, LHSV of 1/h, H_2:oil volume ratio of 500 Nm^3/m^3. This catalyst resulted in conversion of 89%, selectivity of 90%, and yield of 80% [249].

In spite of showing an incredible performance by using the noble metals-assisted solid catalyst, the high-priced noble metals become a drawback to use this method in the large-scale industrial application as reported by Ramírez et al. [171]. Consequently, this triggers authors and researchers to discover other highly attractive and cost-effective components. The transition metals properly answer this challenging and opportunity because they are cheaper components as studied by Mahdi et al. [52]. Therefore, the transition metals have been currently extensively exploited not only as the main catalysts but also as the catalyst supporting doped into the primary catalysts.

Either the noble metals-doped catalysts or the transition metals are compared by Wang [250] who investigated the performance of both catalysts. The authors harnessed ZSM-5 assisted with Ru metal (the noble metal) and NiMo catalyst (the transition metals). Both catalysts generated the green diesel yield of 20–29%. With the same yield, but with the lower metal cost, the transition metals are more superior if compared to the noble metals. Addition of the transition metals on the catalysts was also inspected by Choo et al. [149] using Y zeolite supported by Ni, Cu, Co, Zn, and Mn metals. The highest conversion (76.21%) and selectivity (92.61%) were achieved by the Ni-assisted Y zeolite at 380 °C within 2 h. The most optimum efficiency was demonstrated by Ni-Y zeolite followed by Co-Y zeolite, Cu-Y zeolite, Mn-Y zeolite, and Zn-Y zeolite owing to the synergistic effect between Ni and Y zeolite [149]. The Ni and Co metals were investigated by Kamaruzaman et al. [12] using SBA-15 catalyst. The Ni metal was more superior than Co metal because Co formed a coke causing blockage on the active areas causing reduction of the catalyst activation.

The noble metals, transition metals, and metal sulfides modified into the solid catalysts play an important factor in the improvement of the catalyst capability in consequence of more enormous Bronsted-Lewis acidity, more sizeable active sites, higher pore sites, and larger approachability of the reactant. The developed catalyst performance dramatically contributes to augment the conversion, yield, and selectivity. Nonetheless, the noble metals are known as the expensive components whereas the metal sulfides require the sulfiding agents to promote the MoS_2 active areas, thus it will totally induce the higher process budget. For the time being, utilization of the transition metals is the most promising, potential, and effective method because they are cost-effective, sustainable, and eco-friendly components as well as resulting in the elevated conversion, selectivity, and yield.

4.4 The Role of Nanoparticle Catalysts

Larger active sites, enlarged accessibility of the reactant into the product, high-speed dispersing of the oils and the products over the active areas, more robust Bronsted-Lewis acidity, and higher crystal porosity can be maximally acquired by using micro- and nano-sized catalyst particles [251]. Those factors positively play a highly important role in the replenishment of the catalysts capability and the obtained results such as conversion, selectivity, and yield as well as the higher process efficiency. Additionally, the micro- and nano-sized catalyst particles donate the great catalyst stability [252, 253]. By dint of the extraordinary ability of the micro- and nano-sized particles, these catalysts have been recently extensively explored by authors and researchers.

Baharudin et al. [73] employed NiO-doped ZnO catalyst in nano-sized particles in which the more spacious meso-macro-sized ZnO crystalline particle was treated under the co-precipitation method. The process was carried out at 350 °C within 2 h using PFAD to produce the green diesel. Approximately, 83.4% of biofuel yield was achieved with the green diesel selectivity of 86%. The high result was correlated with higher Bronsted-Lewis acidity using the nano-sized NiO catalyst [73]. Meanwhile, Alsultan et al. [254] exploited Ag-La/AC catalyst in nano-sized particles in which this catalyst was capable of resulting in the higher hydrocarbon yield of 89% with the green diesel selectivity of 93%. Moreover, the catalyst can be reused during 6 batches. The more elevated yield and selectivity were correlated to the synergistic impact of the catalyst and the support leading to enlarged crystallite areas [254]. Furthermore, the high catalytic activity was also demonstrated by Cu-doped Ni/Al$_2$O$_3$ catalyst possessing size of 3–5 nm with the active surface area of 192–285 m^2/g as studied by Gousi et al. [255]. This nano-sized catalyst resulted in high conversion and yield in consequence of the more spacious accessibility of the oils [255].

Not only micro- and nano-sized particle can affect the catalyst capability and the yield, but also meso-sized particle can influence the hydroprocessing technology as studied by Papanikolaou et al. [7]. The mesopores size of Beta catalyst employed was 20–75 nm and the mesopore-sized Beta had higher Bronsted-Lewis acidity (69/187 μmol/g$_{cat}$) if compared to Beta catalyst (60/88 μmol/g$_{cat}$) and Al-SBA-15 catalyst (36/120 μmol/g$_{cat}$). Consequently, the mesopore-sized Beta catalyst resulted in higher conversion (77%) and selectivity (61.9%) rather than the Beta with those of 38% and 3.3% and Al-SBA-15 with the ones of 47% and 12.6%, seriatim [7]. The role of the mesopores catalysts in the green diesel production was also studied by Jin et al. [144] employing Pt/N-AC catalyst in nanoparticles. This catalyst generated the conversion of over 90% with the process efficiency of 76%. On the contrary, Mahdi et al. [52] claimed that the nanoparticle catalysts had a high reusability thus can be reused during up to 5 batches and resulted in high-quality product.

According to the above reports and studies, it seems that the catalyst particle in meso-, micro- and nano-sized demonstrates stronger Bronsted-Lewis acidity, more extensive crystallite areas, higher porosity, more comprehensive active sites, and preferable catalyst stability, which significantly jack up the process efficiency, catalyst reusability, yield, and selectivity, as well as high-grade of green diesel quality.

5 Conclusion

In summary, the role of heterogeneous catalyst is highly on demand for green diesel production. The superior performance of heterogeneous catalyst helps to address many problems related to separation, purification, regeneration, and recyclability of the catalyst. The selectivity of suitable catalyst will mainly affect the selectivity of green diesel and the hydroprocessing techniques are much more preferable compared to other techniques due to the higher conversion of raw materials into its product. The utilization of zeolite heterogeneous catalysts shows a highly favorable catalyst for large-scale green diesel production due to its regenerability, reusability, stability, and high stability catalyst framework during the hydroprocessing reaction. Furthermore, the proper operating parameters, the appropriate oils, and the metal supports modified on the active catalyst sites play a highly crucial role in determining the efficiency of the hydroprocessing technology.

Micro-, meso-, and nano-sized catalyst has become more apparent nowadays since the larger active area and the higher porosity of the catalyst have significantly contributed to the enlarged accessibility of substrates, rapid diffusion for the reactants and products at a faster reaction rate. These donate a perfect conversion (100%), high selectivity (70–95%) and yield (70–97%) as well as the high reusability of the catalysts during several cycles. Nonetheless, the practicability of the catalyst used for large-scale green diesel production needs to be deeply studied in terms of the profit margin which include the capital and investment cost. Eventually, the cost-effective and eco-friendly technology is considered as the ultimate purpose for the clean process and the high company profit.

Acknowledgements N.I.W.A. greatly appreciates the Universiti Teknologi Malaysia (UTM) and the Ministry of Education (MOE) Malaysia through Fundamental Research Grant Scheme (FRGS/1/2020/TK0/UTM/02/8) for the financial and facilities support. L.M. thanks to the Coordination of Superior Level Staff Improvement (CAPES/Brazil), the National Council for Scientific and Technological Development (CNPq/Brazil), and the Alagoas State Research Support Foundation (FAPEAL/Brazil). The findings achieved herein are solely the responsibility of the authors. H.I.M greatly appreciates PT Global Amine Indonesia and PT Wilmar Nabati Indonesia for the support and some of the good advices in order to make the book chapter more perfect.

References

1. By fuel type-Exajoules, C. and C.D. Emissions (2006) bp Statistical Review of World Energy June 2020
2. Nylund N-O, Aakko-Saksa P, Sipilä K (2008) Status and outlook for biofuels, other alternative fuels and new vehicles
3. Pelemo J, Inambao F, Onuh EI, A study to evaluate optimal catalyst properties sourced from biowaste for hydro processing of used cooking oil into green diesel
4. Demirbas A (2007) Importance of biodiesel as transportation fuel. Energy Policy 35(9):4661–4670

5. Ian Tiseo (2021) Annual global emissions of carbon dioxide 1940-2020. Stat: Energy Environ (Assessed on 14th Disember 2021)
6. Chouhan APS, Sarma AK (2013) Biodiesel production from Jatropha curcas L. oil using Lemna perpusilla Torrey ash as heterogeneous catalyst. Biomass Bioenergy 55:386–389
7. Papanikolaou G et al (2020) Highly selective bifunctional Ni zeo-type catalysts for hydroprocessing of methyl palmitate to green diesel. Catal Today 345:14–21
8. Tran QK et al (2020) Hydrodeoxygenation of a bio-oil model compound derived from woody biomass using spray-pyrolysis-derived spherical γ-Al2O3-SiO2 catalysts. J Ind Eng Chem 92:243–251
9. Ho WWS, Ng HK, Gan S (2012) Development and characterisation of novel heterogeneous palm oil mill boiler ash-based catalysts for biodiesel production. Bioresour Technol 125:158–164
10. Zhao C, Brück T, Lercher JA (2013) Catalytic deoxygenation of microalgae oil to green hydrocarbons. Green Chem 15(7):1720–1739
11. Pyra K et al (2019) Desilicated zeolite BEA for the catalytic cracking of LDPE: the interplay between acidic sites' strength and accessibility. Catal Sci Technol 9(8):1794–1801
12. Kamaruzaman MF, Taufiq-Yap YH, Derawi D (2020) Green diesel production from palm fatty acid distillate over SBA-15-supported nickel, cobalt, and nickel/cobalt catalysts. Biomass Bioenergy 134:105476
13. Carrero A et al (2017) Production of renewable hydrogen from glycerol steam reforming over bimetallic Ni-(Cu Co, Cr) catalysts supported on SBA-15 silica. Catalysts 7(2):55
14. Roh H-S et al (2011) The effect of calcination temperature on the performance of Ni/MgO–Al2O3 catalysts for decarboxylation of oleic acid. Catal Today 164(1):457–460
15. Ma B, Zhao C (2015) High-grade diesel production by hydrodeoxygenation of palm oil over a hierarchically structured Ni/HBEA catalyst. Green Chem 17(3):1692–1701
16. Peng B et al (2012) Stabilizing catalytic pathways via redundancy: selective reduction of microalgae oil to alkanes. J Am Chem Soc 134(22):9400–9405
17. Choi I-H et al (2015) The direct production of jet-fuel from non-edible oil in a single-step process. Fuel 158:98–104
18. Lestari S et al (2008) Synthesis of biodiesel via deoxygenation of stearic acid over supported Pd/C catalyst. Catal Lett 122(3):247–251
19. Simakova I et al (2009) Deoxygenation of palmitic and stearic acid over supported Pd catalysts: effect of metal dispersion. Appl Catal A 355(1–2):100–108
20. Immer JG, Kelly MJ, Lamb HH (2010) Catalytic reaction pathways in liquid-phase deoxygenation of C18 free fatty acids. Appl Catal A 375(1):134–139
21. Do PT et al (2009) Catalytic deoxygenation of methyl-octanoate and methyl-stearate on Pt/Al 2 O 3. Catal Lett 130(1):9–18
22. Argüelles A, Amezcua-Allieri MA, Ramírez LF (2021) Life cycle assessment of green diesel production by hydrodeoxygenation of palm oil. Front Energy Res 9:296
23. Tran C-C, Akmach D, Kaliaguine S (2020) Hydrodeoxygenation of vegetable oils over biochar supported bimetallic carbides for producing renewable diesel under mild conditions. Green Chem 22(19):6424–6436
24. de Sousa FP, Cardoso CC, Pasa VM (2016) Producing hydrocarbons for green diesel and jet fuel formulation from palm kernel fat over Pd/C. Fuel Process Technol 143:35–42
25. Shim J-O et al (2015) Optimization of unsupported CoMo catalysts for decarboxylation of oleic acid. Catal Commun 67:16–20
26. Kordulis C et al (2016) Development of nickel based catalysts for the transformation of natural triglycerides and related compounds into green diesel: a critical review. Appl Catal B 181:156–196
27. Shim J-O et al (2016) Bio-diesel production from deoxygenation reaction over Ce0. 6Zr0. 4O2 supported transition metal (Ni, Cu, Co, and Mo) catalysts. J Nanosci Nanotechnol 16(5):4587–4592
28. Jeon K-W et al (2019) Synthesis and characterization of Pt-, Pd-, and Ru-promoted Ni–Ce0. 6Zr0. 4O2 catalysts for efficient biodiesel production by deoxygenation of oleic acid. Fuel 236:928–933

29. Taromi AA, Kaliaguine S (2018) Green diesel production via continuous hydrotreatment of triglycerides over mesostructured γ-alumina supported NiMo/CoMo catalysts. Fuel Process Technol 171:20–30
30. Liu Y et al (2011) Hydrotreatment of vegetable oils to produce bio-hydrogenated diesel and liquefied petroleum gas fuel over catalysts containing sulfided Ni–Mo and solid acids. Energy Fuels 25(10):4675–4685
31. Sazzad B et al (2016) Retardation of oxidation and material degradation in biodiesel: a review. RSC Adv 6(65):60244–60263
32. Priecel P et al (2011) The role of alumina support in the deoxygenation of rapeseed oil over NiMo–alumina catalysts. Catal Today 176(1):409–412
33. Parkhomchuk EV et al (2013) Meso/macroporous CoMo alumina pellets for hydrotreating of heavy oil. Ind Eng Chem Res 52(48):17117–17125
34. Kiatkittipong W et al (2013) Diesel-like hydrocarbon production from hydroprocessing of relevant refining palm oil. Fuel Process Technol 116:16–26
35. Kochaputi N et al (2019) Catalytic behaviors of supported Cu, Ni, and Co phosphide catalysts for deoxygenation of oleic acid. Catalysts 9(9):715
36. Liu Y et al (2015) The production of diesel-like hydrocarbons from palmitic acid over HZSM-22 supported nickel phosphide catalysts. Appl Catal B 174:504–514
37. Phimsen S et al (2016) Oil extracted from spent coffee grounds for bio-hydrotreated diesel production. Energy Convers Manag 126:1028–1036
38. Yenumala SR et al (2020) Hydrodeoxygenation of karanja oil using ordered mesoporous nickel-alumina composite catalysts. Catal Today 348:45–54
39. Scaldaferri CA, Pasa VMD (2019) Hydrogen-free process to convert lipids into bio-jet fuel and green diesel over niobium phosphate catalyst in one-step. Chem Eng J 370:98–109
40. Shim J-O et al (2018) Facile production of biofuel via solvent-free deoxygenation of oleic acid using a CoMo catalyst. Appl Catal B 239:644–653
41. Silva LN et al (2016) Biokerosene and green diesel from macauba oils via catalytic deoxygenation over Pd/C. Fuel 164:329–338
42. Harnos S, Onyestyák G, Kalló D (2012) Hydrocarbons from sunflower oil over partly reduced catalysts. React Kinet Mech Catal 106(1):99–111
43. Šimáček P et al (2010) Fuel properties of hydroprocessed rapeseed oil. Fuel 89(3):611–615
44. Loe R et al (2016) Effect of Cu and Sn promotion on the catalytic deoxygenation of model and algal lipids to fuel-like hydrocarbons over supported Ni catalysts. Appl Catal B 191:147–156
45. Emori EY et al (2017) Catalytic cracking of soybean oil using ZSM5 zeolite. Catal Today 279:168–176
46. Itthibenchapong V et al (2017) Deoxygenation of palm kernel oil to jet fuel-like hydrocarbons using Ni-MoS2/γ-Al2O3 catalysts. Energy Convers Manag 134:188–196
47. Gong S et al (2013) Isomerization of n-alkanes derived from Jatropha oil over bifunctional catalysts. J Mol Catal A: Chem 370:14–21
48. Yasir M et al (2016) Hydroprocessing of crude Jatropha oil using hierarchical structured TiO2 nanocatalysts. Procedia Eng 148:275–281
49. Lee W-S et al (2014) Selective vapor-phase hydrodeoxygenation of anisole to benzene on molybdenum carbide catalysts. J Catal 319:44–53
50. Janarthanam H et al (2020) Performance and emission analysis of waste cooking oil as green diesel in 4S diesel engine. In: AIP conference proceedings. AIP Publishing LLC
51. da Rocha Filho G, Brodzki D, Djéga-Mariadassou G (1993) Formation of alkanes, alkyl-cycloalkanes and alkylbenzenes during the catalytic hydrocracking of vegetable oils. Fuel 72(4):543–549
52. Mahdi HI et al (2021) Catalytic deoxygenation of palm oil and its residue in green diesel production: a current technological review. Chem Eng Res Des
53. Papadopoulos C et al (2021) W promoted Ni-Al2O3 co-precipitated catalysts for green diesel production. Fuel Process Technol 217:106820
54. Hongloi N, Prapainainar P, Prapainainar C (2021) Review of green diesel production from fatty acid deoxygenation over Ni-based catalysts. Mol Catal 111696

55. Lycourghiotis S et al (2021) Transformation of residual fatty raw materials into third generation green diesel over a nickel catalyst supported on mineral palygorskite. Renew Energy 180:773–786
56. Hafeez S et al (2021) Theoretical investigation of the deactivation of Ni supported catalysts for the catalytic deoxygenation of palm oil for green diesel production. Catalysts 11(6):747
57. Fernández-Villamil JM, de Mendoza Paniagua AH (2030) Preliminary design of the green diesel production process by hydrotreatment of vegetable oils
58. Ruangudomsakul M et al (2021) Hydrodeoxygenation of palm oil to green diesel products on mixed-phase nickel phosphides. Mol Catal 111422
59. Chumaidi A et al (2021) Effect of temperature and Mg-Zn catalyst ratio on decarboxylation reaction to produce green diesel from kapok oil with saponification pretreatment using NaOH. In: IOP Conf Ser: Mater Sci Eng. IOP Publishing
60. Setiadi SA, Reaksi Dekarboksilasi Minyak Jarak Pagar untuk Pembuatan Hidrokarbon Setara Fraksi Diesel Dengan Penambahan Ca (OH) 2. Hlm: 1–8. In: Prosiding Seminar Nasional Teknik Kimia Indonesia
61. Pimenta JL et al (2020) Effects of reaction parameters on the deoxygenation of soybean oil for the sustainable production of hydrocarbons. Environ Prog & Sustain Energy 39(5):e13450
62. Zikri A et al (2021) Production of green diesel from Crude Palm Oil (CPO) through hydrotreating process by using zeolite catalyst. In: 4th forum in research, science, and technology (FIRST-T1-T2-2020). Atlantis Press
63. Vázquez-Garrido I et al (2021) Synthesis of NiMo catalysts supported on Mn-Al2O3 for obtaining green diesel from waste soybean oil. Catal Today 365:327–340
64. Attaphaiboon W et al (2021) Potential of vegetable oils for producing green diesel via hydrocracking process. Thai Environ Eng J 35(2):1–11
65. Alisha GD, Trisunaryanti W, Syoufian A (2021) Hydrocracking of Waste Palm Cooking Oil into Hydrocarbon Compounds over Mo Catalyst Impregnated on SBA-15. Silicon, p 1–7
66. Jin M et al (2020) Coproduction of value-added lube base oil and green diesel from natural triglycerides via a simple two-step process. Ind Eng Chem Res 59(19):8946–8954
67. Zikri A, Aznury M (2020) Green diesel production from Crude Palm Oil (CPO) using catalytic hydrogenation method. IOP Conf Ser: Mater Sci Eng. IOP Publishing
68. Aliana-Nasharuddin N et al (2020) Production of green diesel from catalytic deoxygenation of chicken fat oil over a series binary metal oxide-supported MWCNTs. RSC Adv 10(2):626–642
69. Ameen M et al (2020) Parametric studies on hydrodeoxygenation of rubber seed oil for diesel range hydrocarbon production. Energy Fuels 34(4):4603–4617
70. Ameen M et al (2020) Process optimization of green diesel selectivity and understanding of reaction intermediates. Renew Energy 149:1092–1106
71. Asikin-Mijan N et al (2020) Production of renewable diesel from Jatropha curcas oil via pyrolytic-deoxygenation over various multi-wall carbon nanotube-based catalysts. Process Saf Environ Prot 142:336–349
72. Baek H et al (2020) Production of bio hydrofined diesel, jet fuel, and carbon monoxide from fatty acids using a silicon nanowire array-supported rhodium nanoparticle catalyst under microwave conditions. ACS Catal 10(3):2148–2156
73. Baharudin KB et al (2020) Renewable diesel via solventless and hydrogen-free catalytic deoxygenation of palm fatty acid distillate. J Clean Prod 274:122850
74. Gamal MS et al (2020) Effective catalytic deoxygenation of palm fatty acid distillate for green diesel production under hydrogen-free atmosphere over bimetallic catalyst CoMo supported on activated carbon. Fuel Process Technol 208:106519
75. Ochoa-Hernández C, Coronado JM, Serrano DP (2020) Hydrotreating of methyl esters to produce green diesel over Co-and Ni-containing Zr-SBA-15 catalysts. Catalysts 10(2):186
76. de Oliveira Camargo M et al (2020) Green diesel production by solvent-free deoxygenation of oleic acid over nickel phosphide bifunctional catalysts: effect of the support. Fuel 281:118719
77. de Barros Dias Moreira J, de Rezende DB, Pasa VMD (2020) Deoxygenation of Macauba acid oil over Co-based catalyst supported on activated biochar from Macauba endocarp: a potential and sustainable route for green diesel and biokerosene production. Fuel 269:117253

78. Srihanun N et al (2020) Biofuels of green diesel–kerosene–gasoline production from palm oil: effect of palladium cooperated with second metal on hydrocracking reaction. Catalysts 10(2):241
79. Ameen M et al (2019) HY zeolite as hydrodeoxygenation catalyst for diesel range hydrocarbon production from rubber seed oil. Mater Today: Proc 16:1742–1749
80. Scaldaferri CA, Pasa VMD (2019) Production of jet fuel and green diesel range biohydrocarbons by hydroprocessing of soybean oil over niobium phosphate catalyst. Fuel 245:458–466
81. Lycourghiotis S et al (2019) Nickel catalysts supported on palygorskite for transformation of waste cooking oils into green diesel. Appl Catal B: Environ 259:118059
82. Chumachenko V et al (2016) Hydrocracking of vegetable oil on boron-containing catalysts: effect of the nature and content of a hydrogenation component. Catal Ind 8(1):56–74
83. Trisunaryanti W, Suarsih E, Falah II (2019) Well-dispersed nickel nanoparticles on the external and internal surfaces of SBA-15 for hydrocracking of pyrolyzed α-cellulose. RSC Adv 9(3):1230–1237
84. Pattanaik BP, Misra RD (2017) Effect of reaction pathway and operating parameters on the deoxygenation of vegetable oils to produce diesel range hydrocarbon fuels: a review. Renew Sustain Energy Rev 73:545–557
85. Jeon K-W et al (2021) Deoxygenation of non-edible fatty acid for green diesel production: effect of metal loading amount over Ni/MgO–Al2O3 on the catalytic performance and reaction pathway. Fuel 122488
86. Makertihartha IGB et al (2020) Biogasoline production from palm oil: optimization of catalytic cracking parameters. Arab J Sci Eng 45(9):7257–7266
87. Honorato de Oliveira BF et al Renewable diesel production from Palm Fatty Acids Distillate (PFAD) via deoxygenation reactions. Catalysts, 2021. 11(9): p. 1088.
88. binti Ramlee NN, Mahdi HI, Azelee NIW (2022) Biodiesel production using enzymatic catalyst. Biofuels Bioenergy. Elsevier, p 133–169
89. Mahdi HI, Muraza O (2019) An exciting opportunity for zeolite adsorbent design in separation of C4 olefins through adsorptive separation. Sep Purif Technol 221:126–151
90. Quispe CA, Coronado CJ, Carvalho JA Jr (2013) Glycerol: production, consumption, prices, characterization and new trends in combustion. Renew Sustain Energy Rev 27:475–493
91. Putra R et al (2018) Fe/Indonesian natural zeolite as hydrodeoxygenation catalyst in green diesel production from palm oil. Bull Chem React Eng & Catal 13(2):245–255
92. Kalnes TN et al (2009) A technoeconomic and environmental life cycle comparison of green diesel to biodiesel and syndiesel. Environ Prog & Sustain Energy: Off Publ Am Inst Chem Eng 28(1):111–120
93. Mahdi HI et al (2016) Glycerol carbonate production from biodiesel waste over modified natural clinoptilolite. Waste Biomass Valorization 7(6):1349–1356
94. Wibowo AA et al (2020) Green diesel production from waste vegetable oil: a simulation study. In: AIP conference proceedings. AIP Publishing LLC
95. Douvartzides SL et al (2019) Green diesel: biomass feedstocks, production technologies, catalytic research, fuel properties and performance in compression ignition internal combustion engines. Energies 12(5):809
96. Bezergianni S, Dimitriadis A (2013) Comparison between different types of renewable diesel. Renew Sustain Energy Rev 21:110–116
97. Laveille P et al (2021) Sustainable pilot-scale production of a Salicornia oil, its conversion to certified aviation fuel, and techno-economic analysis of the related biorefinery. Biofuels Bioprod Biorefining
98. Kalnes T, Marker T, Shonnard DR (2007) Green diesel: a second generation biofuel. Int J Chem React Eng 5(1)
99. Water C (2008) Green diesel production by hydrorefining renewable feedstocks
100. Papageridis K et al (2020) Effect of operating parameters on the selective catalytic deoxygenation of palm oil to produce renewable diesel over Ni supported on Al2O3, ZrO2 and SiO2 catalysts. Fuel Process Technol 209:106547

101. Papageridis KN et al (2021) Continuous selective deoxygenation of palm oil for renewable diesel production over Ni catalysts supported on Al2O3 and La2O3–Al2O3. RSC Adv 11(15):8569–8584
102. Kukushkin R et al (2020) Deoxygenation of esters over sulfur-free Ni–W/Al2O3 catalysts for production of biofuel components. Chem Eng J 396:125202
103. Haider MS, Castello D, Rosendahl LA (2021) The art of smooth continuous hydroprocessing of biocrudes obtained from hydrothermal liquefaction: hydrodemetallization and propensity for coke formation. Energy & Fuels
104. Palanisamy S, Gevert BS (2014) Hydroprocessing of fatty acid methyl ester containing resin acids blended with gas oil. Fuel Process Technol 126:435–440
105. Palanisamy S, Gevert BS (2016) Study of non-catalytic thermal decomposition of triglyceride at hydroprocessing condition. Appl Therm Eng 107:301–310
106. Palanisamy S, Kandasamy K (2020) Direct hydrogenation and hydrotreating of neat vegetal oil into renewable diesel using alumina binder with zeolite. Rev Chim 71(9):98–112
107. Chen H et al (2018) Tuning the decarboxylation selectivity for deoxygenation of vegetable oil over Pt–Ni bimetal catalysts via surface engineering. Catal Sci Technol 8(4):1126–1133
108. Sotelo-Boyás R, Trejo-Zárraga F, de Jesus Hernández-Loyo F (2012) Hydroconversion of triglycerides into green liquid fuels. Hydrogenation 338:338
109. Satyarthi J et al (2013) An overview of catalytic conversion of vegetable oils/fats into middle distillates. Catal Sci Technol 3(1):70–80
110. Imai H et al (2020) Hydroconversion of methyl laurate over beta-zeolite-supported Ni–Mo catalysts: effect of acid and base treatments of beta zeolite. Fuel Process Technol 197:106182
111. Aatola H et al (2009) Hydrotreated Vegetable Oil (HVO) as a renewable diesel fuel: trade-off between NOx, particulate emission, and fuel consumption of a heavy duty engine. SAE Int J Engines 1(1):1251–1262
112. Karaba A et al (2021) Experimental evaluation of hydrotreated vegetable oils as novel feedstocks for steam-cracking process. Processes 9(9):1504
113. Dimitriadis A et al (2018) Evaluation of a hydrotreated vegetable oil (HVO) and effects on emissions of a passenger car diesel engine. Front Mech Eng 4:7
114. Liu C et al (2013) A cleaner process for hydrocracking of Jatropha oil into green diesel. J Taiwan Inst Chem Eng 44(2):221–227
115. Badoga S et al (2020) Co-processing of hydrothermal liquefaction biocrude with vacuum gas oil through hydrotreating and hydrocracking to produce low-carbon fuels. Energy Fuels 34(6):7160–7169
116. Kimura T et al (2012) Conversion of isoprenoid oil by catalytic cracking and hydrocracking over nanoporous hybrid catalysts. J Biomed Biotechnol 2012
117. Muraza O (2015) Maximizing diesel production through oligomerization: a landmark opportunity for zeolite research. Ind Eng Chem Res 54(3):781–789
118. Bellussi G et al (2012) Oligomerization of olefins from Light Cracking Naphtha over zeolite-based catalyst for the production of high quality diesel fuel. Microporous Mesoporous Mater 164:127–134
119. Panpian P et al (2021) Production of bio-jet fuel through ethylene oligomerization using NiAlKIT-6 as a highly efficient catalyst. Fuel 287:119831
120. Hancsók J et al (2011) Investigation the effect of oxygenic compounds on the isomerization of bioparaffins over Pt/SAPO-11. Top Catal 54(16):1094–1101
121. Glisic SB, Pajnik JM, Orlović AM (2016) Process and techno-economic analysis of green diesel production from waste vegetable oil and the comparison with ester type biodiesel production. Appl Energy 170:176–185
122. Glišić S, Lukic I, Skala D (2009) Biodiesel synthesis at high pressure and temperature: analysis of energy consumption on industrial scale. Bioresour Technol 100(24):6347–6354
123. Glisic SB, Orlovic AM (2012) Modelling of non-catalytic biodiesel synthesis under sub and supercritical conditions: the influence of phase distribution. J Supercrit Fluids 65:61–70
124. Glisic SB, Orlović AM (2014) Review of biodiesel synthesis from waste oil under elevated pressure and temperature: phase equilibrium, reaction kinetics, process design and techno-economic study. Renew Sustain Energy Rev 31:708–725

125. West AH, Posarac D, Ellis N (2008) Assessment of four biodiesel production processes using HYSYS. Plant. Bioresour Technol 99(14):6587–6601
126. Zeng D et al (2021) CuO@ NiO nanoparticles derived from metal–organic framework precursors for the deoxygenation of fatty acids. ACS Sustain Chem & Eng
127. Hanafi SA, Elmelawy MS, Ahmed HA (2021) Solvent-free deoxygenation of low-cost fat to produce diesel-like hydrocarbons over Ni–MoS2/Al2O3–TiO2 heterogenized catalyst. Int J Energy Water Resour 1–13
128. Gorimbo J, Moyo M, Liu X (2022) Oligomerization of bio-olefins for bio-jet fuel. In: Hydrocarbon biorefinery. Elsevier, p 271–294
129. Mirzaei N et al (2021) Flexible production of liquid biofuels via thermochemical treatment of biomass and olefins oligomerization: a process study. Chem Eng Trans 86:187–192
130. Maghrebi R et al (2021) Isomerization of long-chain fatty acids and long-chain hydrocarbons: a review. Renew Sustain Energy Rev 149:111264
131. Glisic S, Skala D (2009) The problems in design and detailed analyses of energy consumption for biodiesel synthesis at supercritical conditions. J Supercrit Fluids 49(2):293–301
132. Gieleciak R, Farooqi H, Chen J (2021) Detailed characterization of diesel fractions from co-hydroprocessing vegetable oil and petroleum heavy vacuum gas oil blends. Energy & Fuels
133. Srifa A et al (2015) Roles of monometallic catalysts in hydrodeoxygenation of palm oil to green diesel. Chem Eng J 278:249–258
134. Li C et al (2018) Catalytic cracking of Swida wilsoniana oil for hydrocarbon biofuel over Cu-modified ZSM-5 zeolite. Fuel 218:59–66
135. Onyestyák G et al (2012) Sunflower oil to green diesel over Raney-type Ni-catalyst. Fuel 102:282–288
136. Yang Y et al (2013) Hydrotreating of C18 fatty acids to hydrocarbons on sulphided NiW/SiO2–Al2O3. Fuel Process Technol 116:165–174
137. Lázaro MJ et al (2015) Carbon-based catalysts: synthesis and applications. C R Chim 18(11):1229–1241
138. Arun N, Sharma RV, Dalai AK (2015) Green diesel synthesis by hydrodeoxygenation of bio-based feedstocks: strategies for catalyst design and development. Renew Sustain Energy Rev 48:240–255
139. Li X et al (2018) Heterogeneous sulfur-free hydrodeoxygenation catalysts for selectively upgrading the renewable bio-oils to second generation biofuels. Renew Sustain Energy Rev 82:3762–3797
140. Hájek M et al (2021) The catalysed transformation of vegetable oils or animal fats to biofuels and bio-lubricants: a review. Catalysts 11(9):1118
141. Nikolopoulos I et al (2021) Cobalt–alumina coprecipitated catalysts for green diesel production. Ind & Eng Chem Res
142. Silva GCR, de Andrade MHC (2021) Simulation of deoxygenation of vegetable oils for diesel-like fuel production in continuous reactor. Biomass Convers Biorefinery 1–15
143. Wang F et al (2018) Hydrotreatment of vegetable oil for green diesel over activated carbon supported molybdenum carbide catalyst. Fuel 216:738–746
144. Jin W et al (2020) Catalytic conversion of palm oil to bio-hydrogenated diesel over novel N-doped activated carbon supported Pt nanoparticles. Energies 13(1):132
145. Jin W et al (2018) Catalytic upgrading of biomass model compounds: novel approaches and lessons learnt from traditional hydrodeoxygenation–a review
146. Zhao X et al (2015) Catalytic cracking of non-edible sunflower oil over ZSM-5 for hydrocarbon bio-jet fuel. New Biotechnol 32(2):300–312
147. Choo M-Y et al (2019) The role of nanosized zeolite Y in the H 2-free catalytic deoxygenation of triolein. Catal Sci Technol 9(3):772–782
148. Choo M-Y et al (2020) Deposition of NiO nanoparticles on nanosized zeolite NaY for production of biofuel via hydrogen-free deoxygenation. Materials 13(14):3104
149. Choo M-Y et al (2020) Deoxygenation of triolein to green diesel in the H2-free condition: effect of transition metal oxide supported on zeolite Y. J Anal Appl Pyrolysis 147:104797

150. Azreena IN et al (2021) A promoter effect on hydrodeoxygenation reactions of oleic acid by zeolite beta catalysts. J Anal Appl Pyrolysis 155:105044

151. Chintakanan P et al (2021) Bio-jet fuel range in biofuels derived from hydroconversion of palm olein over Ni/zeolite catalysts and freezing point of biofuels/Jet A-1 blends. Fuel 293:120472

152. Freitas LNS et al (2021) Study of direct synthesis of bio-hydrocarbons from macauba oils using zeolites as catalysts. Fuel 287:119472

153. Kang Y-H et al (2021) Green and effective catalytic hydroconversion of an extractable portion from an oil sludge to clean jet and diesel fuels over a mesoporous Y zeolite-supported nickel catalyst. Fuel 287:119396

154. Arumugam M et al (2021) Hierarchical HZSM-5 for catalytic cracking of oleic acid to biofuels. Nanomaterials 11(3):747

155. Wang F et al (2021) Promoting hydrocarbon production from fatty acid pyrolysis using transition metal or phosphorus modified Al-MCM-41 catalyst. J Anal Appl Pyrolysis 156:105146

156. Lee C-W et al (2020) Hydrodeoxygenation of palmitic acid over zeolite-supported nickel catalysts. Catal Today

157. Liu Y et al (2020) Rapid and green synthesis of SAPO-11 for deoxygenation of stearic acid to produce bio-diesel fractions. Microporous Mesoporous Mater 303:110280

158. Ghaffar N et al (2020) Catalytic cracking of high density polyethylene pyrolysis vapor over zeolite ZSM-5 towards production of diesel. IOP Conf Ser: Mater Sci Eng. IOP Publishing

159. Sousa FP et al (2018) Simultaneous deoxygenation, cracking and isomerization of palm kernel oil and palm olein over beta zeolite to produce biogasoline, green diesel and biojet-fuel. Fuel 223:149–156

160. Zandonai CH et al (2016) Production of petroleum-like synthetic fuel by hydrocracking of crude soybean oil over ZSM5 zeolite–improvement of catalyst lifetime by ion exchange. Fuel 172:228–237

161. Gomes LC et al (2017) Hydroisomerization of n-hexadecane using Pt/alumina-Beta zeolite catalysts for producing renewable diesel with low pour point. Fuel 209:521–528

162. Hachemi I et al (2017) Sulfur-free Ni catalyst for production of green diesel by hydrodeoxygenation. J Catal 347:205–221

163. El-Sawy MS et al (2020) Co-hydroprocessing and hydrocracking of alternative feed mixture (vacuum gas oil/waste lubricating oil/waste cooking oil) with the aim of producing high quality fuels. Fuel 269:117437

164. French RJ, Hrdlicka J, Baldwin R (2010) Mild hydrotreating of biomass pyrolysis oils to produce a suitable refinery feedstock. Environ Prog Sustain Energy 29(2):142–150

165. Gruia A (2006) Recent advances in hydrocracking. In: Practical advances in petroleum processing. Springer, pp 219–255

166. Kubička D et al (2014) Effect of support-active phase interactions on the catalyst activity and selectivity in deoxygenation of triglycerides. Appl Catal B 145:101–107

167. Fogassy G et al (2010) Biomass derived feedstock co-processing with vacuum gas oil for second-generation fuel production in FCC units. Appl Catal B 96(3–4):476–485

168. Bezergianni S, Dimitriadis A, Meletidis G (2014) Effectiveness of CoMo and NiMo catalysts on co-hydroprocessing of heavy atmospheric gas oil–waste cooking oil mixtures. Fuel 125:129–136

169. Dewi TK (2014) Conversion of waste oil into fuel oil. In: 5th Sriwijaya international seminar on energy and environmental science and technology. Sriwijaya University

170. Hanafi SA et al (2016) Hydrocracking of waste chicken fat as a cost effective feedstock for renewable fuel production: a kinetic study. Egypt J Pet 25(4):531–537

171. Ramírez J, Rana MS, Ancheyta J (2007) Characteristics of heavy oil hydroprocessing catalysts. Hydroprocessing of heavy oils and residua. Taylor & Francis, New York, pp 121–190

172. Task IB (2021) Progress in commercialization of biojet/Sustainable Aviation Fuels (SAF): technologies. Potential Chall

173. Wang F et al (2016) Co-hydrotreating of used engine oil and the low-boiling fraction of bio-oil blends for the production of liquid fuel. Fuel Process Technol 146:62–69

174. Pelemo J, Inambao FL, Onuh EI (2020) Potential of used cooking oil as feedstock for hydroprocessing into hydrogenation derived renewable diesel: a review. IJERT 13:500–519
175. Furimsky E (1998) Selection of catalysts and reactors for hydroprocessing. Appl Catal A 171(2):177–206
176. Marafi M et al (2009) Activity of hydroprocessing catalysts prepared by reprocessing spent catalysts. Fuel Process Technol 90(2):264–269
177. Patel M, Kumar A (2016) Production of renewable diesel through the hydroprocessing of lignocellulosic biomass-derived bio-oil: a review. Renew Sustain Energy Rev 58:1293–1307
178. Choudhary TV, Phillips CB (2011) Renewable fuels via catalytic hydrodeoxygenation. Appl Catal A 397(1–2):1–12
179. Yulia D, Zulys A (2020) Hydroprocessing of kemiri sunan oil (reutealis trisperma (blanco) airy shaw) over NiMoCe/γ-Al2O3 catalyst to produce green diesel. IOP Conf Ser: Mater Sci Eng. IOP Publishing
180. Kaewchada A et al (s2021) Production of bio-hydrogenated diesel from palm oil using Rh/HZSM-5 in a continuous mini fixed-bed reactor. Chem Eng Process-Process Intensif 168:108586
181. Van de Velden M et al (2010) Fundamentals, kinetics and endothermicity of the biomass pyrolysis reaction. Renew Energy 35(1):232–242
182. Carpenter D et al (2014) Biomass feedstocks for renewable fuel production: a review of the impacts of feedstock and pretreatment on the yield and product distribution of fast pyrolysis bio-oils and vapors. Green Chem 16(2):384–406
183. Mousavi-Avval SH, Shah A (2021) Techno-economic analysis of hydroprocessed renewable jet fuel production from pennycress oilseed. Renew Sustain Energy Rev 149:111340
184. McIntosh S et al (2021) Combined pyrolysis and sulphided NiMo/Al2O3 catalysed hydroprocessing in a multistage strategy for the production of biofuels from milk processing waste. Fuel 295:120602
185. Phichitsurathaworn N et al (2021) Techno-economic analysis of co-production of bio-hydrogenated diesel from palm oil and methanol. Energy Convers Manag 244:114464
186. Plazas-González M, Guerrero-Fajardo CA, Modeling and simulation for the hydroprocessing of palm oil components for the green diesel production
187. Wang W-C, Hsieh C-H (2020) Hydro-processing of biomass-derived oil into straight-chain alkanes. Chem Eng Res Des 153:63–74
188. Mangesh V et al (2020) Green energy: hydroprocessing waste polypropylene to produce transport fuel. J Clean Prod 276:124200
189. Jeong H et al (2020) Comparison of activity and stability of supported Ni2P and Pt catalysts in the hydroprocessing of palm oil into normal paraffins. J Ind Eng Chem 83:189–199
190. Varakin A et al (2020) Toward HYD/DEC selectivity control in hydrodeoxygenation over supported and unsupported Co (Ni)-MoS2 catalysts. A key to effective dual-bed catalyst reactor for co-hydroprocessing of diesel and vegetable oil. Catal Today 357:556–564
191. Carrara N et al (2015) Selective hydrogenation by novel composite supported Pd egg-shell catalysts. Catal Commun 61:72–77
192. Liu C et al (2014) Improvement of methane production from waste activated sludge by on-site photocatalytic pretreatment in a photocatalytic anaerobic fermenter. Bioresour Technol 155:198–203
193. Xu J et al (2019) Integrated catalytic conversion of waste triglycerides to liquid hydrocarbons for aviation biofuels. J Clean Prod 222:784–792
194. Banu JR et al (2020) Impervious and influence in the liquid fuel production from municipal plastic waste through thermo-chemical biomass conversion technologies-a review. Sci Total Environ 718:137287
195. Faussone GC (2018) Transportation fuel from plastic: two cases of study. Waste Manag 73:416–423
196. Mangesh V et al (2020) Experimental investigation to identify the type of waste plastic pyrolysis oil suitable for conversion to diesel engine fuel. J Clean Prod 246:119066

197. Mangesh V et al (2020) Combustion and emission analysis of hydrogenated waste polypropylene pyrolysis oil blended with diesel. J Hazard Mater 386:121453
198. Adu-Mensah D et al (2019) A review on partial hydrogenation of biodiesel and its influence on fuel properties. Fuel 251:660–668
199. Kemp K et al (2013) An exploration of the follow-up up needs of patients with inflammatory bowel disease. J Crohn's Colitis 7(9):e386–e395
200. Šimáček P et al (2009) Hydroprocessed rapeseed oil as a source of hydrocarbon-based biodiesel. Fuel 88(3):456–460
201. Zhao X et al (2016) Development of hydrocarbon biofuel from sunflower seed and sunflower meat oils over ZSM-5. J Renew Sustain Energy 8(1):013109
202. Patil SJ, Vaidya PD (2018) On the production of bio-hydrogenated diesel over hydrotalcite-like supported palladium and ruthenium catalysts. Fuel Process Technol 169:142–149
203. Srifa A et al (2014) Production of bio-hydrogenated diesel by catalytic hydrotreating of palm oil over NiMoS2/γ-Al2O3 catalyst. Bioresour Technol 158:81–90
204. Tiwari R et al (2011) Hydrotreating and hydrocracking catalysts for processing of waste soya-oil and refinery-oil mixtures. Catal Commun 12(6):559–562
205. Yang H et al (2017) The effects of contact time and coking on the catalytic fast pyrolysis of cellulose. Green Chem 19(1):286–297
206. Chen N et al (2013) Effects of Si/Al ratio and Pt loading on Pt/SAPO-11 catalysts in hydroconversion of Jatropha oil. Appl Catal A 466:105–115
207. Bezergianni S et al (2011) Toward hydrotreating of waste cooking oil for biodiesel production. Effect of pressure, H2/oil ratio, and liquid hourly space velocity. Ind & Eng Chem Res 50(7):3874–3879
208. Lambert N (2012) Study of hydrogenation derived renewable diesel as a renewable fuel option in North America. Nat Resour Canada. Montreal
209. Naguran M (2015) Sustainability issues and strategies of biofuel development in Southeast Asia. In: Sustainability matters: environmental and climate changes in the Asia-Pacific. World Scientific, pp 331–369
210. Wertz J-L, Bédué O (2013) Lignocellulosic biorefineries. EPFL Press Lausanne, Switzerland
211. Milbrandt A, Kinchin C, McCormick R (2013) Feasibility of producing and using biomass-based diesel and jet fuel in the United States. National Renewable Energy Lab (NREL), Golden, CO (United States)
212. Kraemer KL, Dedrick J, Yamashiro S (2000) Refining and extending the business model with information technology: dell computer corporation. Inf Soc 16(1):5–21
213. Suopajärvi H, Fabritius T (2013) Towards more sustainable ironmaking—an analysis of energy wood availability in Finland and the economics of charcoal production. Sustainability 5(3):1188–1207
214. Qian J, Shi H, Yun Z (2010) Preparation of biodiesel from Jatropha curcas L. oil produced by two-phase solvent extraction. Bioresour Technol 101(18):7025–7031
215. Mohammad M et al (2013) Overview on the production of paraffin based-biofuels via catalytic hydrodeoxygenation. Renew Sustain Energy Rev 22:121–132
216. Mamaghani AH et al (2016) Techno-economic feasibility of photovoltaic, wind, diesel and hybrid electrification systems for off-grid rural electrification in Colombia. Renew Energy 97:293–305
217. Abomohra AE-F et al (2016) Microalgal biomass production as a sustainable feedstock for biodiesel: current status and perspectives. Renew Sustain Energy Rev 64:596–606
218. de Souza T et al (2022) Biodiesel in South American countries: a review on policies, stages of development and imminent competition with hydrotreated vegetable oil. Renew Sustain Energy Rev 153:111755
219. Kuznetsov NI et al (2017) Economic research of transfer of technologies for manufacturing high-tech production in Russia: bio-fuel. J Environ Manag & Tour 8(3(19)):606–611
220. Tarkowski M (2021) Towards a more sustainable transport future—the cases of ferry shipping electrification in Denmark, Netherland, Norway and Sweden. In: Innovations and traditions for sustainable development. Springer, pp 177–191

221. Irwan S et al (2012) Biodiesel progress in Malaysia. Energy Sources Part A: Recover Util Environ Effects 34(23):2139–2146
222. Aziz HA et al (2016) Production of palm-based esteramine through heterogeneous catalysis. J Surfactants Deterg 19(1):11–18
223. Mukherjee I, Sovacool BK (2014) Palm oil-based biofuels and sustainability in southeast Asia: a review of Indonesia, Malaysia, and Thailand. Renew Sustain Energy Rev 37:1–12
224. Khatiwada D, Palmén C, Silveira S (2021) Evaluating the palm oil demand in Indonesia: production trends, yields, and emerging issues. Biofuels 12(2):135–147
225. Yasinta T, Karuniasa M (2021) Palm oil-based biofuels and sustainability In Indonesia: assess social, environmental and economic aspects. IOP Conf Ser: Mater Sci Eng. IOP Publishing.
226. Plazas-González M, Guerrero-Fajardo CA, Sodré JR (2018) Modelling and simulation of hydrotreating of palm oil components to obtain green diesel. J Clean Prod 184:301–308
227. Yaakob Z et al (2012) Utilization of palm empty fruit bunch for the production of biodiesel from Jatropha curcas oil. Bioresour Technol 104:695–700
228. Šimáček P et al (2011) Premium quality renewable diesel fuel by hydroprocessing of sunflower oil. Fuel 90(7):2473–2479
229. Mazubert A et al (2014) Intensification of waste cooking oil transformation by transesterification and esterification reactions in oscillatory baffled and microstructured reactors for biodiesel production. Green Process Synth 3(6):419–429
230. Hazrat M et al (2019) Emission characteristics of waste tallow and waste cooking oil based ternary biodiesel fuels. Energy Procedia 160:842–847
231. Meng X, Chen G, Wang Y (2008) Biodiesel production from waste cooking oil via alkali catalyst and its engine test. Fuel Process Technol 89(9):851–857
232. Falade AO, Oboh G, Okoh AI (2017) Potential health lmplications of the consumption of thermally-oxidized cooking oils–a review. Polish J Food Nutr Sci 67(2)
233. Karamé I (2012) Hydrogenation. BoD–Books on Demand
234. Ardiyanti A et al (2011) Hydrotreatment of wood-based pyrolysis oil using zirconia-supported mono-and bimetallic (Pt, Pd, Rh) catalysts. Appl Catal A 407(1–2):56–66
235. Bejblová M et al (2005) Hydrodeoxygenation of benzophenone on Pd catalysts. Appl Catal A 296(2):169–175
236. Bunch AY, Ozkan US (2002) Investigation of the reaction network of benzofuran hydrodeoxygenation over sulfided and reduced Ni–Mo/Al2O3 catalysts. J Catal 206(2):177–187
237. Yang YQ, Tye CT, Smith KJ (2008) Influence of MoS2 catalyst morphology on the hydrodeoxygenation of phenols. Catal Commun 9(6):1364–1368
238. Laurent E, Delmon B (1994) Study of the hydrodeoxygenation of carbonyl, car□ylic and guaiacyl groups over sulfided CoMo/γ-Al2O3 and NiMo/γ-Al2O3 catalysts: I. Catalytic reaction schemes. Appl Catal A: Gen 109(1):77–96
239. Whiffen VM, Smith KJ (2010) Hydrodeoxygenation of 4-methylphenol over unsupported MoP, MoS2, and MoO x catalysts. Energy Fuels 24(9):4728–4737
240. Alvarez-Galvan MC et al (2018) Metal phosphide catalysts for the hydrotreatment of non-edible vegetable oils. Catal Today 302:242–249
241. Mahdi HI, Muraza O (2016) Conversion of isobutylene to octane-booster compounds after methyl tert-butyl ether phaseout: the role of heterogeneous catalysis. Ind Eng Chem Res 55(43):11193–11210
242. de Miguel Mercader F, Groeneveld M (2011) SR a. Kersten, C. Geantet, G. Toussaint, NWJ Way, CJ Schaverien and KJ a. Hogendoorn. Energy Environ. Sci. 4:985
243. Pinheiro A et al (2011) Impact of the presence of carbon monoxide and carbon dioxide on gas oil hydrotreatment: investigation on liquids from biomass cotreatment with petroleum cuts. Energy Fuels 25(2):804–812
244. Kumar R et al (2010) Hydroprocessing of Jatropha oil and its mixtures with gas oil. Green Chem 12(12):2232–2239
245. Philippe M et al (2013) Transformation of dibenzothiophenes model molecules over CoMoP/Al2O3 catalyst in the presence of oxygenated compounds. Appl Catal B 132:493–498

246. Varakin A et al (2018) Comparable investigation of unsupported MoS2 hydrodesulfurization catalysts prepared by different techniques: advantages of support leaching method. Appl Catal B 238:498–508

247. Asikin-Mijan N et al (2016) Waste clamshell-derived CaO supported Co and W catalysts for renewable fuels production via cracking-deoxygenation of triolein. J Anal Appl Pyrol 120:110–120

248. Asikin-Mijan N et al (2017) Catalytic deoxygenation of triglycerides to green diesel over modified CaO-based catalysts. RSC Adv 7(73):46445–46460

249. Kovács S et al (2011) Fuel production by hydrotreating of triglycerides on NiMo/Al2O3/F catalyst. Chem Eng J 176:237–243

250. Wang H (2012) Biofuels production from hydrotreating of vegetable oil using supported noble metals, and transition metal carbide and nitride

251. Tago T et al (2012) Size-controlled synthesis of nano-zeolites and their application to light olefin synthesis. Catal Surv Asia 16(3):148–163

252. Santillan-Jimenez E et al (2014) Catalytic deoxygenation of triglycerides and fatty acids to hydrocarbons over Ni–Al layered double hydroxide. Catal Today 237:136–144

253. Santillan-Jimenez E et al (2013) Catalytic deoxygenation of triglycerides and fatty acids to hydrocarbons over carbon-supported nickel. Fuel 103:1010–1017

254. Abdulkareem-Alsultan G et al (2019) Pyro-lytic de-oxygenation of waste cooking oil for green diesel production over Ag2O3-La2O3/AC nano-catalyst. J Anal Appl Pyrol 137:171–184

255. Gousi M et al (2020) Green diesel production over nickel-alumina nanostructured catalysts promoted by copper. Energies 13(14):3707

Chapter 6
Green Diesel: Integrated Production Processes, Future Perspectives and Techno-Economic Feasibility

Jaspreet Kaur, Mohammad Aslam, M. K. Jha, and Anil K. Sarma

Abstract All over the world, an increase in biofuels consumption, e.g. green diesel, reduces the cost impact and dependence on petroleum and detrimental environmental consequences. Green diesel, next-generation fuel, an alternative energy product, has a similar molecular structure as petroleum diesel but provides better diesel properties. It has exceptional storage stability and is completely compatible for blending with the standard mix of petroleum-derived diesel fuels. The green diesel has been produced by hydrotreating triglycerides or vegetable oils with hydrogen. It is produced using the same feedstocks as biodiesel (mainly animal fats or vegetable oil) but the production process for both differs significantly. It is an optimum biocomponent for blending into mineral diesel. The high quality of green diesel is determined by its higher heating value and energy density, high cetane number and outstanding cold flow properties. Also, the low density of green diesel makes it a very good blending component for refiners, which are usually limited in accepting heavy gas oil bases in the diesel blend. In addition, the low aromatic content benefits from blending with other petroleum diesel bases. The various oxygenates produced that may be considered as additives for diesel fuel are various alcohols, ethers, esters, acetals and carbonates. These fuel additives also lead to the production of specific products that meet international and regional standards allowing the fuels trade to take place. The future production of

J. Kaur · A. K. Sarma (✉)
Chemical Conversion Division, Sardar Swaran Singh National Institute of Bio-Energy, Kapurthala, Punjab, India
e-mail: anil.sarma@nibe.res.in

J. Kaur
e-mail: jassi0713@gmail.com

J. Kaur · M. K. Jha
Department of Chemical Engineering, DR B R Ambedkar National Institute of Technology, Jalandhar, India
e-mail: jhamk@nitj.ac.in

M. Aslam
Department of Chemistry, National Institute of Technology, Srinagar (J&K), India
e-mail: maslam@nitsri.ac.in

these additives needs the development of the integrated production process or reactive separation technologies, which helps in the reduction of energy consumption and capital costs. The production of these green diesel additives provides clean and efficient technology. The technical and cost analysis of these production technologies depends upon the unit plant capacity and feedstock price. Unit capacities of the investigated processes that are below 100,000 tonnes/year are likely to result in negative net present values after 10 years of the project lifetime.

Keywords Biofuels · Biomass · Green Diesel · Life Cycle Assessment · Ecofining

1 Introduction

The fuels used in transportation such as gasoline, diesel and jet fuel are the primary energy consuming sources in this world and their contribution towards world's energy consumption is 28% [1]. Though, petroleum diesel has dominated the transport sector, which globally contributes 50% towards energy sector, but its significant contribution to the environmental pollution and climate change derives the need for the production of fuels from renewable sources for sustainable environment [2, 3]. As there is an increase in demand for energy in the developing countries, which is expected to hike by 90%, so, one-third part of the energy would come from renewable energy sources [4]. Due to the presence of oxygen in biofuel's structure, they are considered as lower fuel mileage compared to petroleum-derived transportation fuels [1]. Also, biofuels are incompatible with internal combustion engines due to their various properties such as cold flow properties, vapor pressure, miscibility with water, etc. [1]. Therefore, new biomass processing concepts should be developed for the production of hydrocarbon biofuels that are chemically similar to the petroleum-derived transportation fuels [5, 6]. The biomass is the most common and only renewable carbon source that has great potential in producing various bio-based liquid transportation fuels [1] and more than 80% of the energy in the world is supplied by it. Biomass is the renewable organic material that comes from plants and animals, which includes agricultural crops and residues, forest wastes and residues and municipal and industrial wastes. On the basis of the chemical nature, biomass is broadly classified into three categories namely triglycerides, sugar and starchy biomass and lignocellulosic biomass [1]. A triglyceride is an ester derived from glycerol and three fatty acids. Triglycerides are the main constituents of body fat in humans and other vertebrates, as well as vegetable fat. The sugar is a disaccharide of glucose and fructose. Starch is composed of a mixture of linear polysaccharide, amylose (homopolymer of d-glucose linked via a α-1,4 glycosidic bond) and branched polysaccharide, amylopectin (homopolymer of d-glucose linked via a linear α-1,4 glycosidic bond and branched α-1,6 glycosidic bond). Lignocellulosic biomass is an abundant renewable resource primarily composed of cellulose (40–50%), lignin (10–25%) and hemicelluloses (20–30%) that form a complex composite structure. Hence, on the basis of the nature of the biomass, there are various processes that are generally developed for the conversion of biomass [1]. Therefore, hydro-carbon biofuels are commonly known as called as green transportation or green diesel or green fuels, for example, as green gasoline, green diesel and green jet fuel. The processing of biomass into these fuels involves (a) thermochemical conversion processes—processes such as liquefaction, gasification and pyrolysis that involve the conversion of the biomass into fuels using heat energy, (b) chemical conversion processes—processes that involve the conversion of triglycerides in the presence of catalysts into green diesel, (c) biochemical conversion processes—processes that involve the conversion of biomass via biological means such as fermentation and (d) platform chemical-based conversion processes—processes that are based on the conversion of platform chemicals 5-hydroxymethylfurfural (HMF) and furfural into renewable fuels [1]. Amongst the first-generation, second-generation and third-generation fuels, third-generation biofuels mainly produced from microalgae so called "third-generation" biodiesel and green diesel fuels, have received the highest emphasis in the last 3 years [2, 7]. The different types of biofuels produced from various kinds of biomasses are shown in Fig. 1, and they are advantageous in terms of both in reducing the effect of global warming as well energy dependency [2].

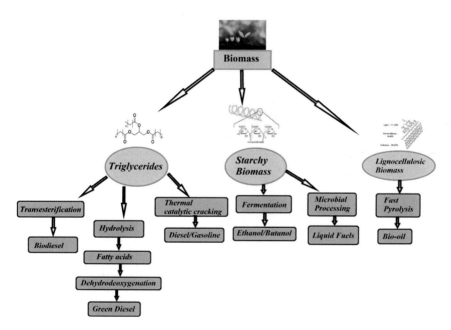

Fig. 1 Different fuels produced from different biomasses

2 Production Processes

Green diesel, also known as "renewable diesel", "second-generation diesel", "bio-hydrogenated diesel", "Hydrogenated Esters and Fatty Acids (HEFA)", "Bio-Hydrogenated Diesel (BHD)", "Hydrogenation Derived Renewable Diesel (HDRD)", "Hydrotreated Vegetable Oil" or "Hydrogenated Vegetable Oil", is a new generation biofuel with biological origin [2]. It is a product obtained from biological matter and is a mixture of straight chain and branched saturated hydrocarbons (C_{15} to C_{18}). Also, it is free from aromatics or naphthenes and oxygen and is, therefore, stable in nature and provides similar heating as compared to other diesel fuels. Green diesel is of a higher quality than biodiesel and has similar properties both in terms of composition and combustion properties as to syndiesel. Also, it is independent of feed origin and the process of green diesel formation is more flexible than biodiesel production with respect to feedstock selection and plant location [3]. The comparison of the green diesel properties with other kind of fuels is presented in Table 1.

There are numerous technological pathways that are being studied, explored and reviewed in the literature for producing renewable or green biofuels such as transesterification, fermentation, hydroprocessing, Fischer Tropsch synthesis, oligomerization, aldol-condensation, hydroxyalkylation alkylation (HAA), pyrolysis, catalytic cracking, etc. The four technologies that are widely involved in the transformation of biomass to green diesel are hydroprocessing, catalytic upgrading, thermal conversion and thermochemical processes.

- **Transesterification**—It is the chemical process for the conversion of the biomass lipids, in which triglycerides are transformed into Fatty Acid Methyl Esters

Table 1 Comparison of properties of different types of fuels produced [2–7]

Fuel type property	Mineral ULSD	Biodiesel	Petroleum diesel	Renewable diesel	FT diesel	Green diesel
Specific gravity	0.84	0.88	0.85	–	0.77	0.78
Sulphur content (ppm)	<10	<1	High	–	<1	<1
Heating value (MJ/kg)	43	38	43.2	–	43.5	44
Cloud point (°C)	−5	−5 to −15	<−34	–	–	−10 to −20
Distillation (°C)	200–350	340–355	–	–	–	265–320
Cetane number	40	50–65	44.5–67	~70	> 75	70–90
Stability	Good	Marginal	–	–	–	Good

Fig. 2 Hydrotreatment of typical glycerides

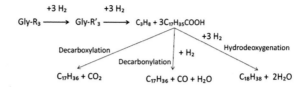

(FAME or biodiesel) and glycerol usually with methanol in the presence of a homogeneous alkaline catalyst such as KOH or NaOH, at a temperature of about 60 °C and atmospheric pressure [8, 9]. The supercritical transesterification has environmental benefits regarding wastewater and spent chemicals generation, as well as energy consumption, high production efficiency, high feedstock flexibility while comparing it with the conventional biodiesel production [10].

- **Hydrotreating**—This process is used in petroleum refineries, hydrotreating involves reaction of the feedstock (lipids) with hydrogen under elevated temperatures and pressures in the presence of a catalyst. The hydrotreating process of vegetable oils leads to C_{15}–C_{18} primary hydrocarbon products within the temperature range of 300–450 °C and pressure above 30 bar [11]. Commercial plants currently use this technology. The hydrotreating process mainly consists of three reactions, i.e. hydrodeoxygenation, decarbonylation and decarboxylation as shown in Fig. 2 [11].

Hydrotreating has become a more common way to produce renewable diesel (RD), known as hydro-processed esters and fatty acids (HEFA) [12]. The hydrodeoxygenation process is expected to produce better quality green diesel by reducing oxygen content as shown in figure 3 [13]. This process is carried out under elevated temperature and pressure. Molecular sieve catalysts and supported metal catalysts in sulfided or reduced form are generally used as they favor the conversion of bio-oil feedstocks into hydrocarbon fuels by decarboxylation [14]. The hydrodeoxygenation process is highly favored because of the formation of hydrocarbons in a narrow range of molecular weights. Hydrotreatment of vegetable oils could be used as first approach to remove sulfur from refinery products in comparison to the conventional hydrotreatment units [15]. Neste Oil, Petrobras, SK-Innovation and UOP/Eni companies have been producing and commercializing RD. Neste oil is operating two units with a combined capacity of 170,000 tonnes/year of green diesel in Finland as well as a production in Singapore and Rotterdam with 800,000 tonnes/year capacity [16].

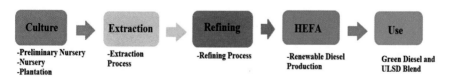

Fig. 3 Hydrodeoxygenation of oil for green diesel production

- **Biological sugar upgrading**—This pathway uses a biochemical deconstruction process, similar to that used with cellulosic ethanol with the addition of organisms that convert sugars to hydrocarbons. Ethanol can be obtained from various carbon sources by engineering or by exploiting native fermentation pathways of various microbial hosts. Production of higher alcohols as alternatives to ethanol with better fuel properties has been demonstrated by engineering fermentative pathways, non-fermentative keto-acid pathways and isoprenoid pathways [17]. In addition to higher alcohols, fatty acid-derived biofuels and isoprenoid-derived biofuels have also been proposed as good diesel alternatives (Fig. 4).
- **Catalytic conversion of sugars, Starches and alcohols**—This pathway involves a series of catalytic reactions to convert a carbohydrate stream into hydrocarbon fuels such as aqueous phase reforming (APR). The reactions that are considered to be involved during APR are (a) the reforming of the sugars that produce H_2, (b) the dehydrogenation of alcohols, (c) the hydrogenation of carbonyls, (d) deoxygenation reactions, (e) hydrogenolysis and (f) cyclization [2]. During APR, the hydrolysate slurry reacts with water, which converts the sugars and sugar alcohols

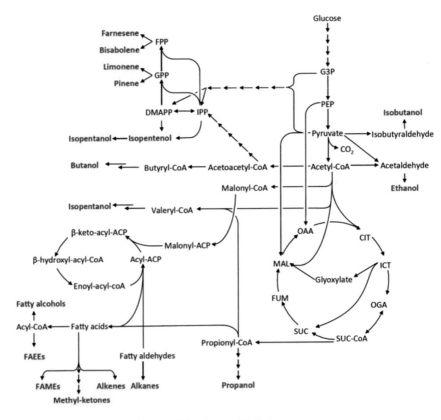

Fig. 4 Pathways for the production of the advanced biofuels

into H_2, CO_2, lower saturated hydrocarbons and condensable chemical intermediates at temperatures about 175–300 °C and pressures of 10–90 bar over supported metal or metal alloy catalysts such as Pt or Ni–Sn. Hydrogenation takes place in the presence of hydrogen at 100–150 °C and 10–30 bar causing the saturation of the C = C, C = O and C–O–C bonds over metal catalysts such as Ru, Pt, Pd and Ni. Hydrogenolysis occurs in the presence of hydrogen at 125–230 °C and 14–300 bar resulting from the selective cleavage of C–C or C–O bonds of specific molecules like glycerol for the production of more valuable polyols or diols, which are useful in the production of chemical polymers. This takes place under alkaline conditions over supported metal catalysts such as Ru, Pd, Pt, Ni and Cu. Catalytic dehydration is an alternative pathway for the production of alkenes from oxygenated feedstocks. The disadvantages associated with these processes are the matter that is susceptible to the existed lignin pretreatment methods.

- **Biomass to Liquid (BTL) Thermochemical Processes (Gasification)**—During this process, biomass is thermally converted to syngas and catalytically converted to hydrocarbon fuels [8]. Gasification is the exothermal reaction, which converts biomass or carbonaceous fuels into synthesis gas (syngas) by partial oxidation [2]. The green diesel produced by the Fischer–Tropsch method is sometimes referred as FT green diesel [2, 18]. One competitive technology for the production of hydrocarbon fuel is the gasification of oil followed by Fisher–Tropsch (FT) synthesis but, the major disadvantages of this process are the high cost and unfavorable carbon dioxide emissions [1New]. Fischer–Tropsch synthesis is a highly exothermic surface polymerization reaction that converts H_2 and CO into saturated or unsaturated hydrocarbons of 1–40 carbon atoms via heterogeneous catalysts such as iron (Fe) or cobalt (Co). The reactions that proceed with the surface polymerization of these groups into larger chains according to produce a crude oil mixture of saturated hydrocarbons, unsaturated hydrocarbons, alcohols and carbonyls are as follows [19, 20]:

$$2nH_2 + nCO \rightarrow (CH_2)_n + nH_2O$$

$$(2n + 1)H_2 + nCO \rightarrow C_nH_{2n+2} + nH_2O$$

$$2nH_2 + nCO \rightarrow C_nH_{2n} + nH_2O$$

$$2nH_2 + nCO \rightarrow H(CH_2)_n OH + (n - 1)H_2O$$

$$nCO + (2n - 1)H_2 \rightarrow (CH_2)_n O + (n - 1)H_2O$$

- **Thermal conversion (Pyrolysis)**—This pathway involves the chemical decomposition of organic materials at elevated temperatures in the absence of oxygen. The process produces liquid pyrolysis oil that can be upgraded to hydrocarbon fuels, either in a standalone process or as a feedstock for co-feeding with crude oil into a standard petroleum refinery.

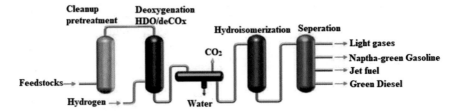

Fig. 5 Hydroprocessing plant

- **Hydrothermal processing**—This process uses very high temperature and pressure to initiate chemical decomposition of biomass or wet waste materials to produce oil that may be catalytically upgraded to hydrocarbon fuels. Hydrothermal processing technology is being commercialized in the United States by Changing World Technologies (CWT) [9].
- **Hydroprocessing**—It is the catalytic process that involves the reaction of petroleum distillates for the removal of impurities by treating with hydrogen pressure for the formation of transportation/hydrocarbon fuels. Hydro-processing yields different liquid fuels such as green diesel (C_{15}–C_{18}), green jet fuel (C_{11}–C_{13}) and green naphtha (C_5–C_{10}) based on how hydrogen saturates the double bonds as presented in Fig. 5 [2]. Hydroprocessing has been explored widely and commercially made the availability of renewable diesel. Nestle in Finland has commercialized NExBTL, the renewable diesel by Nestle.
- **Ecofining process**—This process basically is an integration form of two-stage hydrorefining process in which feedstock is mixed with recycle hydrogen, which is then sent to a multi-stage adiabatic, catalytic hydrodeoxygenation reactor for saturation and complete deoxygenating [3]. This product is then mixed with additional hydrogen gas and then forwarded to an integrated catalytic hydro-isomerization reactor to produce paraffin-rich diesel fuel. The advantages of the Ecofining process over tranesterification process are that it is robust to high concentrations of free fatty acids and allow the use of other lower-cost materials as feedstocks. The schematic representation of ecofining process is given in Fig. 6. The commercial unit of green diesel based on UOP/Eni Ecofining process technology is to become operational in Italy [21]. Diamond Green Diesel facility in Norco, Louisiana, is

Fig. 6 UOP's ecofining process

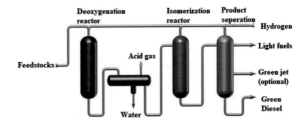

also installed operating on the basis of this technology with a capacity more than 130 million gallons/year [10].

3 Process Economics

3.1 The Energy Optimization of a Green Diesel Production Process

The cost of the process can be reduced by the efficient utilization of the energy of the process. To minimize the energy consumption of the green diesel production process, the heat exchanger network (HEN) should be considered [11]. Energy optimization strategies can also keep the quality of green diesel product.

Life Cycle Assessment (LCA) is a methodology commonly used for the evaluation of the effect on the environment caused by industrial processes and services, from acquisition, manufacture, use and maintenance of the raw material, until the final disposal of the product or service. It is one of the suitable tools for environmental decision-making [22]. LCA studies were performed to determine the impacts of biofuel feedstock, allocation method and other study assumptions on biofuel production.

Finally, production of green diesel by hydro-processing of triglycerides produces propane as a byproduct which is a gaseous fuel of high market value. This makes the green diesel production more attractive in economic terms while comparing with the production of biodiesel and other production processes [23, 24]. There are two options that can be considered for hydro-processing: co-processing in a distillate hydro-processing unit or building a stand-alone unit. The co-processing route emerges out as an attractive option due to resulting in a lower cost implementation [25].

3.2 Cost Estimation

Sensitivity analysis based on the Net Present Value (NPV) method was performed by the variation of key variables affecting process unit economic performance, such as the feedstock price, unit capacity and the by-product price. To understand the effects of uncertain variables on green diesel production, sensitivity analysis by accounting the Net Present Value (NPV) using impact analysis was reported [26]. The NPV method represents the comparison of the present worth of all cash inflows to the present worth of all cash outflows associated with the investment project and positive NPV implies a profitable investment.

The formula for NPV could be as follows:

$$NPV = -C_0 + \sum_{i=0}^{n} \frac{C_i}{(1+r)^i},$$

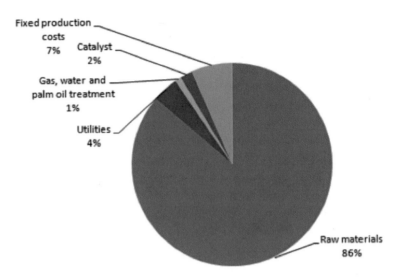

Fig. 7 Operating costs associated with a plant

where C_0 is the total investment, C_i is the net cash flow at the end of period n and r is the minimum acceptable rate of return of 0.08.

Calculations for all costs as well as estimation details and the proportion of production costs are depicted in Fig. 7.

The average cost of green diesel in Indian market (Indiamart.com) is Rs. 51/- excluding GST whereas petroleum diesel cost rises to Rs. 90/- due to the imposition of local taxes (2021). Green diesel's major advantage over FAME is the deoxygenation that occurs during processing, so that green diesel does not display a cleaning effect and should exhibit improved storage stability [21]. It also has a comparable cloud point to diesel, making it compatible with the existing oil pipeline in the opinions of several of its current producers. This potential, coupled with the current lack of extensive research on its compatibility and fungibility, makes green diesel a good candidate for future study. Research, development and demonstration of cost-effective technologies would be essential, especially for the integrated biorefineries that can take advantage of waste streams as throughput feedstocks for cost offsets for the production of different by-products. Toward commercialization of biofuels for reducing fossil fuel usage, both options would have to be aggressively pursued viz. integrating the biofuels with existing petroleum-based refineries via implementation of efficient processing pathways and development of new, integrated biorefineries, preferably colocated plants.

4 Conclusion

From the perspective of production technology, green diesel has a very similar production route from lipids to cracking, hydrotreatment or hydroreforming of wax type heavy products of petroleum distillates. Biological sugar upgradation to green diesel and other biofuels is an environmentally benign process, if the economics is made favorable by process integration. Moreover, these can be synthesized from CO and H_2 by using suitable Fischer Tropsch catalysts. Hydrodeoxygenated products are characterized by a desired viscosity, low or no oxygen content, enhanced atomization and good lubricity. Also, hydrotreatment can be executed in currently existing petroleum refineries with a low capital cost as stated in Eco refining. As compared to biodiesel green diesel has favorable economy in the LCA study. Fossil energy consumption over the life cycle is expected to be reduced by 84–90% for green diesel produced from soybean oil or palm oil, respectively, when H_2 is produced internally from byproducts other than from fossil resources. Thus, green diesel has the potential to displace more petroleum resources per energy content in the fuel compared to biodiesel. Larger reductions in greenhouse gas emissions for green diesel relative to biodiesel were predicted by this study for soybean feedstocks, but lack of verifiable data on palm oil prevented any conclusions to be made for this feedstock. Overcoming this omission and inclusion of other environmental impacts will be the subject of future research in green diesel production and use.

References

1. Kumar P, Varkolu M, Mailaram S, Kunamalla A, Maity SK (2018) Biorefinery polyutilization systems: production of green transportation fuels from biomass. For Chemical and Energy Hubs. Elsevier Inc., In Polygeneration with Polystorage. https://doi.org/10.1016/B978-0-12-813306-4.00012-4
2. Douvartzides SL, Charisiou ND, Papageridis KN, Goula MA (2019). Green diesel: biomass feedstocks, production technologies, catalytic research, fuel properties and performance in compression ignition internal combustion engines. Energies 12(5). https://doi.org/10.3390/en12050809
3. Kalnes, T., Shonnard, D. R., & Marker, T (2007). Green diesel: a second generation biofuel green diesel: a second generation biofuel. Int. J. Chem Reactor Eng. www.eere.energy
4. Vignesh P, Pradeep Kumar AR, Ganesh NS, Jayaseelan V, Sudhakar K (2021) Biodiesel and green diesel generation: an overview. Oil Gas Sci Technol 76. https://doi.org/10.2516/ogst/2020088
5. Maity SK (2015) Opportunities, recent trends and challenges of integrated biorefinery: part i. Renew Sustain Energy Rev 43:1427–1445. https://doi.org/10.1016/j.rser.2014.11.092
6. Maity SK (2015) Opportunities, recent trends and challenges of integrated biorefinery: part II. Renew Sustain Energy Rev 43:1446–1466. https://doi.org/10.1016/j.rser.2014.08.075
7. Zhou J, Xiong Y, Gong Y, Liu X (2017) Analysis of the oxidative degradation of biodiesel blends using FTIR, UV–Vis, TGA and TD-DES methods. Fuel 202:23–28. https://doi.org/10.1016/j.fuel.2017.04.032

8. Ma Y, Gao Z, Wang Q, Liu Y (2018) Biodiesels from microbial oils: opportunity and challenges. Biores Technol 263(May):631–641. https://doi.org/10.1016/j.biortech.2018.05.028

9. Dahiya A (2020) Cutting-edge biofuel conversion technologies to integrate into petroleum-based infrastructure and integrated biorefineries. Elsevier, Bioenergy (Second Edition). https://doi.org/10.1016/b978-0-12-815497-7.00031-2

10. Glisic SB, Pajnik JM, Orlović AM (2016) Process and techno-economic analysis of green diesel production from waste vegetable oil and the comparison with ester type biodiesel production. Appl Energy 170:176–185. https://doi.org/10.1016/j.apenergy.2016.02.102

11. Kittisupakorn P, Sae-ueng S, Suwatthikul A (2016). Optimization of energy consumption in a hydrotreating process for green diesel production from palm oil. In Computer Aided Chemical Engineering, vol. 38. Elsevier Masson SAS. https://doi.org/10.1016/B978-0-444-63428-3.50130-2

12. Arguelles-Arguelles A, Amezcua-Allieri MA, Ramírez-Verduzco LF (2021) Life cycle assessment of green diesel production by hydrodeoxygenation of palm oil. Frontiers in Energy Res 9(July):1–10. https://doi.org/10.3389/fenrg.2021.690725

13. Murti SDS, Yanti FM, Sholihah A, Juwita A R, Prasetyo, J., Thebora, M. E., Pramana, E., & Saputra, H. (2020) Synthesis of green diesel through hydrodeoxygenation reaction of used cooking oil over NiMo/Al2O3 catalyst. AIP Conference Proceedings, 2217(April). https://doi.org/10.1063/5.0000604

14. Mohammad M, Kandaramath Hari T, Yaakob Z, Chandra Sharma Y, Sopian K (2013) Overview on the production of paraffin based-biofuels via catalytic hydrodeoxygenation. Renew Sustain Energy Rev 22(X) 121–132. https://doi.org/10.1016/j.rser.2013.01.026

15. Fernández-Villamil JM, Paniagua AHDM (2018) Preliminary design of the green diesel production process by hydrotreatment of vegetable oils. Eurecha, 15. https://web.fe.up.pt/~fgm/eurecha/scp_2018/eurecha2018_mainreport_1stprize.pdf

16. Neste Oil. NExBTL diesel. (2009). http://www.nesteoil.com/default.asp?path=1,41,11991,12243,12335,12337.

17. Kang A, Lee TS (2015) Converting sugars to biofuels: ethanol and beyond. Bioengineering 2(4):184–203. https://doi.org/10.3390/bioengineering2040184

18. Gousi M, Andriopoulou C, Bourikas K, Ladas S, Sotiriou M, Kordulis C, Lycourghiotis A (2017) Green diesel production over nickel-alumina co-precipitated catalysts. Appl Catal A 536:45–56. https://doi.org/10.1016/j.apcata.2017.02.010

19. Ail SS, Dasappa S (2016) Biomass to liquid transportation fuel via Fischer Tropsch synthesis—Technology review and current scenario. Renew Sustain Energy Rev 58:267–286. https://doi.org/10.1016/j.rser.2015.12.143

20. Mahmoudi H, Mahmoudi M, Doustdar O, Jahangiri H, Tsolakis A, Gu S, LechWyszynski M (2018) A review of Fischer Tropsch synthesis process, mechanism, surface chemistry and catalyst formulation. Biofuels Engineering 2(1):11–31. https://doi.org/10.1515/bfuel-2017-0002

21. Honeywell UOP renewable fuel technology. http://www.uop.com/processingsolutions/renewables/green-diesel/#commercial-production

22. Curran MA (2006). Data from: scientific aapplications international corporation. Life cycle assessment: principles and practice. Cincinnati, OH: U.S. Environmental Protection Agency. (EPA), EPA/600/R-06/060. Available at: https://cfpub.epa.gov/si/si_public_record_report.cfm?Lab_NRMRL&dirEntryId_155087.

23. Kordouli E, Sygellou L, Kordulis C, Bourikas K, Lycourghiotis A (2017) Probing the synergistic ratio of the NiMo/Γ-Al2O3 reduced catalysts for the transformation of natural triglycerides into green diesel. Appl Catal B 209:12–22. https://doi.org/10.1016/j.apcatb.2017.02.045

24. Kordouli E, Kordulis C, Lycourghiotis A, Cole R, Vasudevan PT, Pawelec B, Fierro JLG (2017) HDO activity of carbon-supported Rh, Ni and Mo-Ni catalysts. Molecular Catalysis 441:209–220. https://doi.org/10.1016/j.mcat.2017.08.013

25. Holmgren J, Gosling C, Marker T, Kokayeff P, Faraci G, Perego C (2007) Green diesel production from vegetable oil. 10th Topical Conference on Refinery Processing 2007, Held at the 2007 AIChE Spring National Meeting, September, 61–67
26. Birgisson S. Feasibility study of converting rapeseed to biodiesel for use on a fishing vessel. MSc thesis, Reykjavík Energy Graduate School of Sustainable Systems, REYST Bæjarháls 1, Reykjavík; 2011

Chapter 7
Technological Advancements in the Production of Green Diesel from Biomass

Sudhakara Reddy Yenumala, Baishakhi Sarkhel, and Sunil K. Maity

Abstract Diesel-range hydrocarbons derived from biomass have similar chemical compositions and physicochemical properties with petroleum-derived diesel, known as green diesel. Green diesel also meets the American Society for Testing and Materials (ASTM) specification ASTM D975. Green diesel is thus compatible with existing petroleum pipelines, storage tanks, fueling stations, and diesel engines. Currently, green diesel is produced from non-edible tree-borne oils and lignocellulose biomass. Hydroprocessing of oils and fats, biomass-to-liquid, and a combination of fast pyrolysis and hydrodeoxygenation of bio-oil, are some of the thermochemical processes used to produce green diesel. While commercial technologies are available for producing green diesel from oils and fats, these technologies are suffering from the challenges of the dearth and high cost of feedstock. On the contrary, lignocellulosic biomass is abundant and inexpensive. However, the technologies for converting lignocellulosic biomass to green diesel are in the developing stage. This chapter presents the existing and upcoming hydropyrolysis technologies for converting lignocellulosic biomass to green diesel.

Keywords Hydroprocessing · Green diesel · Pyrolysis · Hydrodeoxygenation · Non-edible oils · And Biomass

S. R. Yenumala (✉) · B. Sarkhel
Thermo-Catalytic Processes Area, Material Resource Efficiency Division, CSIR-Indian Institute of Petroleum, Dehradun 248005, UK
e-mail: sudha.reddy06@gmail.com

S. K. Maity
Department of Chemical Engineering, Indian Institute of Technology Hyderabad, Kandi, Sangareddy, Telangana 502284, India
e-mail: sunil_maity@che.iith.ac.in

© The Author(s), under exclusive license to Springer Nature Singapore Pte Ltd. 2022   219
M. Aslam et al. (eds.), *Green Diesel: An Alternative to Biodiesel and Petrodiesel*,
Advances in Sustainability Science and Technology,
https://doi.org/10.1007/978-981-19-2235-0_7

Abbreviations

CFHP   Catalytic fast hydropyrolysis
CFP   Catalytic fast pyrolysis
FFA   Free fatty acids
F-TS   Fischer–Tropsch synthesis
HDO   Hydrodeoxygenation
LHSV   Liquid hourly space velocity
MMT   Million metric tons
WHSV   Weight hourly space velocity

1   Introduction

Fossil fuels are the principal sources of energy and provide about 80% of the world's energy, which is equivalent to 617 EJ [39]. The distribution of primary energy supply by fossil carbon is shown in Fig. 1. In India, fossil carbon provides around 77% of the primary energy. This excessive dependency on fossil carbon has detrimental consequences on the earth's environment, human health, and sustainable development. Fossil carbon emits greenhouse gases into our environment that will ultimately lead to more than 6 °C global average temperature rise by the end of this century [99]. The annual CO_2 emission is approximately 36.4 Gigatons, with 0.9% rise from 2010–19 [25]. The Paris agreement proposed to hold the global average temperature rise to less than 2 °C above the pre-industrial level [95]. This target can be achieved only by decarbonization, i.e., by developing alternative technologies for harvesting energy from carbon–neutral renewable sources, such as solar, wind, geothermal, biomass, etc.

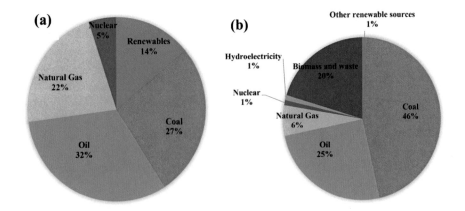

Fig. 1 Primary energy supply in (**a**) the world in 2017 and (**b**) India in 2020 [40, 98]

Transportation fuels (petrol, jet fuel, and diesel) play a crucial role in modern society and are largely obtained from fossil carbons. Further, the demand for transportation fuels is growing rapidly due to the enhanced standard of living, growing world population, and industrialization. In 2021, the transportation sector consumed around 17,653 trillion BTU energy, which was approximately 27% of the world's primary energy consumption [20]. US Energy Information Administration reported that roughly 100 million barrels petroleum and other liquid fuels were consumed per day in the world in 2018 [21]. The consumption of petroleum products in India was approximately 211 million metric tons (MMT) in 2019, with an annual growth rate of 7.9 MMT from 2019 to 2021 [65, 86]. India imports around 350 MMT of fresh carbon to its fuel import pool to meet the increasing energy demand [76]. Sustainable and carbon–neutral sources of transportation fuels are thus necessary to avert dependency on fossil carbon and preserve the clean environment of our planet.

Various renewable energy sources have been emphasized in different parts of the world to meet the future energy demand. However, excluding biomass, all other renewable energy sources are not carbon-based and can merely serve heat and electric energy. On the contrary, biomass is the lone carbon-containing renewable energy source and can provide both transportation fuels and organic chemicals, besides heat and electric energy. The European Renewable Energy Council has projected to substitute 48% of the global energy from renewable sources by 2040 [22]. Among these, biomass will continue to supply 50% of total renewable energy. In 2018, National biofuels policy, India proposed to blend 20% biofuels with existing transportation fuels for sustainable development and to supplement the rapidly increasing energy demand [63].

The transportation fuels obtained from biomass are known as biofuel. Bioethanol, biobutanol, biodiesel, etc., are some of the examples of biofuels. Bioethanol and biobutanol are obtained via microbial fermentation of sugar, which is derived from either sugar, starch, or lignocellulosic biomass. In contrast, biodiesel is obtained from vegetable oils, animal fats, microalgal oils, and used cooking oils by the transesterification reaction with methanol. However, these are oxygen-containing biofuels. The properties of these biofuels thus do not allow their direct use in the existing internal combustion engines. For instance, the unfavorable cold flow properties limit the application of biodiesel in diesel to the extent of 20 vol% only. In contrast, hydrocarbon biofuels have similar chemical compositions, calorific value, fuel mileage, and physicochemical properties with current transportation fuels. These biofuels are thus compatible with existing internal combustion engines and petroleum refinery infrastructures. This approach further reduces the burden of creating capital-intensive new infrastructures. This novel manufacturing concept is known as hydrocarbon biorefinery [59, 60]. The manufacturing of hydrocarbon biofuels has thus been emphasized in recent times. Hydrocarbon biofuels are also referred to as green gasoline, green diesel, and green jet fuel.

Green diesel is currently produced from triglyceride feedstock and lignocellulosic biomass. The catalytic hydrodeoxygenation (HDO) is generally used for the conversion of vegetable oils into green diesel [52, 100, 103] the green diesel is also produced from lignocellulosic biomass by thermochemical conversion processes, such as gasification and fast pyrolysis. However, the gasification of biomass in the presence of air produces synthesis gas with low hydrogen content (8–14 vol%) and is inappropriate for producing synthetic liquid fuels by Fischer–Tropsch synthesis (F-TS) [57]. Fast pyrolysis is a promising technology for thermal disintegration of biomass in the absence of oxygen or in the presence of hydrogen at 623–773 K to produce a liquid product, known as bio-oil. The bio-oil is subsequently subjected to catalytic up-gradation to obtain green diesel. In this context, the present book chapter is focused on the technological advancements in green diesel production by HDO of oils and fats and hydropyrolysis of lignocellulosic biomass.

2 Feedstock for Green Diesel

Biomass is the organic matter obtained from living organisms, such as plants and animals. Biomass is generally three types depending on their chemical structures: triglyceride, lignocellulosic biomass, and starch & sugar biomass. However, in this chapter, the chemical structure of triglycerides and lignocellulosic biomass is briefly described.

2.1 Oils and Fats

Oils and fats are the promising feedstock for the production biodiesel and green diesel owing to their simple chemical structure, consistent chemical composition, and lesser oxygen content than lignocellulosic biomass. The triglycerides are composed of linear C_8–C_{24} fatty acids with mostly C_{16} and C_{18} fatty acids (Fig. 2) [102]. Therefore, the removal of oxygen from triglycerides produces hydrocarbon biofuels in the carbon range of C_8–C_{20}.

Non-edible oils. Non-edible oils are obtained from tree born oil seeds, such as Jatropha (Jatropha curcas), Karanja (Pongamia pinnata), Wild Apricot (Prunus armeniaca), Cheura (Diploknema butyracea), Simarouba (Simarouba glauca), Mahua (Madhuca indica), Kokum (Garcinia indica), Jojoba (Simmondsia chinensis), Neem (Azadirachta indica), Tung (Aleurites fordii), rubber, etc. In India, the production

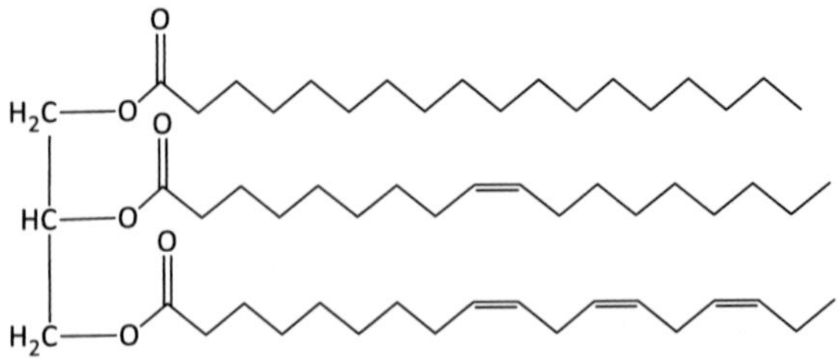

Fig. 2 Structure of triglyceride containing both saturated and unsaturated fatty acids linked to a glycerol backbone

potential of natural non-edible oilseeds (jatropha, karanja, neem, mahua, etc.) is about 20 MMT per annum, of which only 20–25% is consumed currently. These non-edible oils can meet only around 6% of the country's transportation fuels. The government of India thus proposed jatropha cultivation in non-agricultural lands to improve biofuel availability [60]. The jatropha oil has low free fatty acids (FFA) and needs minimal maintenance. In recent times, microalgae have been emerged as a promising source of triglycerides due to their rapid growth rate and high oil content. However, microalgae are facing major economic challenges due to the expensive harvesting and extraction of oil. The physicochemical properties of some of these oils are presented in Table 1.

Animal fats. Animal fats are the lipids obtained from animal organs and secretions. Oil is extracted from the rendered tissue fats from livestock animals, such as aquatic species, pigs, chickens, and cows. The global animal fat market (tallow and lard) is around 10% of oil production, equivalent to nearly 27.1 MMT in 2020, and is anticipated to be 31.8 MMT by 2026 with compounded annual growth rate of 2.8% [70]. Animal fats are used in the production of oleochemical, animal feed, pet food, and other food applications. However, the production and use of animal fats for food applications are restricted in India. Further, the use of animal fats for food applications are declined during COVID-19 due to its origin. The animal fats can be used for the production of renewable diesel because of their lower cost than vegetable oils. Animal fats are classified as tallow, lard, chicken fat, and grease. Tallow is a waste product generated in slaughterhouses from the processing or rendering operations, and lard is extracted from swine slaughter residues [5]. The typical composition of animal fats is shown in Table 2.

Used cooking oils. The global cooking oil production is about 200 MMT, of which around 10% is generated as used cooking oil. The cooking oils vary widely depending on the country. Some examples of cooking oils are oil palm, soybean oil, rapeseed oil, sunflower oil, peanut (groundnut) oil, corn oil, castor oil, mustard oil, linseed oil, etc. Food Safety and Standards Authority of India estimated that India has an annual availability of 2 MMT used cooking oils, considering the generation of 5% and 15% used cooking oil in domestic and industrial usage, respectively [26].

Pretreatment of triglycerides. The composition of used cooking oils varies depending on the type of oils, usage, logistics of collection, storage, location, etc. So, it is very important to pretreat the used cooking oils before processing for biofuels to achieve the desired fuel standards. Further, the tree-borne oils contain gums, phospholipids, metals (sodium, potassium, calcium, and magnesium), and undesirable organic compounds. Microbial oils contain polar lipids, FFA, sterols, terpenes, carotenoids, chlorophyll, and some unidentified compounds. First of all, these impurities consume a large quantity of hydrogen during hydroprocessing due to a high degree of unsaturation. Secondly, they deposit on the active metal surface (phospholipid, sulfolipid, and trace metals), deactivating the catalyst. These impurities also contribute to poor cold flow properties depending on the process and catalyst used. The combinations of degumming, adsorption, acid bleaching, hydrolysis, and FFA distillation are used to remove these impurities from tree-borne oils and microbial oils [12, 49, 55].

Table 1 Physicochemical properties of non-edible oils [2, 28, 49, 81, 91, 92, 104]

	Jatropha	Karanja	Polanga	Rubber	Mahua	Neem	Jojoba	Z. moelleri
Palmitic ($C_{16:0}$)	14.1–15.3	3.7–7.9	12.01–14.6	10.2	16.0–28.2	14.9	3–16	27.3–29.5
Stearic ($C_{18:0}$)	3.7–9.8	2.4–8.9	12.95–19.96	8.7	20.0–25.1	20.6	0.5–6.5	14.4–16.6
Oleic ($C_{18:1}$)	34.3–45.8	44.5–71.3	34.09–37.57	24.6	41.0–51.0	43.9	43.5–66	23.5–25.9
Linoleic ($C_{18:2}$)	29–44.2	10.8–18.3	–	39.6	8.9–13.7	17.9	25.2–34.4	12.8–14.2
Linolenic ($C_{18:3}$)	0.80	–	26.33–38.26	16.3	–	0.4	–	13.3–15.5
Cetane number	52.3	58	–	37	43.5	51	–	
Lipid content, wt%	50	25–40	65–75	50–60	35–50	20–30	40–50	18.1–22
Viscosity 40 °C (mm^2/s)	18.2	4.85	4–5.34	5.81	24.58	20.5–48.5	26.6	–
Flash point, °C	174	180	151–170	130	232	34–285	292	–
Calorific value (MJ/kg)	38.2	34–38	39.25–41.3	36.5	36	33.7–39.5	42.761	–

Table 2 Fatty acid composition of animal fats [5, 24, 70]

Fatty acids	Lard rendered pork fat	Premier jus tallow	Fish fat	Chicken fat	Yellow grease
Palmitic (C16:0)	20–30	20–30	15.97–31.04	24	23.24
Stearic (C18:0)	8–22	15–30	2.79–11.20	5.8	12.96
Palmitoleic (C16:1)	2–4	1–5	1.48–19.61	5.8	3.79
Oleic (C18:1)	35–55	30–45	2.44–28.97	38.2	44.32
Linoleic (C18:2)	4–12	1–6	0.06–3.48	23.8	6.97
Linolenic (C18:3)	<1.5	<1.5	–	1.9	0.67

2.2 Lignocellulosic Biomass

Lignocellulosic biomass is obtained from forest management operations, crop residues, and industrial wastes. The primary components of lignocellulosic biomass are cellulose, hemicellulose, lignin, and extractives. The cellulose, a polysaccharide, is a straight-chain biopolymer of D-glucose bonded by β1,4 glycosidic bond and has around 10,000 degrees of polymerization. The hemicellulose is a branched polymer of pentose and hexose sugars. These sugars are linked by β1,3 glycosidic and β1,4 glycosidic bonds. The lignin is a three-dimensional aromatic biopolymer formed from three different building blocks: sinapyl, coumaryl, and coniferyl alcohol. The composition of lignocellulosic biomass varies depending on environmental conditions, geographical location, soil conditions, crop types, etc. It has an estimated annual supply potential of 97 EJ–147 EJ by 2030 with about 40% share from agricultural residues and waste (37–66 EJ) [41]. The remaining biomass are from energy crops (33–39 EJ), and forest products, including forest residues (24–43 EJ). India has an estimated biomass generation potential of 683 MMT in the Rabi, Kharif, and summer seasons. Approximately 178 MMT surplus biomass is available for value addition. The major crop residues in the kharif season are rice, sugarcane, cotton, and soybean. Wheat, gram, rice, and mustard are the surplus crop residues in the rabi season. The typical composition of selected lignocellulosic biomass is shown in Table 3.

3 Biodiesel

The global biofuels (comprising of bioethanol, biodiesel, and hydrotreated vegetable oils) production was 144 billion liters in 2020, which was equivalent to 2480 thousand

Table 3 Physicochemical characteristics of selected lignocellulosic biomass [10, 16, 38, 42, 64, 83, 105]

	Chemical composition, wt%						Elemental composition, wt%				HHV, MJ/kg
	CL	HC	Lignin	VM	FC	Ash	C	H	N	S	
Rice straw	38	25	12	65.5	15.9	18.7	38.2	5.2	0.9	0.2	14.1–15.1
Wheat straw	30	22	17	82.1	11	6.9	43	5.4	0.0	–	18.0
Cotton	89.7	1.0	2.7	72.8	20.6	6.6	47.0	6.0	1.8	0.19	16.9
Sugarcane bagasse	41.6	25.1	20.3	83.7	13.1	3.2	45.5	6.0	45.2	–	18.7
Soybean	44–8	5–14	–	69.3	14.9	6.3	41.7	6.2	7.1	–	–
Mustard	48.3	29.6	24.6	82.7	6.4	0.9	55.7	4.8	0.4	0.2	17.6

CL = cellulose, HC = hemicellulose, VM = volatile matter, FC = fixed carbon, HHV = higher heating value

barrels per day [59]. Global biofuels production is expected to reach 182 billion liters by 2023–25. Currently, bioethanol is the dominant biofuel in the world and blended with gasoline. At present, 42 billion liters of biodiesel and green diesel are produced with a major share of biodiesel [59]. Biodiesel, fatty acid methyl esters, can be used directly in engines (neat, B100) or blended with diesel (B20, 20 vol% biodiesel with 80 vol% diesel). The presence of oxygen (around 15 wt%) in biodiesel improves the combustion characteristics. The emissions of CO, particulate matters, and hydrocarbons are thus lesser for biodiesel–diesel blends. Indonesia, US, and Brazil are the world's leading biodiesel producers, producing 8.0, 6.5, and 5.9 billion liters in 2019, respectively [87]. Biodiesel is mainly produced from soybean, rapeseed, sunflower, and palm oils worldwide (Fig. 3). The major obstacle for the commercialization of biodiesel is the dearth and high cost of feedstock and food *vs.* fuel issues. The economic analysis shows that the feedstock contributes approximately 70–90% of biodiesel production cost. Non-edible oils are cheap with no food *vs.* fuel conflicts and readily available in developing countries like India, Bangladesh, etc. Non-edible oils are thus gaining importance for biodiesel production. In India, National Biofuel

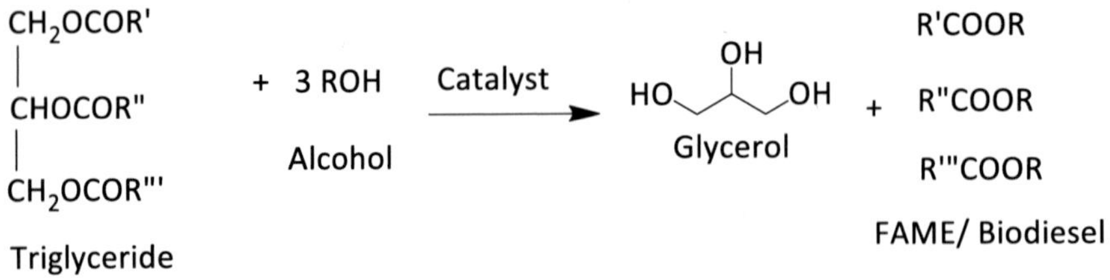

Fig. 3 Transesterification reaction of oils and fats

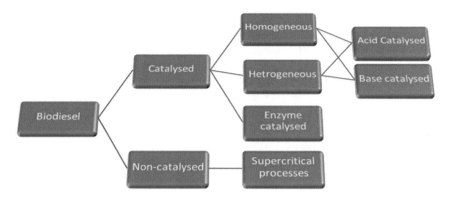

Fig. 4 Classification of biodiesel production process

Policy in 2018 planned to blend 5 vol% biodiesel with diesel by 2030. Currently, India has an installed biodiesel production capacity of 4223 tons per day [26].

The commonly used alcohols in the transesterification reaction are methanol and ethanol. However, the methanol is most frequently used in the transesterification process because of its low cost and ease of separation from biodiesel after the reaction. The transesterification reaction produces glycerol as a by-product and is carried out in the presence of a catalyst. Three different types of catalysts are generally used in this reaction: alkali, acid, and enzyme (Fig. 4). The alkali and acid catalysts are either homogeneous or heterogeneous. The transesterification reaction is also carried out under supercritical conditions in the absence of any catalyst, known as non-catalytic or supercritical transesterification. The commonly used homogeneous alkali catalysts are sodium hydroxide (NaOH) and potassium hydroxide (KOH). The H_2SO_4 and HCl are common homogeneous acid catalysts in the transesterification reaction.

The base-catalyzed transesterification reaction is most frequently used due to the rapid reaction rate, which allows a short reaction time and lesser alcohol to oils ratio. The low alcohol to oil ratio further reduces the cost of separation of excess alcohol from the reaction mixture. However, the base-catalyzed transesterification reaction can not be used for oils with high FFA content as the catalyst reacts with FFA to form soap. The soap formation consumes alkali catalyst and renders difficulty in the separation of biodiesel from glycerol. Similarly, the alcohol used in base-catalyzed transesterification reaction should be anhydrous as water hydrolyzes esters, forming FFA. The oils with high FFA content are thus converted to biodiesel by a two-step transesterification process. In the first step, a predetermined quantity of methanol and sulfuric acid are added to oil depending on the FFA content. In this step, the FFA is converted to fatty acid methyl esters under atmospheric pressure at 60–70 °C. In the second step, a transesterification reaction is carried out by a base catalyst. In contrast, acid-catalyzed transesterification reaction is very slow compared to base-catalyzed transesterification. Therefore, a long reaction time and high alcohol to oils ratio are required to drive the reaction towards completion. The acid catalysts alone are thus rarely used in the transesterification reaction.

However, the processing of feedstock with high FFA content in the presence of homogenous catalysts is facing numerous challenges, such as the complexity of cleaning processes, poor quality of glycerol, loss of yield, and inconsistency in product quality. The heterogeneous and enzyme catalysts or non-catalytic transesterification reactions are thus used to overcome these problems. The heterogeneous transesterification reaction uses solid catalyst materials, such as metal hydroxides, oxides (alkaline earth, transition, and mixed) and complexes, ion exchange resins, sulfated oxides, and carbon. These include calcium, magnesium, and zirconium oxide, zeolites, hydrotalcite, and supported catalysts. However, the heterogeneous transesterification reaction is slow that requires a high reaction temperature (60–200 °C), high alcohol to oils ratio (up to 40), and long reaction time (6–20 h).

Conversely, the reaction completes within 30 min reaction time in non-catalytic or supercritical transesterification. The advantages of supercritical transesterification are the handling of unrefined and high FFA feedstock. Moreover, in supercritical transesterification, the separation of biodiesel from the reaction mixture is quite easy with low cost as the reaction is carried out in the absence of a catalyst. However, the supercritical transesterification is carried out at the critical temperature and pressure of the solvents. The critical temperature and pressure of methanol are 240 °C and 80 bar. Supercritical transesterification is thus highly energy-intensive and involves a high cost of equipment with increased process complexity.

The enzyme-catalyzed transesterification is also insensitive to FFA. However, choice of the enzyme is very important in this reaction. The immobilized lipases from different origins, such as Candida antractica, Mucor miehei, Burkholderia cepacia Novozym 435, and Lipozyme TLIM are most frequently used in transesterification reactions. The activity of enzymes varies based on the water content, concentration of alcohol, and reaction temperature. However, the slow reaction rate and high cost of the enzymes are the main bottlenecks in the commercialization of enzyme-catalyzed transesterification. The biodiesel produced by different processes should meet the various fuel standards, such as IS 15607:2005, ASTM D6751, or EN 14,214.

4 Green Diesel by Hydroprocessing of Oils and Fats

Hydroprocessing refers to transforming oils and fats into long-chain hydrocarbons in the range of green diesel and green jet fuel. The process is carried out in the presence of hydrogen and supported metal catalysts. The process requires moderate temperature (300–350 °C) and high hydrogen pressure (up to 150 bar). The most common catalysts for the hydroprocessing of oils and fats are sulfided NiMo, CoMo, and NiW supported on alumina. The economic analysis further showed that the green diesel produced by hydroprocessing of oils and fats is quite competitive with the current diesel price [58]. For an annual plant capacity of 0.12 MMT of oils, the production cost per kg of green diesel was reported to be US $0.84. The minimum selling price per kg of green diesel was reported to be US $1.19 for 8.5% return on investment and five years as the payback period.

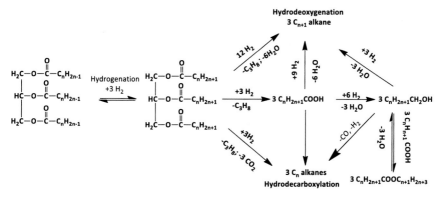

Fig. 5 Reaction mechanism for HDO of oils and fats

4.1 Reaction Mechanism

The hydroprocessing of oils and fats begins with the saturation of the double bonds by hydrogenation reaction over metal sites of the catalyst. The saturated oils are further converted to fatty acids through the monoglycerides and diglycerides intermediates. This reaction step converts glycerol backbone into propane. The resulting fatty acids are further converted to fuel-range hydrocarbons via various oxygen elimination reactions, such as HDO, decarbonylation, decarboxylation, and catalytic cracking (Fig. 5). Decarboxylation and decarbonylation are jointly referred to as deCOx reactions. In deCOx reaction, the oxygen present in the triglycerides is removed as CO_2 (decarboxylation) and CO (decarbonylation). In comparison, the HDO route eliminates oxygen in the form of water. The fuel-range hydrocarbons formed by the HDO reaction have the same number of carbon atoms as in the parent fatty acids [51, 101]. In contrast, deCOx reactions are associated with the loss of one carbon atom in the form of CO/CO_2 from the parent fatty acids. The HDO route dominates over acidic catalysts and bimetallic NiMo and CoMo catalysts, while decarbonylation reaction is predominant over monometallic Ni and Co catalysts. The catalytic cracking is also significant over acidic catalysts, especially at high reaction temperatures.

4.2 Role of Catalysts and Process Conditions

The extent of deoxygenation and product distribution depends on the active metals, metal concentration, acidity, and various process parameters. HDO of sunflower oil was carried out over NiMo/γ-Al$_2$O$_3$, Ni/γ-Al$_2$O$_3$, and Pd/γ-Al$_2$O$_3$ catalysts [37]. While NiMo catalyst followed the HDO pathway, the decarboxylation/decarbonylation route was dominant over Ni and Pd catalysts. HDO of rapeseed oil was carried out over NiMo/Al$_2$O$_3$, Ni/Al$_2$O$_3$, and Mo/Al$_2$O$_3$ catalysts [50]. The NiMo catalyst displayed higher catalytic activity than Ni and Mo catalysts. While

Ni catalyst showed the catalytic activity towards the formation of fatty acids at lower temperatures, Mo catalyst demonstrated the catalytic activity towards esters, aldehydes, and alcohols.

NiMo/alumina showed the higher deoxygenation activity and selectivity towards the formation of octadecane than CoMo/Al$_2$O$_3$ and NiW/SiO$_2$-Al$_2$O$_3$ catalysts [53], whereas PtPd/Al$_2$O$_3$ showed activity towards C$_{15}$ and C$_{17}$ hydrocarbon only. HDO of Jatropha curcas oil was carried out using commercial hydrodesulfurization catalysts, CoMo/Al$_2$O$_3$ and NiMo/Al$_2$O$_3$, under hydrodesulfurization operating conditions [27]. The yield of liquid products was higher over NiMo/Al$_2$O$_3$ catalysts. The effect of co-metal (Ni and Co) doping on MoS$_2$ was studied for this reaction [107]. The Ni-doped MoS$_2$ catalyst was reported to promote HDO reaction, while the Co-promoted MoS$_2$ catalyst favored decarbonylation/decarboxylation reaction. The Ni and Pd catalysts tend to activate decarboxylation and decarbonylation reaction, whereas the CoMoSx and NiMoSx catalysts activate the HDO route for removing oxygen atoms from triglyceride [46]. The monometallic catalysts favored the decarboxylation route, and bimetallic catalysts favored the HDO route [16].

At a relatively low reaction temperature (523 K), the acidity of the catalyst improves the catalytic activity. In comparison, the strong acidity favors C–C bond cracking at high reaction temperature (573 K), decreasing the yield of green diesel [18]. Highly acidic HY and HZSM-5 supported NiMo catalysts produced mainly green gasoline, while NiMo/SiO$_2$-Al$_2$O$_3$ (SiO$_2$/Al$_2$O$_3$ = 8.0) with appropriate acid strength led to proper extent of isomerization/cracking, forming green diesel with cold flow properties comparable to conventional diesel [56]. The shape selectivity of the HZSM-5 zeolite may have prevented the cracking of multi-branched paraffins by limiting the introduction of these molecules into the inner micropores of the zeolite. The HY catalyst promotes the formation of multi branched iso-paraffins that have the advantage of lowering the pour point but the disadvantage of being susceptible to cracking, resulting in a yield loss of diesel-range hydrocarbons [85].

The CoMo/Al$_2$O$_3$ catalyst showed higher olefin yields than NiMo/Al$_2$O$_3$ and NiW/Al$_2$O$_3$, and the trends in isoparaffin yield were as follows: NiMo/B$_2$O$_3$–Al$_2$O$_3$ (B$_2$O$_3$/Al$_2$O$_3$ = 15/85 wt/wt%) > NiMo/Al$_2$O$_3$ [94]. During HDO of sunflower oil, the ratio of C$_{17}$/C$_{18}$ was increased with an increase in reaction temperatures from 573 to 633 K with a simultaneous decrease in the yield of hydrocarbons [48]. The n-C$_{17}$/n-C$_{18}$ ratio was decreased with increasing temperatures over sulfided NiMo/γ-Al$_2$O$_3$ at 70 bar H$_2$. The relative rate of decarboxylation/decarbonylation vs. HDO decreases as the hydrogen pressure increases [37, 47]. For HDO of rapeseed oil over commercial NiMo/alumina catalyst at 633 K and 70 bar, iso-alkanes were observed to increase with increasing reaction temperature and pressure [84]. HDO results of waste cooking oil were compared with fresh cooking oil [7]. The green diesel selectivity did not show any appreciable difference for these two feedstock under the hydroprocessing temperatures.

The process parameters affect the deoxygenation activity and product distribution. Increasing the reaction temperatures improves the formation of iso-paraffin [6]. The cold flow properties of the resulting fuel were observed to improve for two-stage hydroprocessing of waste cooking oil than single-stage hydroprocessing [6]. The free fatty acid and water contents in waste cooking oil improved the yield of green diesel over Ni/Al$_2$O$_3$ catalyst that was more selective to the decarbonylation route [43]. Ni/Meso-Y exhibited a high yield of jet fuel-range alkanes (40.5%) and a low yield of jet fuel-range aromatic hydrocarbons (11.3%) from waste cooking oil at 673 K compared to Ni/HY and Ni/SAPO-34 [54].

HDO of microalgae oil was studied over supported (ZrO$_2$, SiO$_2$, Al$_2$O$_3$, and HBeta) metal (Ni, Pt, Rh, Fe, and NiMo) catalysts. The NiMo catalyst showed lesser selectivity to methane and coke formation compared to Pt and Rh [108]. Peng et al. reported that an increase in reaction temperature and pressure led to a decrease in n-C$_{17}$ and iso-C$_{18}$ yield and an increase in n-C$_{18}$ yield over Ni/HBeta (Si/Al = 180) [73]. Zhou et al. reported the optimum conditions for HDO of microalgae oil: 360 °C, 500 psig hydrogen pressure, 1000 SmL/mL hydrogen/feed ratio, and 1 s residence time [109]. Kandel et al. reported that the decarboxylation or decarboxylation products selectivity over Fe/Meso-SiO$_2$ could be controlled by tuning the degree of oxidation of Fe in the catalyst [45]. The hydrogen consumption was increased in the order HDO > deCO > deCO$_2$. The end product from the hydroprocessing of oils and fats should match the properties of the fuel as per the ASTM or BIS, or EN.

4.3 Commercial Processes

The commercial hydroprocessing technologies for the conversion of oil and fats were developed by Neste NEXBTL, UOP/Eni Ecofining™, MaxFlu™ hydrotreating technology, Bio-Synfining, Vegan® technology HydroFlex™, and UPM BioVerno. The worldwide installed plant capacities of these hydroprocessing technologies are listed in Table 4. The fuel properties of green diesel obtained by different technologies are compared with ASTM D975 and EN 590 specifications, as shown in Table 5.

NEXBTL™ Technology. Neste developed a commercial hydroprocessing technology, called NEXBTL™, to convert vegetable oils and animal fats with high FFA content into green diesel (Fig. 6). The oil is first sent to a distillation column, which is operated under vacuum (0.5–5 kPa). In this distillation column, the FFA is separated as distillate with mono-, di-, and triglycerides as the bottom product. The bottom product further undergoes pretreatment to remove nitrogen, phosphorous, solids, metal (Ca, Na, Mg, Fe, and K) salts, and water. The fatty acid-rich stream undergoes ketoniation reaction using TiO$_2$ catalyst, where oxygen is removed as CO$_2$ and H$_2$O at 340–360 °C, 15–20 bar, and 1.0–1.5 h^{-1} weight hourly space velocity (WHSV) with gas/feed ratio of 0.1–1.5 (w/w). The product stream undergoes mild HDO in a fixed-bed reactor using NiMo/Al$_2$O$_3$ catalyst at 300–330 °C, 40–50 bar, 1.0–2.0 h^{-1} WHSV, and 350–500 Nm^3 H$_2$/m^3 feed. The product stream is sent to the stripper for gas–liquid separation. The liquid stream is sent to the hydroisomerization reactor to yield a renewable base oil and naphtha as a product. The hydroisomerization reaction

Table 4 Various hydroprocessing technologies in the world [8, 17, 32–36, 62, 69, 72, 78, 93, 106]

Technology/company	Feedstock	Location	Capacity, million gallons/Year	Process and catalysts
NEXBTL/ Neste	Vegetable oils and animal fats	Rotterdam, Netherlands	23	• Single-stage hydroprocessing, followed by isomerization • Hydroprocessing catalyst: NiMo/Al$_2$O$_3$ or CoMo/Al$_2$O$_3$ • Isomerization catalyst: Pt/ZSM-22/Al$_2$O$_3$, Pt/SAPO-11/Al$_2$O$_3$, Pt/SAPO-11/SiO$_2$, or Pt/ZSM-23/Al$_2$O$_3$
		Porvoo, Finland	11	
		Singapore	2.8	
Honeywell's Ecofining technology, Diamond green diesel	Vegetable oils, animal fats, and used cooking oils	Norco, Louisiana	675	• Single-stage hydroprocessing, followed by isomerization • Hydroprocessing catalyst: sulfided NiMo, CoMo, NiW, or CoW supported on alumina • Isomerization catalyst: Pt, Pd, Ir, Ru, Rh, and Re supported on zeolites, sulfonated oxides, SAPOs, or micro-mesoporous silica-alumina
Honeywell's Ecofining technology, Diamond green diesel	Vegetable oils, animal fats, and used cooking oils	Port Arthur, Texas	400	
Honeywell's Ecofining technology, Next renewable fuels	White and brown grease, animal fat, soy oil, and variety of vegetable oils	Port Westward, Oregon	575	

HydroFlex™, Haldor Topsoe, HollyFrontier Corp	Soybean oil, corn oil, and animal fats	Artesia, New Mexico	110	• Two-stage hydrotreating and hydroisomerization
				• Hydroprocessing catalyst
				  – First stage: Mo/γ-Al$_2$O$_3$ (TK 335, 337, 339)
				  – Second stage: NiMo/alumina (TK 340, 341, 359)
				• Isomerization catalyst: NiW catalyst (TK- 920, 930, 935, 928, 932)

HydroFlex™, Haldor Topsoe, HollyFrontier Corp		Cheyenne, Wyoming	90	

(continued)

Table 4 (continued)

Technology/company	Feedstock	Location	Capacity, million gallons/Year	Process and catalysts
HydroFlex™, Haldor Topsoe, Grön Fuels LLC		Baton Rouge, Louisiana	990	
HydroFlex™, Haldor Topsoe, Marathon Petroleum		Dickinson, North Dakota	184	
HydroFlex™, Haldor Topsoe, Marathon Petroleum		Martinez, California	736	
HydroFlex™, Haldor Topsoe, Phillips 66		Rodeo, California	680	

Bio-Synfining, dynamic fuels LLC, REG Geismar LLC	High and low free fatty acid feedstock	Geismar, Louisiana	340	• Two-stage hydroprocessing, followed by isomerization
				• Hydroprocessing catalyst
				  – First stage: Mo/γ-Al$_2$O$_3$
				  – Second stage: NiMo/γ-Al$_2$O$_3$
				• Isomerization catalyst: Pt, Pd, and Ni on alumina, fluorided alumina, silica, ferrierite, ZSM-12, ZSM-21, SAPO-11, or SAPO-31

MaxFlux™ hydrotreating technology/ bioflux®, Ryze Renewables	Distiller's corn oil, esters, and fatty acids	Las Vegas, Nevada	100	Single-stage process with dual beds in the reactor

(continued)

Table 4 (continued)

Technology/company	Feedstock	Location	Capacity, million gallons/Year	Process and catalysts
Vegan® technology by Axens	Vegetable oil, animal fats, and algal oil	France	14.3	Single-stage hydroprocessing, followed by isomerization Hydroprocessing catalyst: NiMo/γ-Al$_2$O$_3$ Isomerization catalyst: Pt–Pd/MSA
UPM BioVerno	Crude tall oil	Finland	2.8	Single-stage hydroprocessing, followed by hydrocracking/isomerization

Table 5 Fuel properties of green diesel by different technologies

	ASTM D975	EN 590	A	B	C	D	E
Density at 15 °C, Kg/m³	–	820–845	770–790	780	800–830	–	773–779
Kinematic viscosity, mm²/s	1.9–4.1	2–4.5	3	–	–	1.9–4.1	–
Cetane number	≥40	≥51	>70	70–90	Min 51	Min 65	>75
Sulfur, mg/kg	≤15	≤10	≤5	<1		2	–
Flash point, °C	≥52	>55	>70	–	–	Min 52	–
Cloud point and cold filter plugging point, °C	–	≤−10 to ≤ −34, ≤ −5 to ≤ −44,	−5/ −15/ −22/−34, close to cloud point	−20 to + 20	Max. −16, max −26	–	−20 to 0
Aromatic content, (w/w)%	≤ 35	–	≤1.1	–	Max 5.0	2	–
Heating value, MJ/Kg	–	43.1	44.1	44	–	–	–

A. NEXBTL [67], B. UOP Ecofining™ [44], C. Hydroflex™ [30], D. Bio-Synfining® [77], and E. Vegan™ [61]

is performed in the presence of isomerization catalysts: Pt/SAPO-11/Al₂O₃, Pt/ZSM-22/Al₂O₃, Pt/ZSM-23/Al₂O₃, or Pt/SAPO-11/SiO₂ at 300–350 °C, 25–45 bar, 0.5–1.0 h⁻¹ WHSV, and 300–500 Nm³ H₂/m³ feed. The fatty acid-depleted stream undergoes HDO over NiMo/Al₂O₃ catalyst at about 310 °C, 50 bar, 1.0–1.5 h⁻¹ WHSV, and H₂/oil ratio of 900 Nm³/m³ to yield the deoxygenated product. The paraffin product is then sent for hydroisomerization using a hydroisomerization catalyst at about 300–350 °C, 20–40 bar, and 0.8–1.0 h⁻¹ WHSV to yield renewable diesel and naphtha [66]. The EN590 diesel standard limits the density of winter qualities to a minimum of 800 kg/m³, limiting the blending of Neste Renewable Diesel (Table 5). However, the ASTM D976 standard in the USA and CAN/CGSB 3.517 standard in Canada allow the use of 100% Neste Renewable Diesel [67, 68].

Honeywell UOP Ecofining™ Technology. Honeywell UOP developed the Ecofining™ process to produce green diesel and green jet fuel from vegetable oils and fats. The process consists of a series of steps, such as hydrotreating, deoxygenation, and hydrocracking/hydroisomerization (Fig. 7). The hydrotreating and deoxygenation reactions are performed in a single reactor with two different layers of the

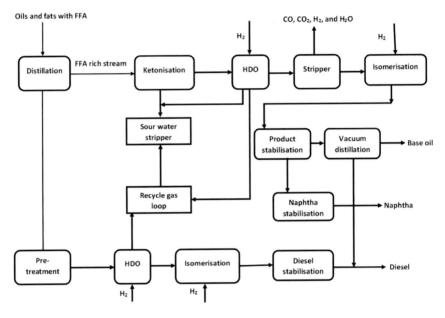

Fig. 6 NEXBTL process for green diesel production

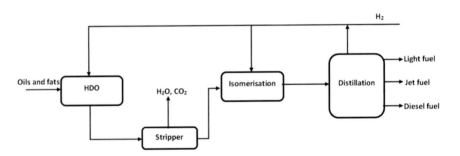

Fig. 7 Honeywell UOP Ecofining™ process for green diesel production [75]

catalyst bed. Hydrotreating reaction happens over Ni and Mo supported on alumina or Pt and Pd supported on Al_2O_3/C at 200–300 °C, 15–50 bar, and 1–4 h^{-1} WHSV. The decarboxylation reaction occurs at 288–345 °C, 50–70 bar, and 1–4 h^{-1} WHSV. The reactor product is sent to a high-pressure stripper to remove gases and water. The linear paraffin obtained from the stripper is further isomerized in a hydroisomerizartion/hydrocracking reactor to obtain the green diesel with improved cold flow properties [9]. The final product is distilled to produce green diesel. The green diesel can be blended in any proportion with petroleum-derived fuels and is suitable as a blending component for EN590 or ASTM 975 diesel (Table 5) [44].

Haldor Topsoe Hydroflex™ Technology. Haldor Topsoe developed the Hydroflex™ technology to manufacture green diesel by co-processing vegetable oils with light gas oil. This process includes two-stage hydroprocessing and isomerization reactor. Hydrocracking is also employed to obtain flexible end products, green diesel or green jet fuel [96]. A mixture of light gas oil and vegetable oils (15–25 vol%) is used as feedstock. The mixture was sent to the HDO reactor loaded with hydrotreating (20 vol% Mo/Al_2O_3) and deoxygenation (80 vol% $NiMo/Al_2O_3$) catalysts. The reaction is carried out at 300–400 °C, 30 bar H_2, 0.1–10 h^{-1} liquid hourly space velocity (LHSV), and H_2/oil ratio of 200–300 Nm^3/m^3. The product from the reactor is sent to the hydrocracking/dewaxing unit to convert n-paraffin into branched isomers. The reaction conditions employed for the conversion of n-paraffin into isoparaffin are 250–400 °C, 60 bar, 0.1–10 h^{-1} LHSV, and H_2/oil ratio of 200–300 Nm^3/m^3 [19]. The Haldor Topsoe process can be adapted in an existing refinery unit with only minor modifications. Several refineries have already planned to expand their facility to convert vegetable oils to green diesel, as listed in Table 4. Haldor Topse further improved their processes and catalysts for converting neat vegetable oils into green diesel by sour and sweet process (Fig. 8). These processes are also flexible to produce either green jet fuel or green diesel. The range of catalysts used in this process is listed in Table 4. In the recent development, Haldor Topsoe developed H2bridgre™ process [31] and integrated it with the Hydroflex process to bridge the required hydrogen (Haldor Topsoe). The green diesel meets the ASTM D975 and EN 590 specifications (Table 5) [30].

Bio-Synfining® technology. The Bio-Synfining® technology involves a two-step hydrotreating of pretreated feedstock in two different beds of catalyst in series (Fig. 9). In the first stage, TK 709 catalyst is used as catalyst, while NiMo, CoMo, and NiW supported on alumina, aluminum phosphate, or silica is employed as the catalyst in the second stage. The pretreated vegetable oil first undergoes partial HDO (first stage), where the fatty acids are converted into paraffin, and triglycerides are

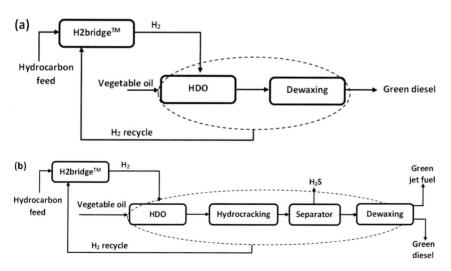

Fig. 8 Hydroflex™ (**a**) sour, (**b**) sweet process [29, 96]

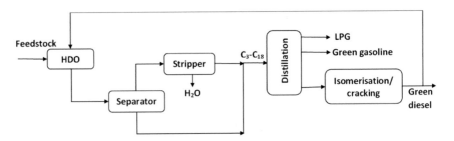

Fig. 9 Bio-Synfining® process [74]

transformed into fatty alcohol and small ester. The product stream from this reactor is cooled and sent to a three-phase separator, where water and gases are separated from hydrocarbons. The partially hydrogenated product from the first stage is sent to the second reactor, where complete deoxygenation is achieved to obtain paraffinic hydrocarbons. The paraffin product is then separated in a distillation column. The C_{15}-C_{18} paraffinic product is subjected to hydrocracking/hydroisomerization to obtain green diesel. Ni, Pd, and Pt dispersed on fluorided alumina, amorphous or crystalline alumina, silica, ZSM-21, ZSM-12, SAPO-31, SAPO-11, or ferrierite, are used as isomerization catalysts. The hydrotreating/HDO is carried out at 371 °C, 83 bar, 0.5–5 h^{-1} LHSV, and H$_2$/feed ratio of 350–2670 m^3/m^3. The hydrocracking is operated at 300–400 °C, 69–140 bar, 0.2–4 h^{-1} LHSV, and H$_2$/feed ratio of 180–1800 m^3/m^3 [1]. This process is developed to obtain green gasoline as the targeted product, with green diesel as a byproduct. The green diesel meets the ASTM D975 specifications (Table 5).

Vegan™ technology. The Vegan™ technology is Axens's HVO technology, in which pretreated vegetable oils are processed in a fixed-bed hydroprocessing reactor, followed by isomerization of paraffinic product, to obtain green diesel (Fig. 10) [106]. In this process, the feedstock first undergoes a pretreatment step to remove the impurities. The pretreated feedstock is then subjected to prehydrotreatment for hydrogenation of the double bonds to avoid undesirable side reactions. The hydrogenation reaction is carried out at 150–200 °C and 1–100 bar over Pd, Pt, NiMo, or CoMo supported on alumina/silica. The product from the prehydrotreatment reactor undergoes hydrotreatment over NiMo/Al$_2$O$_3$ catalyst at 220–310 °C, 10–40 bar, 0.1–10 h^{-1} LHSV, and 150–450 Nm^3/m^3 H$_2$/oil ratio. The gases present in the hydrotreatment product are removed by a stripper. The water is also stripped off from the liquid product before isomerization. The isomerization reaction is carried out at 320–420

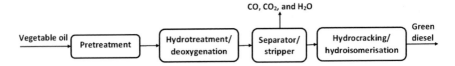

Fig. 10 Simplified block flow diagram of Vegan™ process [82]

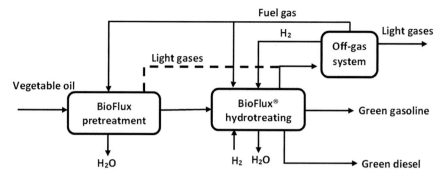

Fig. 11 Bioflux® process [3]

°C, 10–90 bar, 0.5–5 h^{-1} WHSV, and 150–1500 Nm^3/Nm^3 as H$_2$/oil ratio over Pt, Pd, or Pt–Pd/amorphous silica-alumina catalyst [11]. The fuel properties of green diesel, obtained by this technology, matches with ASTM specification, and it can be either blended or directly used in internal combustion engines (Table 5) [4].

Bioflux® technology. The Bioflux® process combines the pretreatment and Maxflux™ hydrotreating technology to obtain green diesel (Fig. 11) [3, 62]. In this process, a thermal degradation technique is used to remove the contaminants present in vegetable oils. The pretreated oils are hydrotreated to remove oxygen. A part of the product stream is recycled to preheat the fresh feed. The product stream undergoes hydroisomerization to convert *n*-paraffinic hydrocarbons into iso-paraffin to meet the desired cold flow properties of green diesel. The green diesel yield is 95–99 wt%, and it meets ASTM D975, EN590, and EN15940 specifications [15].

5 Green Diesel by Hydropyrolsis of Lignocellulosic Biomass

Lignocellulosic biomass can be converted to hydrocarbon biofuels by hydropyrolysis and hydroconversion. Hydropyrolysis is similar to the fast pyrolysis process, in which hydrogen is additionally used as a fluidising agent at slightly high pressure in the presence of a catalyst. The liquid product obtained from hydropyrolysis has low or no oxygen content. Fast hydropyrolysis refers to the thermal treatment of biomass under the hydrogen environment with high heating rates (500 °C/s) [79]. Pyrolysis can be performed either in the presence of a catalyst or without a catalyst. However, the high oxygen content, high acid number, high viscosity, and poor thermal stability restrict the use of pyrolysis oil as transportation fuels. Therefore, the vapors produced from pyrolysis must be treated further using a suitable catalyst for upgrading into hydrocarbon biofuels.

5.1 Mechanism

The cellulose component of biomass first decomposes into furfural, levoglucosan, small aldehydes, ketones, and light gases. However, hemicellulose decomposes more easily than cellulose at a relatively lower temperature. The lignin, which binds cellulose and hemicellulose together, is most difficult to decompose and has a high carbonizing characteristics. The decomposition of lignin produces phenol, catechol, guaiacol, cresol, and other oxygenates. These oxygenates make bio-oil acidic and immiscible with petroleum fractions. The acidity of bio-oil leads to corrosion of commercial process equipment [23]. Therefore, the bio-oils require chemical pretreatment for integrating into a petroleum refinery.

5.2 Catalytic Fast Hydropyrolysis (CFHP)

Catalytic fast pyrolysis (CFP) and HDO of biomass pyrolysis vapor have drawn the wide attention of the scientific community in the last few decades. The catalyst can be incorporated either in the reactor (in-situ CFHP) or downstream of the reactor with a close-coupled vapor upgrading reactor (ex-situ CFHP). The in-situ process is much simpler with a short vapor residence time and gives desired product selectivity. However, the contamination of catalysts by biomass ash can permanently deactivate the catalyst's active site. In the case of ex-situ CFHP, direct contact of catalyst and biomass ash is avoided, which lowers the risk of catalyst poisoning. Another benefit of ex situ CFHP is that the temperature of pyrolysis reactor and catalytic reactor can be operated independently at their optimum temperature [23].

The biomass hydropyrolysis generates two liquid phases: aqueous and organic phases. It contains a mixture of aromatic and aliphatic hydrocarbons, linear ketones, furans, furfurals, and phenols naphthalene, etc., along with char and non-condensable gases [97]. Choice of catalyst, temperature, pressure, WHSV, and time-on-stream are the major factors that impact the liquid yield and quality of product [13, 88, 90]. Tummann et al. reported that temperature and pressure significantly affect char and gas yields. The char yield was declined, and gas yield was enhanced by rising the hydropyrolysis reaction temperature. However, the yield of condensable organics was not affected much. The yield of aqueous phase was enhanced, and the yield of CO and CO_2 was diminished at elevated pressure with negligible impact on the yields of condensable organics and char [90]. Dayton et al. reported optimum C and O content (26.4 mol% C & 6.2 wt% O) in bio-oil at 450 °C. At higher temperatures, the gas yield was increased at the cost of liquid yield. Conversely, at a low temperature, the amount of oxygen in the bio-oil was increased. At higher temperatures, the cracking of pyrolysis vapor leads to CO, CO_2, CH_4, and C_2-C_3 gases. The solid yield was decreased at higher temperatures due to the devolatilization of the biomass [13].

The organic bio-oil yield depends on WHSV slightly, but the yield of solid, gaseous, and aqueous products depends largely on WHSV. With increasing WHSV, the solid and carbon yield in the aqueous phase increases, but gas yield decreases with increasing WHSV. The oxygen content of bio-oil increases with increasing WHSV. At higher WHSV, some of the catalytic sites are fully occupied, which lowers its activity. Overall, it can be concluded that low WHSV is preferred for hydropyrolysis. With increasing biomass feed rate, more amount of pyrolysis vapor is generated, which in turn lowers the residence time. With increasing pressure, it has been observed that the carbon yield of bio-oil increases slightly, but the oxygen content decreases significantly [13].

5.3 Commercial Processes

The major players on hydropyrolysis for conversion of lignocellulosic biomass to hydrocarbon biofuels are RTI International, GTI/IH^2, and Haldor Topsøe. The process conditions and yield and properties of biofuels by these technologies are show in Table 6. IH^2 established the pilot-scale facilities, but commercial plants are yet to be installed. The liquid product obtained from the process must be upgraded to achieve the desired fuel standards.

GTI/IH^2 technology. In GTI/IH^2 process, the biomass is converted to liquid fuels by a two-stage catalytic conversion process (Fig. 12). In the first stage, biomass is fed to the fluidized-bed reactor in the presence of a catalyst under a pressurized hydrogen environment. In the fast hydropyrolysis step, biomass is converted to bio-oil, gas, and

Table 6 Hydropyrolysis of biomass

	IH^2 (GTI) [80]			Haldor Topsoe [89]	RTI [14]
Hydropyrolysis catalyst	CR-4211			Sulfided CoMo	NiMo
Hydroconversion catalyst	CR-4202			Sulfided NiMo	–
Temperature (fluidized-bed/HDO), °C	340–470/ 370–400			450/370	425
Pressure, bar	20–22			24–26	20–22
Biomass rate, g/h	300			250–350	180–250
Feedstock	Bagasse	Corn stover	Microalgae	Beech wood	Loblolly pine
C$_4$ + liquid yield, wt%	28.6	20.6	46.3	21.5	24.1
Oxygen, wt%	<0.5	0.3	0.3	0.0003	2.1
Sulfur, wt%	0.21	0.07	0.03	0.117	–
Density, g/mL at 40 °C	0.81	0.82	0.79	0.827	–
Diesel, wt%	25	30	47	–	–
Gasoline, wt%	75	70	53	–	–

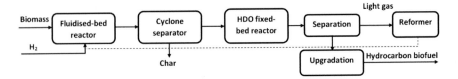

Fig. 12 Schematic of GTI/IH^2 process [80]

char. The hot pyrolysis vapor is sent to a cyclone separator, where the solid char and catalyst particles are separated. Catalyst make-up is also done in the fluidized-bed reactor to maintain a consistent amount of catalysts. In the second stage, the vapor stream, consisting of bio-oil and non-condensable gases, is upgraded in a fixed-bed catalytic reactor using a suitable HDO catalyst, in which oxygen is removed to produce green gasoline and green diesel. The product from the fixed-bed reactor consists of non-condensable gases and liquids (water and organic liquids), which are separated to organic liquid and gases. The light gases obtained from the separator are reformed to meet the hydrogen requirement in the process [80]. The hydropyrolysis reaction is carried at 340–470 °C, 14–23 bar, and 0.5–2.0 h^{-1} WHSV over the CRI 4211 catalyst. The HDO is performed at 370–430 °C over the CRI-4202 catalyst. A maximum of 48 wt% hydrocarbons was observed from microalgae, while only 28.6 wt% hydrocarbons was obtained from woody biomass (Table 6). IH^2 has developed a 5 tons/day demonstration facility at Shell Tech Center, Bangalore [71].

Haldor Topsøe technology. The Haldor Topsøe process is quite similar to the IH^2 process. The CoMo/MgAl$_2$O$_3$ and NiMo/Al$_2$O$_3$ catalysts are used in the hydropyrolysis reactor and pyrolysis vapor upgrading reactor, respectively. The catalysts are sulfided at 350 °C and 26 bar for 3 h using 1.8 mol% H$_2$S and 11 mol% N$_2$ in H$_2$ gas. Biomass enters the fluidized-bed reactor by a screw feeder. In the fluidized-bed reactor, the biomass is converted to gas, liquid, and solid in the presence of a catalyst under the hydrogen environment. The product from this reactor is filtered to separate the solid biochar. The pyrolysis vapor is upgraded by HDO to remove oxygen. The product is sent to a three-stage condensation unit operated at 20 °C, 2 °C, and −40 °C, and the non-condensable gases are flared. The final product consists of only 0.003% of oxygen. The hydropyrolysis (365–511 °C) and HDO (345–400 °C) reactors are operated at 3–36 bar hydrogen pressure [90]. The maximum hydrocarbons yield is 21.5 wt% at fluidization temperature of 450 °C and HDO temperature of 370 °C with 24–26 bar hydrogen pressure [89].

RTI technology. RTI process consists of a single-stage hydropyrolysis, where both biomass decomposition and HDO happen in the fluidized-bed reactor (Fig. 13). In this process, the catalyst is loaded in the fluidized-bed reactor and reduced by hydrogen for 2 h. The biomass particles (180–300 μ) are fed to a fluidized-bed reactor at 375−475 °C and 3–35 bar hydrogen pressure. The solid biochar and catalyst particles are separated from the product by a cyclone separator. The hydropyrolysis vapor is sent to a heat exchanger and impinger to remove the aqueous phase. The outlet of the impinger is connected to an electrostatic precipitator to collect organic liquid [14]. The maximum liquid yield of 24.1 wt% with 2.1 wt% oxygen content

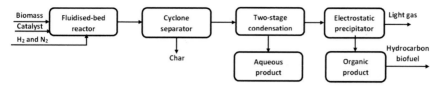

Fig. 13 Schematic of RTI process [14]

is reported at 425 °C and 20.7 bar (40% hydrogen) using hydroprocessing NiMo catalyst.

6 Conclusions

This chapter presents the various commercial technologies for converting lignocellulosic biomass and oils & fats into hydrocarbon biofuels, especially green diesel. Neste NEXBTL, UOP/Eni Ecofining™, MaxFlu™, Bio-Synfining, Vegan® technology HydroFlex™, and UPM BioVerno are some of the commercial technologies for producing green diesel from oils and fats. These technologies employ a combination of catalytic HDO, hydroisomerization, and hydrocracking for removal of oxygen from oils and fats and isomerization of n-parafins to iso-paraffins. The green diesel obtained by these technologies meets the ASTM D975 and EN 590 specifications. The green diesel can be blended or used in pure form in internal combustion engines. Green diesel is also economically competitive with the current diesel price. However, the dearth and high cost of feedstock are the major challenges in these technologies. Hydropyrolysis is an important technology for the disintegration of lignocellulosic biomass into hydrocarbon biofuels in the presence of high hydrogen pressure and catalyst. RTI International, GTI/IH^2, and Haldor Topsøe are the main players in this area. These technologies are, however, in the demonstration stage.

References

1. Abhari R, Tomlinson HL, Roth G (2013) Biorenewable naphtha. US8558042B2
2. Ashraful AM, Masjuki HH, Kalam MA et al (2014) Production and comparison of fuel properties, engine performance, and emission characteristics of biodiesel from various non-edible vegetable oils: a review. Energy Convers Manag 80:202–228. https://doi.org/10.1016/j.enconman.2014.01.037
3. AsianDownstreamInsights (2021) BioFlux®: The solution for challenges of renewable feeds. https://asiandownstreaminsights.com/petrochemicals/bioflux-the-solution-for-challenges-of-renewable-feeds/. Acessed 21 Dec 2021
4. Axens (2021) Renewable Diesel and Jet. https://www.axens.net/markets/renewable-fuels-bio-based-chemicals/renewable-diesel-and-jet. Acessed 21 Dec 2021

5. Banković-Ilić IB, Stojković IJ, Stamenković OS et al (2014) Waste animal fats as feedstocks for biodiesel production. Renew Sustain Energy Rev 32:238–254. https://doi.org/10.1016/J.RSER.2014.01.038

6. Bezergianni S, Dimitriadis A, Chrysikou LP (2014) Quality and sustainability comparison of one-vs. Two-step catalytic hydroprocessing of waste cooking oil. Fuel 118:300–307. https://doi.org/10.1016/j.fuel.2013.10.078

7. Bezergianni S, Voutetakis S, Kalogianni A (2009) Catalytic hydrocracking of fresh and used cooking oil. Ind Eng Chem Res 48:8402–8406. https://doi.org/10.1021/ie900445m

8. Biodiesel Magazine (2021) Renewable Diesel's Rising Tide. http://www.biodieselmagazine.com/articles/2517318/renewable-diesels-rising-tide. Acessed 21 Dec 2021

9. Brady JP, Tom N. Kalnes, Marker TL (2012) US 8198492 B2: production of transportation fuel from renewable feedstocks

10. Channiwala SA, Parikh PP (2002) A unified correlation for estimating HHV of solid, liquid and gaseous fuels. Fuel 81:1051–1063. https://doi.org/10.1016/S0016-2361(01)00131-4

11. Chapus T, Dupassieux N (2011) US 7880043 B2: method of converting feedstocks coming from renewable sources into high-quality gas-oil fuel bases

12. Chen SY, Terananont N, Sonthisawate T et al (2015) A cost-effective acid degumming process produces high-quality Jatropha oil in tropical monsoon climates. Eur J Lipid Sci Technol 117:1079–1087. https://doi.org/10.1002/ejlt.201400293

13. Cross P, Wang K, Weiner J et al (2020) Reactive catalytic fast pyrolysis of biomass over molybdenum oxide catalysts: a parametric study. Energy Fuels 34:4678–4684. https://doi.org/10.1021/acs.energyfuels.0c00320

14. Dayton DC, Hlebak J, Carpenter JR et al (2016) Biomass hydropyrolysis in a fluidized bed reactor. Energy Fuels 30:4879–4887. https://doi.org/10.1021/acs.energyfuels.6b00373

15. Digital Refining (2020) Versatile performance with bioFlux renewable diesel (ERTC). https://www.digitalrefining.com/article/1002540/versatile-performance-with-bioflux-renewable-diesel-ert#.YbctY5FByUk. Acessed 21 Dec 2021

16. Dorez G, Ferry L, Sonnier R et al (2014) Effect of cellulose, hemicellulose and lignin contents on pyrolysis and combustion of natural fibers. J Anal Appl Pyrolysis 107:323–331. https://doi.org/10.1016/j.jaap.2014.03.017

17. Douvartzides SL, Charisiou ND, Papageridis KN et al (2019) Green diesel: biomass feedstocks, production technologies, catalytic research, fuel properties and performance in compression ignition internal combustion engines. Energies 12:1–42. https://doi.org/10.3390/en12050809

18. Duan J, Han J, Sun H, Chen P, Lou H, Zheng X (2012) Diesel-like hydrocarbons obtained by direct hydrodeoxygenation of sunflower oil over Pd/Al-SBA-15 catalysts. Catal Commun 17:76–80. https://doi.org/10.1016/j.catcom.2011.10.009

19. Egeberg RG, Knudsen KG, Blom NJ et al (2014) Hydroconversion process and catalyst. US 8912375:B2

20. EIA (2021) Consumption & efficiency. https://www.eia.gov/consumption/. Acessed 22 Dec 2021

21. EIA (2019) The strait of hormuz is the world's most important oil transit chokepoint. https://www.eia.gov/todayinenergy/detail.php?id=39932. Acessed 22 Dec 2021

22. EREC Rnewable energy scenario to 2040.

23. Eschenbacher A, Fennell P, Jensen AD (2021) A review of recent research on catalytic biomass pyrolysis and low-pressure hydropyrolysis. Energy Fuels 35:18333–18369. https://doi.org/10.1021/acs.energyfuels.1c02793

24. FAO Codex Standard for Fats and Oils from Animal Sources. https://www.fao.org/3/y2774e/y2774e05.htm. Acessed 22 Dec 2021

25. Friedlingstein P, O'Sullivan M, Jones M et al (2020) Global carbon budget 2020. Earth Syst Sci Data Discuss 1–3. https://doi.org/10.5194/essd-2020-286

26. Fssai Repurpose Used Cooking Oil

27. García-Dávila J, Ocaranza-Sánchez E, Rojas-López M et al (2014) Jatropha curcas L. oil hydroconversion over hydrodesulfurization catalysts for biofuel production. Fuel 135:380–386. https://doi.org/10.1016/j.fuel.2014.07.006

28. Ghadge SV, Raheman H (2005) Biodiesel production from mahua (Madhuca indica) oil having high free fatty acids. Biomass Bioenerg 28:601–605. https://doi.org/10.1016/j.biombioe.2004. 11.009
29. Green Car Congress (2020) Grön fuels selects Haldor Topsoe technologies for multi-billion-dollar renewable fuels complex in Louisiana. https://www.greencarcongress.com/2020/12/20201215-topsoe.html. Accessed 22.12.2021
30. Grubb M (2019) HydroFlex ^{TM}—the solution to production of renewable jet fuel and diesel. Present. Adv. Biofuels Conf. Svebio 2019
31. Haldor-Topse H2bridge^{TM}. https://renewables.topsoe.com/h2bridge. Acessed 22 Dec 2021
33. Haldor-Topsoe (2020a) HollyFrontier opts for Topsoe's HydroFlex^{TM} solution to reduce cost of compliance with renewable fuels blending requirements. https://blog.topsoe.com/hollyf rontier-opts-for-topsoes-hydroflex-solution-to-reduce-cost-of-compliance-with-renewable-fuels-blending-requirements. Acessed 22.12.2021
34. Haldor-Topsoe (2020b) Grön Fuels, LLC select Haldor Topsoe's HydroFlex^{TM} renewable fuels and H2bridge^{TM} bio-hydrogen technologies with bio-CCS option for a multi-billion low carbon intensity complex in Louisiana. https://blog.topsoe.com/grön-fuels-llc-select-hal dor-topsoes-hydroflex-renewable-fuels-and-h2bridge-bio-hydrogen-technologies-with-bio-ccs-option-for-a-multi-billion-low-carbon-intensit-1607702386153. Accessed 22.12.2021
32. Haldor-Topsoe (2021a) CVR Energy, Inc. subsidiary selects Haldor Topsoe's HydroFlex^{TM} technology for revamp to renewable diesel production. https://blog.topsoe.com/cvr-energy-inc.-subsidiary-selects-haldor-topsoes-hydroflex-technology-for-revamp-to-renewable-die sel-production. Acessed 22 Dec 2021
35. Haldor-Topsoe (2021b) Haldor Topsoe HydroFlex^{TM} technology results in successful test run at Marathon Petroleum Corp. facility producing 100% renewable diesel. https://blog. topsoe.com/marathon-petroleum-corporation-confirms-successful-test-run-for-us-refinery-producing-100-renewable-diesel-based-on-topsoes-hydroflex-technology. Acessed 22 Dec 2021
36. Haldor-Topsoe (2021c) Topsoe HydroFlex^{TM} being used in renewable diesel production at Phillips 66 refinery. https://blog.topsoe.com/phillips-66-confirms-start-up-of-renewable-die sel-production-using-topsoe-hydroflex. Acessed 22 Dec 2021
37. Harnos S, Onyestyák G, Kalló D (2012) Hydrocarbons from sunflower oil over partly reduced catalysts. React Kinet Mech Catal 106:99–111. https://doi.org/10.1007/s11144-012-0424-6
38. Hung N Van, Maguyon-detras MC, Migo MV, et al (2020) Rice straw overview: availability, properties, and management practices. In: Sustainable Rice Straw Management, pp 1–13
39. IEA (2021) World energy. https://www.iea.org/world. Acessed 22 Dec 2021
40. IEA (2020) Primary energy data: India. https://www.eia.gov/international/analysis/countr y/IND. Acessed 22 Dec 2021
41. IRENA (2014) Global bioenergy supply and demand projections: a working paper for REmap 2030
42. Jacob GA, Prabhakaran SPS, Swaminathan G et al (2022) Thermal kinetic analysis of mustard biomass with equiatomic iron–nickel catalyst and its predictive modeling. Chemosphere 286. https://doi.org/10.1016/j.chemosphere.2021.131901
43. Kaewmeesri R, Srifa A, Itthibenchapong V et al (2015) Deoxygenation of waste chicken fats to green diesel over Ni/Al_2O_3: effect of water and free fatty acid content. Energy Fuels 29:833–840. https://doi.org/10.1021/ef5023362
44. Kalnes TN, Marker T, Shonnard DR, et al (2008) Green diesel production by hydrorefining renewable feedstocks. Biofuels Technol, 7–11
45. Kandel K, Anderegg JW, Nelson NC et al (2014) Supported iron nanoparticles for the hydrodeoxygenation of microalgal oil to green diesel. J Catal 314:142–148. https://doi.org/10.1016/j.jcat.2014.04.009
46. Kim SK, Han JY, Lee HS et al (2014) Production of renewable diesel via catalytic deoxygenation of natural triglycerides: comprehensive understanding of reaction intermediates and hydrocarbons. Appl Energy 116:199–205. https://doi.org/10.1016/j.apenergy.2013.11.062

47. Kovács S, Kasza T, Thernesz A et al (2011) Fuel production by hydrotreating of triglycerides on NiMo/Al$_2$O$_3$/F catalyst. Chem Eng J 176–177:237–243. https://doi.org/10.1016/j. cej.2011.05.110
48. Krár M, Kovács S, Kalló D et al (2010) Fuel purpose hydrotreating of sunflower oil on CoMo/Al$_2$O$_3$ catalyst. Bioresour Technol 101:9287–9293. https://doi.org/10.1016/j.biortech. 2010.06.107
49. Kruger JS, Knoshaug EP, Dong T et al (2021) Catalytic hydroprocessing of single-cell oils to hydrocarbon fuels: converting microbial lipids to fuels is a promising approach to replace fossil fuels. Johnson Matthey Technol Rev 65:227–246. https://doi.org/10.1595/205651321 x16013966874707
50. Kubička D, Kaluža L (2010) Deoxygenation of vegetable oils over sulfided Ni, Mo and NiMo catalysts. Appl Catal A Gen 372:199–208. https://doi.org/10.1016/j.apcata.2009.10.034
51. Kumar P, Maity SK, Shee D (2019) Role of NiMo alloy and Ni species in the performance of NiMo/alumina catalysts for hydrodeoxygenation of stearic acid: a kinetic study. ACS Omega 4:2833–2843. https://doi.org/10.1021/acsomega.8b03592
52. Kumar P, Yenumala SR, Maity SK et al (2014) Kinetics of hydrodeoxygenation of stearic acid using supported nickel catalysts: effects of supports. Appl Catal A Gen 471:28–38. https:// doi.org/10.1016/j.apcata.2013.11.021
53. Kumar R, Rana BS, Tiwari R et al (2010) Hydroprocessing of jatropha oil and its mixtures with gas oil. Green Chem 12:2232–2239. https://doi.org/10.1039/c0gc00204f
54. Li T, Cheng J, Huang R, Zhou J et al (2015) Conversion of waste cooking oil to jet biofuel with nickel-based mesoporous zeolite Y catalyst. Bioresour Technol 197:289–294. https:// doi.org/10.1016/j.biortech.2015.08.115
55. Liu KT, Gao S, Chung TW, Huang C, ming, et al (2012) Effect of process conditions on the removal of phospholipids from Jatropha curcas oil during the degumming process. Chem Eng Res Des 90:1381–1386. https://doi.org/10.1016/j.cherd.2012.01.002
56. Liu Y, Sotelo-Boyás R, Murata K et al (2011) Hydrotreatment of vegetable oils to produce bio-hydrogenated diesel and liquefied petroleum gas fuel over catalysts containing sulfided Ni-Mo and solid acids. Energy Fuels 25:4675–4685. https://doi.org/10.1021/ef200889e
57. Lv PM, Xiong ZH, Chang J, et al (2004) An experimental study on biomass air–steam gasification in a fluidized bed. Bioresour Technol 95:95–101 . https://doi.org/10.1016/j.biortech. 2004.02.003
58. Mailaram S, Maity SK (2019) Techno-economic evaluation of two alternative processes for production of green diesel from karanja oil: a pinch analysis approach. J Renew Sustain Energy 11. https://doi.org/10.1063/1.5078567
59. Maity S, Gayen K, Bhowmick T (2022) Hydrocarbon biorefinery: sustainable processing of biomass for hydrocarbon biofuels. Elsevier
60. Maity SK (2015) Opportunities, recent trends and challenges of integrated biorefinery: Part I. Renew Sustain Energy Rev 43:1427–1445. https://doi.org/10.1016/j.rser.2014.08.075
61. Maniatis K (2019) 2nd EU-INDIA conference on advanced biofuels
62. Michael D. Ackerson, Byars MS (2017) Hydroprocessing method with high liquid mass flux. US 2017/0037325 A1
63. MNRE (2018) National policy on biofuels.
64. Montero G, Coronado MA, Torres R et al (2016) Higher heating value determination of wheat straw from Baja California, Mexico. Energy 109:612–619. https://doi.org/10.1016/J. ENERGY.2016.05.011
65. MoPNG (2020) Energizing India's Progress
66. MYLLYOJA J, NURMI P, KANERVO, TOPPINEN, MAKKONEN, KETTUNEN (2020) Process for the production of renewable base oil, diesel and naphtha. US 2020/0181504 A1
67. Neste (2020) Neste Renewable Diesel Handbook
68. NESTE (2015) High performance from NEXBTL renewable diesel
69. NEXT Renewable fuels (2021) Frequently asked questions. https://nextrenewables.com/faq/. Acessed 22 Dec 2021

70. Özogul Y, Özogul F, Çi˙çek E, et al (2009) Fat content and fatty acid compositions of 34 marine water fish species from the mediterranean sea. Int J Food Sci Nutr 60:464–475. https://doi. org/10.1080/09637480701838175
71. Paggio A Del (2019) Shell in advanced biofuels
72. Pekka K (2010) Integrated process for producing desel fuel from biological material and products, uses and equipment relating to said process. US 2010/0317903 A1
73. Peng B, Yao Y, Zhao C et al (2012) Towards quantitative conversion of microalgae oil to diesel-range alkanes with bifunctional catalysts. Angew Chemie Int Ed 51:2072–2075. https://doi. org/10.1002/anie.201106243
74. Pyl SP, Schietekat CM, Reyniers MF et al (2011) Biomass to olefins: cracking of renewable naphtha. Chem Eng J 176–177:178–187. https://doi.org/10.1016/j.cej.2011.04.062
75. Ray A, Anumakonda A (2011) Production of green liquid hydrocarbon fuels. In: Biofuels. Elsevier, pp 587–608
76. Ray A, Sinha AK, Atray N, Kumar S (2019) Converting indigenous waste carbon sources to useful products. Spec issue Environ 205:204
77. Renewable Energy Group (2019) REG renewable diesel fact sheet
78. Renewable Energy Group (2021) Geismar Biorefinery. https://www.regi.com/find-fuel/pro duction-facilities/geismar. Acessed 22 Dec 2021
79. Resende FLP (2016) Recent advances on fast hydropyrolysis of biomass. Catal Today 269:148–155. https://doi.org/10.1016/j.cattod.2016.01.004
80. Roberts M, Marker TL (2012) Biomass to gasoline and diesel using integrated hydropyrolysis and hydroconversion
81. Sánchez M, Avhad MR, Marchetti JM et al (2016) Jojoba oil: a state of the art review and future prospects. Energy Convers Manag 129:293–304. https://doi.org/10.1016/j.enconman. 2016.10.038
82. Scharff Y, Asteris D, Fédou S (2013) Catalyst technology for biofuel production: conversion of renewable lipids into biojet and biodiesel. OCL—oilseeds fats, crop lipids 20:2–5. https:// doi.org/10.1051/ocl/2013023
83. Şensöz S, Kaynar İ (2006) Bio-oil production from soybean (Glycine max L.); fuel properties of Bio-oil. Ind Crops Prod 23:99–105. https://doi.org/10.1016/j.indcrop.2005.04.005
84. Šimáček P, Kubička D, Šebor G et al (2010) Fuel properties of hydroprocessed rapeseed oil. Fuel 89:611–615. https://doi.org/10.1016/j.fuel.2009.09.017
85. Sotelo-boy R, Liu Y, Minowa T (2011) Renewable diesel production from the hydrotreating of rapeseed oil with Pt / zeolite and NiMo/Al$_2$O$_3$ catalysts. Ind Eng Chem Res 50:2791–2799. https://doi.org/10.1021/ie100824d
86. Statista (2021a) Consumption volume of petroleum products in India from financial year 2012 to 2020, with an estimate for 2021. https://www.statista.com/statistics/715241/india-consum ption-volume-of-petroleum-products/. Acessed 22 Dec 2021
87. Statista (2021b) Leading biodiesel producers worldwide in 2019, by country. https://www. statista.com/statistics/271472/biodiesel-production-in-selected-countries/. Acessed 22 Dec 2021
88. Stummann MZ, Hansen AB, Hansen LP et al (2019a) Catalytic hydropyrolysis of biomass using molybdenum sulfide based catalyst. Effect of Promoters. Energy and Fuels 33:1302–1313. https://doi.org/10.1021/acs.energyfuels.8b04191
89. Stummann MZ, Høj M, Hansen AB et al (2019b) New insights into the effect of pressure on catalytic hydropyrolysis of biomass. Fuel Process Technol 193:392–403. https://doi.org/10. 1016/j.fuproc.2019.05.037
90. Stummann MZ, Høj M, Schandel CB et al (2018) Hydrogen assisted catalytic biomass pyrolysis. Effect of temperature and pressure. Biomass Bioenerg 115:97–107. https://doi.org/10. 1016/j.biombioe.2018.04.012
91. Swami SB, Thakor NJ, Haldankar PM et al (2012) Jackfruit and its many functional components as related to human health: a review. Compr Rev Food Sci Food Saf 11:565–576. https:// doi.org/10.1111/j.1541-4337.2012.00210.x

92. Takase M, Zhao T, Zhang M et al (2015) An expatiate review of neem, jatropha, rubber and karanja as multipurpose non-edible biodiesel resources and comparison of their fuel, engine and emission properties. Renew Sustain Energy Rev 43:495–520. https://doi.org/10.1016/j.rser.2014.11.049

93. TheDigest (2017) The renewing of Nevada renewable fuels and the rise of Ryze renewables. https://www.biofuelsdigest.com/bdigest/2017/08/03/the-renewing-of-nevada-renewable-fuels-and-the-the-rise-of-ryze-renewables/. Acessed 22 Dec 2021

94. Toba M, Abe Y, Kuramochi H et al (2011) Hydrodeoxygenation of waste vegetable oil over sulfide catalysts. Catal Today 164:533–537. https://doi.org/10.1016/j.cattod.2010.11.049

95. United Nations climate change (2015) Paris agreement

96. Verdier S, Alkilde OF, Chopra R, Gabrielsen J, Grubb M (2018) Hydroprocessing of renewable feedstocks—challenges and solutions. Topsoe white Pap

97. Wang K, Dayton DC, Peters JE et al (2017) Reactive catalytic fast pyrolysis of biomass to produce high-quality bio-crude. Green Chem 19:3243–3251. https://doi.org/10.1039/c7gc01088e

98. World Bioenergy Association (2019) Global bioenergy statistics 2019

99. WorldResourceInstitute (2019) Six ways the climate changed over the past decade. https://www.wri.org/insights/6-ways-climate-changed-over-past-decade. Acessed 22 Dec 2021

100. Yenumala SR, Kumar P, Maity SK et al (2020) Hydrodeoxygenation of karanja oil using ordered mesoporous nickel-alumina composite catalysts. Catal Today 348. https://doi.org/10.1016/j.cattod.2019.08.040

101. Yenumala SR, Kumar P, Maity SK, et al (2019) Production of green diesel from karanja oil (Pongamia pinnata) using mesoporous NiMo-alumina composite catalysts. Bioresour Technol Reports 7:100288

102. Yenumala SR, Maity SK (2011) Reforming of vegetable oil for production of hydrogen: a thermodynamic analysis. Int J Hydrogen Energy 36:11666–11675. https://doi.org/10.1016/j.ijhydene.2011.06.055

103. Yenumala SR, Maity SK, Shee D (2016a) Reaction mechanism and kinetic modeling for the hydrodeoxygenation of triglycerides over alumina supported nickel catalyst. React Kinet Mech Catal 120:109–128. https://doi.org/10.1007/s11144-016-1098-2

104. Yenumala SR, Maity SK, Shee D (2016b) Hydrodeoxygenation of karanja oil over supported nickel catalysts: influence of support and nickel loading. Catal Sci Technol 6:3156–3165. https://doi.org/10.1039/x0xx00000x

105. Yu ZT, Xu X, Hu YC et al (2011) Unsteady natural convection heat transfer from a heated horizontal circular cylinder to its air-filled coaxial triangular enclosure. Fuel 90:1128–1132. https://doi.org/10.1016/j.fuel.2010.11.031

106. Zhang B, Wu JJ, Yang C et al (2018) Recent developments in commercial processes for refining bio-feedstocks to renewable diesel. Bioenergy Res 11:689–702. https://doi.org/10.1007/s12155-018-9927-y

107. Zhang H, Lin H, Zheng Y (2014) The role of cobalt and nickel in deoxygenation of vegetable oils. Appl Catal B Environ 160–161:415–422. https://doi.org/10.1016/j.apcatb.2014.05.043

108. Zhou L, Lawal A (2016) Hydrodeoxygenation of microalgae oil to green diesel over Pt, Rh and presulfided NiMo catalysts. Catal Sci Technol 6:1442–1454. https://doi.org/10.1039/C5CY01307K

109. Zhou L, Lawal A (2015) Evaluation of presulfided NiMo/γ-Al$_2$O$_3$ for hydrodeoxygenation of microalgae oil to produce green diesel. Energy Fuels 29:262–272. https://doi.org/10.1021/ef502258q

Chapter 8
Characterization of Green Diesel: Existing Standards and Beyond

Uplabdhi Tyagi, Mohammad Aslam, and Anil K. Sarma

Abstract Green diesel is a second-generation biofuel exhibiting comparable physicochemical structure as petroleum diesel but better diesel properties and is considered as an alternative energy product. It is produced from the hydro-processing of fatty acids in a temperature–pressure range of 600–700 °F and 400–100 atm pressure. Green diesel is made of linear and branched pure paraffin in different proportions, and due to its chemical composition and significant variation in the degree of isomerization occurs, it is considered an effective biocomponent for blending into mineral diesel. However, the emissions from diesel causing sustainability issues and contribute towards severe health and environmental impact. Therefore, the appropriate fuel properties and their characterization are essential to maintain the quality of green diesel and to meet the desired requirements at affordable, feasible, and variable process cost, thus facilitating the integration with existing petroleum refineries. The quality of green diesel is determined by its high heating value, energy density, cold flow properties, and high cetane number. The low density and aromatic content of green diesel are the promising features that make it a good blending material as other petroleum diesel blends in the refineries. Quantification and analysis of organic and elemental carbon fractions are performed by a thermal-optical transmission analysis method, acid value of the green diesel is measured by AOCS method, other fuel properties including viscosity, iodine value, flash point, and copper strip corrosion are measured by the ASTM D6751 and EN 14,214 standard protocols. The properties of vegetable oil-derived green diesel such as loading facilities, storage, and logistics are similar to fossil diesel but have a very high cetane number. Application standards

U. Tyagi · A. K. Sarma (✉)
Chemical Conversion Division, Sardar Swaran Singh National Institute of Bio-Energy,
Kapurthala, Punjab, India
e-mail: anil.sarma@nibe.res.in

U. Tyagi
e-mail: uplabdhityagi200@gmail.com

M. Aslam
Department of Chemistry, National Institute of Technology, Srinagar (J&K), India
e-mail: maslam@nitsri.ac.in

© The Author(s), under exclusive license to Springer Nature Singapore Pte Ltd. 2022   249
M. Aslam et al. (eds.), *Green Diesel: An Alternative to Biodiesel and Petrodiesel*,
Advances in Sustainability Science and Technology,
https://doi.org/10.1007/978-981-19-2235-0_8

of green diesel in CI engine with respect to efficiency parameters, emission patterns as per EURO norms, etc. are some additional aspects that have been addressed.

Keywords Vegetable oils · Green diesel · Fuel properties · ASTM standards · EN Standards

1 Introduction

A growing global population is driving an exponential increase in the demand for energy. There has been an increase of 17-fold in global energy consumption in the last century. Fossil fuel combustion is a primary source of atmospheric pollution due to the emissions of CO_2, SO_2, and NOx. As a result of rise in the crude oil prices, dwindling of petroleum resources and health risks associated with fossil fuel emissions, the outlook and motivation towards creating and using cleaner and renewable sources of energy are growing. A growth in the population and national economy directly affects the production and consumption of energy. Globally, the biofuel industry is experiencing tremendous growth due to several factors including geopolitical issues, oil price fluctuations, and capital interests. According to the International Energy Agency (IEA), conventional biofuels should be able to replace up to 6% of petroleum fuels using the conventional fuels developed for the United States and the European Union, assuming adequate cropland availability [1]. In the EU, producing ethanol requires 5% of the available cropland, whereas in the USA, it requires 8%. To produce 5% of diesel fuel, USA cropland must account for 13% and EU cropland must account for 15%. The variations in the statistics are due to the shortage of reserves (crude oil, coal and natural gas). The requirement for national energy security, and the unfavorable environmental and climatic consequences of conventional utilization of fossil fuels. Green Diesel is also known as "renewable diesel", "second-generation diesel", "bio-hydrogenated diesel", "hydrogenation-derived renewable diesel", "hydrotreated vegetable oil", or "hydrogenated vegetable oil". Green diesel is composed of straight chain and branched saturated hydrocarbons with a carbon atom content ranging from 15 to 18 per molecule (C15 to C18). Figure 1 depicts the production of green fuels from different feedstocks and processes. In addition to green diesel, recently biomass-based diesel has also been produced from fish and animal fats while the terminology HVO is frequently used in the industries and in the fuel standards and European regulations [2, 3]. This composition is similar to that of fossil petroleum diesel; therefore, CI engines can utilize green diesel in pure form or blended according to desired blending ratios without modifying the engine. Table 1 summarizes the detailed comparison of biodiesel and HVO or green diesel. Unlike petroleum diesel, green diesel is a product of biological origin that releases less CO_2 into the atmosphere than biodiesel, and its combustion is cleaner than petroleum diesel due to its lack of aromatics and naphthenes. While biodiesel contains oxygen, green diesel, on the other hand, does not, and therefore have greater

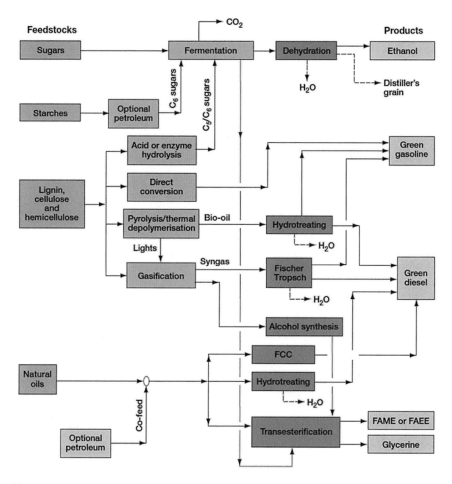

Fig. 1 Production of green fuels from different feedstocks and processes

stability, noncorrosivity, and a similar heating value to petroleum diesel [4]. Additionally, green diesel is better at withstanding cold weather conditions than biodiesel due to its lower emission of NOx and higher cetane number (CN), making it easier to ignite in CI engines. Green diesel produced via hydroprocessing of triglycerides is a byproduct of propane, a gas that holds high market value [5, 6]. As a consequence, the production of green diesel is more economical than the biodiesel production. In general, four technologies can be used to produce green diesel from biomass (i) thermal conversion and upgradation of bio-oil, (ii) hydroprocessing, (iii) biomass to liquid process, (iv) catalytic upgradation of sugars, starch, and alcohols. The hydro-processing of oils and fats involves the catalytic processing of hydrogen to achieve saturated hydrocarbons from triglycerides. The majority of urban carbonaceous aerosol is composed of diesel engine particles, which typically contribute over 70% of the total particulate elemental carbon. Recent epidemiological studies have

Table 1 Detailed comparison of biodiesel with HVO/Green diesel

Biodiesel	HVO/Green diesel
More denser and viscous	Less dense and viscous
Blends well with petrodiesel	Blends well with petrodiesel
Utilizes acids or alkalis as catalyst and requires high strength materials of construction	Utilizes catalyst supports and normal hydrocrackers
High oxygen content	Less oxygen content
Algal oil or vegetable oil can have different properties depending on the feedstock	As a result of hydrocracking, algae oil may be converted into green diesel using a variety of feedstocks and properties may vary with feedstocks
Low-pressure production	High temperature–pressure production
Further development of the technology is possible	Hydrocracking reactions can be carried out using synthesis gas
Good lubricity	A lubricant or another compound must be added to improve lubricity
A new process for transalkylation of petroleum refinery units	The hydrocracking process is easy to integrate with petroleum refineries, which are well acquainted and accustomed to this process
Utilizes coal, natural gas, biomass, petroleum, and derived alcohols	Utilizes biomass/petroleum/coal-derived hydrogen
Widely popular	Emerging as a potential alternative or as a blender of petrodiesel
Different from petrodiesel in terms of chemical composition	Similar in chemical composition with petrodiesel
Low calorific value	High calorific value
Alcohols may be used to improve properties	Produced from cracking or catalytic cracking of vegetable/algal oil under mild conditions
Pretreatment is required for the removal of fatty acids	Produces water, ammonia, and hydrogen sulfide
Marketing challenges for glycerol	Uses propane or other smaller hydrocarbons
Emerges due to the development of heterogeneous catalysts	Emerges due to the development of carbon-based catalysts
Less but sufficient cetane number	High but sufficient cetane number
Good pour point	Low pour point
Moderate emission performance	Good emission performance

established a correlation between fine atmospheric particles and excess morbidity and mortality [7]. These studies have made fine atmospheric particles a growing interest. Both global climate and local air quality are affected by diesel particles, in addition to public health. Carbonaceous particles play a key role in many chemical processes in the atmosphere due to their catalytic properties and high specific surface area. Solid elemental carbon (EC) forms a layer on diesel particles, which are believed to consist of volatile organic carbon (OC) [8, 9]. As a result, the core usually shows a highly agglomerated chainlike structure, containing roughly spherical primary particles with diameters ranging from 10 to 50 nm. Additionally, diesel particles contain ions that can dissolve in water, such as sulfate derived from fuel, and trace elements due to the use of organometallic additives in lubricating oil as well as engine abrasion [10, 11]. Engine design and exhaust after treatment can reduce diesel engine emissions. In addition to reducing carbon monoxide and gaseous hydrocarbons, an oxidation catalytic converter can also reduce the organic fraction of particles in diesel exhaust. Over the course of the next few decades, countless researchers claimed that vegetable oil could be used as an alternative fuel for CI engines [12]. In spite of this, the high availability and superior characteristics of petroleum products created an obstacle to the use of vegetable oils. As a result of the Kyoto Protocol (1991), the Paris Convention (2015), and the rapid depletion of global petroleum reserves, there has been a flurry of interest in the development of alternative fuel sources that would produce less pollution and provide some renewable energy. The biofuel industry has the potential to address these environmental concerns [13, 14]. Due to its similar physicochemical and fuel properties, biofuel can be synthesized in a number of ways and is a possible aspirant to petroleum. A simple lipid such as straight vegetable oil (SVO), especially the non-edible version, can be a valuable feedstock because it has a high level of availability, a renewable nature, and is environmentally friendly. However, high product viscosity results in the prolonged combustion time, low injection pressure and poor fuel atomization as compared to petrodiesel [15]. Furthermore, high density, low calorific value, and low cetane number prohibit its direct application to CI engines. Vegetable oil is prone to fouling injectors, stricken oil rings, gum formation decreased lubricant life, and carbon deposits on injector orifices as outlined above. Straight chain vegetable oil is, therefore, considered unworkable for CI engines and is generally considered unacceptable. Therefore, the challenges associated with the physicochemical properties of straight chain vegetable oil, several techniques such as preheating before the injection of straight chain vegetable oil, pyrolysis, mixing of straight chain vegetable oil with petrodiesel, micro emulsification of straight chain vegetable oil with suitable surfactant and viscosity modifier, transesterification and hydrotreatment of straight chain vegetable oil plays an important role during the production of green diesel [16].

By analyzing and characterizing green diesel properly as summarized in Table 2, it offers several advantages over other petroleum products, such as the fact that it does not exhibit a cleaning effect and exhibits enhanced stability in storage. Moreover, green diesel has a cloud point that is similar to diesel but its pour point is lower, making it highly compatible with existing oil pipelines. This makes green diesel an attractive alternative energy product.

The following properties significantly affect the quality of green diesel.

Cetane Number

Cetane number, an index number, defines the quality of ignition of a diesel fuel and can be calculated according to Eq. 1. Fuels with high cetane numbers ignite more

Table 2 Summary of green diesel fuel properties

	Mineral ultra low sulfur diesel	Bio-diesel	Green diesel
Cetane number	40	50 to 65	70 to 90
Sulfur content (PPM)	<10	<1	<1
Cloud point, °C	23 °F	23 to 59 °F	14 to 68 °F
Specific gravity	0.84	0.88	0.78
Distillation, °F	392 to 662 °F	644 to 671 °F	509 to 608 °F
Heating value, MJ/kg	44	39	45
Stability	Good	Medium	Good

Benefits of green diesel blending

Diesel-pool components	**Barrels in pool**	**Cetane index**
Green diesel	2346	74
Hydrotreated straight run diesel	7500	52
Blended product cetane	NA	50
Kerosene	500	41
Hydrotreated LCO	2000	20
	Base case	**W/green diesel**
Euro IV diesel yield + jet fuel	Base	+30%
Green diesel percentage in diesel pool	0	5.75
Refinery margin, $/bbl crude	Base	+0.6
Capacity of refinery, kBPD	150	150

readily. The ignition delay is not only a measure of speed but also of the delay in the ignition. Upon injection into a diesel engine, it measures how fast the fuel can self-ignite. It is the volume percent of hexadecane in a blend of hexadecane and 1-methylnaphthalene that has characteristics similar to those of a diesel fuel to be tested. Cetane index between 40 and 55 is the range at which diesel engines operate well. Since different vegetable oils are considered as feedstocks, the cetane number of biodiesel is not fixed. Biodiesel is produced by using various kinds of plant oils, including palm oil, peanut oil, karanja oil, soybean oil, jatropha oil, sunflower oil, and rapeseed oil [17]. The cetane number is in the range of 68–94.8 and for animal fat, it ranges between 56 and 60. Cetane number of biodiesel is determined by a number of factors, including the number of C atoms in a long C chain, unresolved bonds, and ester yield. Renewable diesels (RD) are n-paraffinic hydrocarbons that do not contain sulfur or oxygen and ranges from 80 to 99%. Renewable diesels have a high cetane number due to the absence of sulfur, allowing the engine to run at higher speed. In addition to the length of the hydrocarbon chain, the cetane number of the renewable diesels depends on the number of short and isomerized hydrocarbon chains. The renewable diesels produced low HC emissions at low load and with high cetane number. This also contributes to a shorter ignition lag thus results in a lower temperature and a reduced NOx output in the end. Due to the high cetane index of renewable diesels, diesel engine hydrocarbons emissions and combustion noise were lower at low loads than diesel engines [18]. A high load at 200 N-m resulted in a similar ignition delay to the diesel. As the temperature of the combustion chamber increases at high loads, the difference between the ignition of diesel and renewable diesel is reduced due to the increasing inside temperature, thereby reducing HC. Additionally, the nucleation and accumulation modes resulted in a decline in accumulation particle size, and an increase in nucleation particle size accounting biodiesel blends ranging from 10 to 100% in volume. This may be due to the fact that biodiesel contains oxygen atom that lowers the particulate matter (PM) size. In contrast, due to high cetane number, renewable diesels show decline in the "nucleation mode" and "accumulation mode" of the particulate matter for all blend variations. There are diesel fuels available in some markets with higher cetane numbers, normally "premium" diesel fuels with additional cleaners and some synthetics. When burned, diesel fuel has a heating value of 43.1 MJ/kg, compared to 43.2 MJ/kg for gasoline and about 86.1% of diesel fuel mass is made up of carbon. However, diesel fuels offer high energy density, i.e. EN590 diesel fuel density: 0.820 to 0.845 kg/L, which is more than the density of EN 228 gasoline, i.e. 0.720–0.775 kg/L at 15 °C. Also, emissions of carbon dioxide from diesel fuel are 73.25 g/MJ, which is lower than gasoline, i.e. 73.38 g/MJ [19]. Petroleum can be refined into diesel fuel much more easily than gasoline and contains hydrocarbons with boiling points ranging between 180 °C and 360 °C. In order to remove sulfur, more refining often has to be done, which contributes to a higher price. Some areas of the United States as well as the United Kingdom and Australia have higher diesel fuel prices than petrol. The reasons for this may include several factors, including the shutdown of some refineries in the Gulf of Mexico, the diversion of mass refining capacity to gasoline production, and the recent switch from high-sulfur diesel to ultra-low-sulfur diesel (ULSD). In Sweden, a diesel fuel with a lower aromatic content is also available; it is designated as MK-1 (class 1 environmental diesel) and is slightly more expensive than regular ULSD [20]. Moreover, compared to petrol fuel taxes, diesel fuel taxes in Germany are about 28% lower.

$$CetaneNumber = \text{Low cetane number}$$
$$+ \frac{(\text{Difference of two known cetane number} * \text{difference of hand wheel reading between known low cetane and the sample})}{\text{Difference of hand wheel reading of high and low cetane}}$$

Density

Density is a quantity that matches the amount of matter an object has to its volume. The density of a substance is temperature dependent. Some papers measure the density of diesel at 40 °C and others at 15 °C. Density is closely related to some fuel characteristics, primarily cetane number, arene structure, viscosity, boiling point, and volatility of the fuel. With the exception of a few substances, the density and volume decrease when the temperature increases. In general, biodiesel made from different types of oils such as jatropha oil, rapeseed oil, palm oil and mahua oil having the density between 880 and 890 kg/m^3 while the density for renewable diesel ranges between 775 and 816 kg/m^3. In accordance with EN590, the density of renewable diesel is lower than that of diesel fuel and biodiesel. This results in a pour speed that is about 4–5% faster than diesel or biodiesel. The Bernoullis equation can be used to explain this phenomenon, which states that the flow rate is inversely proportional to the square root of the density of the fluid. Renewable diesel fuel has a lower density than biodiesel fuel, so a given quantity of BD fuel will give a smaller volume of combustion in the CI engine [21]. As a consequence, it also shows that the renewable diesel has less energy that accounts for less power and torque. If a constant volume injection system is used to conduct experiments, then variations in the fuel density will be experienced, which, in turn, will affect the variation of the energy content. The amount of energy contained in fuel is determined by its mass infused into the cylinder. It is necessary to increase the fuel quantity to compensate for this loss, which is approximately 3–6% more than diesel, resulting in greater fuel consumption. Fuel injection systems might need to be calibrated due to lower density in order to obtain maximum benefit from renewable diesel [22]. The density of the renewable diesel can be increased to 820 kg/m^3 for summer grade fuel and 800 kg/m^3 for winter-grade fuel by using renewable diesel as a fuel blend in the ratio of 30:70% according to EN590.

Cloud point (CP)

Knowing the cloud point of your fuel is crucial for understanding its working condition in lower temperatures regions. CP is the temperature of crystallization at which fuel behaves like a cloud when the temperature is lowered. In addition to thickening the oil, the crystals block the fuel filter and may even cause injector damage. As a result, diesel engines are difficult to operate. Transesterification and hydroprocessing can further reduce the cloud point of each vegetable oil. The biodiesel can be easily operational up to − 5 °C. RD has a lower temperature range than biodiesel, which fluctuates between −30 °C and −5 °C. Isomerization can further reduce the cloud point of the renewable diesel to −55 °C and at the same time cetane number decreases with an increase in cloud point. The flow properties of isomerized or short HC chains are superior to those of long HC chains [23]. Renewable diesel better operation ability is due to its chemical structure that contains saturated bonds, while in biodiesel, the presence of saturated fatty acids declines the cloud point. Besides the pour point (CP), other properties also determine whether a fuel is capable of being used at low temperatures, such as the cold flow pour point (CFPP). The pour point shows the free pouring of the liquid fuel at low temperature while the CFPP indicates the oil flow through the filter at low temperature in an allocated residence time.

Kinematic viscosity

The viscosity is a measure of friction between adjacent layers of flowing matter such as liquids and gases, which resist motion. Viscous fluids provide resistance between two adjacent layers, for instance, toothpaste is more viscous than oil. Biodiesel has a viscosity ranging from 4 to 4.8 mm²/s, while RD has a viscosity ranging from 3 to 4 mm²/s, the same as petro-diesel (3.6 mm²/s). The suitable fuel viscosity is highly essential for the appropriate design of an engine fuel injector. Due to the higher viscosity than the standard value, biodiesel blends have better lubrication, but insufficient flow and poor atomization due to their viscosity. As a result, large size vapors are generated on injection, resulting in improper mixing, black smoke, and incomplete combustion [24]. Moreover, due to the high viscosity and low volatile characteristics of biodiesel than conventional diesel fuel, exhaust emissions such as particulate matter, hydrocarbons, and carbon monoxide also increase with the rapid consumption of biodiesel as shown in Fig. 2. In addition, under similar process conditions, renewable diesel possesses low emissions on New European Driving Cycle. Besides this, low viscosity leads to malformation of the fuel pump that enhances the probability of leakage within the pump and insufficient supply of fuel and hence destroys the engine capability and performance.

Oxygen content/stability

The presence of oxygen in the molecular arrangement of a fuel plays a crucial role in determining its ability to last a long time. FAME is the biofuel, which contains oxygen atoms and is aromatic. In contrast, RD is a saturated aliphatic straight HC chain with no oxygen atoms, similar to petro-diesel. The presence of oxygen in any

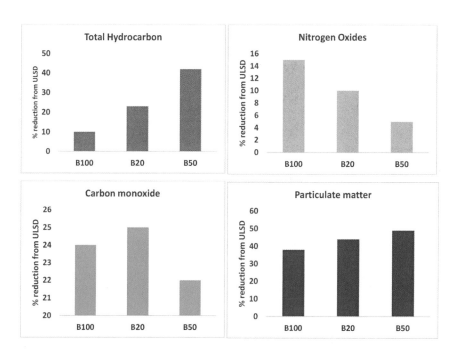

Fig. 2 Percentage of emission reduction estimate for B20, B50, and B100 fuel. *B20 = Blended fuel with 20% biodiesel and 80% petroleum diesel, *B50 = Blended fuel with 50% biodiesel and 50% petroleum diesel, *B100 = 100% biodiesel, *ULSD = Ultra low sulfur diesel

fuel makes the molecule unstable, since oxygen is prone to reacting hence biodiesel's stability is therefore very challenging. In RD, there are no oxygen molecules, while in biodiesel, the oxygen content ranges from 6 to 12%. Biodiesel has the advantage of being more stable and long-lasting than RD. When biodiesel is stored for an extended period of time, it is prone to oxygen deterioration because of the presence of several fatty acids with double bonds in the oil. Various properties of the fuel can be affected by oxygen stability. Biodiesel must meet several standards in order to meet the criteria of EN14112 at 110 °C. According to ASTM D 6751, EN14213, and EN14214 standards, the oxidation stability is reported for minimum 3, 4, and 6 h, respectively. Various antioxidants can also increase oxidation stability, including tertbutylated hydroxy toluene, pyrogallol, tert-butylated hydroxyanisole, and propyl galate. Among all these, propyl galate is found to be more effective. In order to maintain long-term stability, biodiesel must be protected against oxygen, light, and water vapors in the air. Following is the order for the removal of heteroatoms by the hydroprocessing technique; hydrodesulfurization > hydrodeoxygenation > hydro-denitrogenation [25]. In contrast, using vanadium nitride as a catalyst, the affinity for HDO was almost 10 times more in comparison to HDS while huge amount of hydrogen is needed at higher temperature of around 350 °C for HDO. Utilization of noble metals as catalysts for hydroprocessing is more effective as it easily activates the hydrogen molecule at mild temperature and pressure. Several noble metals such as Rh, Pd, Ru, Pt, and Zr show prominent results during hydroprocessing. HDO usually is used in conjunction with noble metal catalysts; the hydrogen is then adsorbed and activated on the surface of the noble metal and further split. In either case, the oxygen-carrying compounds are activated by noble metals or metal supports. By reacting with oxygen, the hydrogen splits, breaking the C–O bond and creating water.

Flash point

Flash point is the lowest temperature at which an organic compound gives off enough vapors to ignite on contact with a flame. It provides a clearer understanding of the fuel's flammability properties, as well as its volatility. A fuel's flash point is very important, as it enables an operator to take precautions when handling it. There is a definite vapor pressure for all fluids, which is a function of the temperature of the fluid and is governed by Boyle's Law. Whenever a fuel's temperature rises, vapor pressure also rises, resulting in an increase in vapor concentration. This shows that temperature plays a significant role in the development of vapor concentrations [26]. In a flash point situation, the amount of vapors produced is insufficient to sustain combustion. The fuel will only ignite once it comes in contact with a source of ignition or a flame. Fuel is more volatile at lower flash points, and that's why they are inversely related. According to ASTM D9751, different vegetable oils have flash points ranging from 210 °C to 290 °C, making them safe. It is considered easier and safer to handle waste cooking oil and canola oil as feedstocks. In order to reduce the flash point of BD-diesel blends and avoid prolonged ignition delays, alcohols such as methanol or ethanol should be used as additives. However, lower carbon chain alcohols reduce the flash point to even lower levels than the ASTM D9751 standard of 52 °C. This is extremely hazardous and poses a fire hazard. As a result, higher C-chain alcohols are preferred to avoid this risk. Biodiesel and renewable diesel flash points range from 96 to 188 °C and 52 to 136 °C, respectively. The range is due to the variations in the properties of vegetable oils used as a feedstock. The flash point of diesel is 76 °C. Based on these values, it appears that the range of renewable diesel is lower than that of biodiesel, indicating less and short ignition delay and suitable combustion with renewable diesel as compared with biodiesel [2, 27] (Table 3).

Calorific value (CV)

Calorific value corresponds to the amount of energy in any substance or heat released during combustion. The calorific value of BD ranges between 33.6 and 43.2 MJ/kg as compared to the renewable diesel, which is 42–47 MJ/kg. Since the oil contains oxygen atoms, it reduces calorific value by about 10–12%. On the other hand, renewable diesel removes the oxygen by hydroprocessing, which results in a higher calorific value than biodiesel. As the calorific value increases, more energy is released for work and less fuel is used thus producing more brake thermal efficiency. The calorific value of renewable diesel is comparable to that of petrodiesel. Table 4 summarizes the different properties of some vegetable oil. Literature reports that rapeseed oil exhibit the highest calorific value, i.e. 39.95 MJ/kg as compared to others including castor oil, waste cooking oil and jatropha oil. Also, the Crambe possesses the highest calorific value, i.e. 40.6 MJ/kg followed by diesel, i.e. 45.58 MJ/Kg and renewable diesel, i.e. 46.83 MJ/Kg. The calorific value for biodiesel is 37 MJ/Kg as it contains a high percentage of oxygen, its molecular structure makes it very light [28]. In addition, calorific value waste date pits derived renewable diesel using two different catalysts, i.e. Pd/Carbon and Pt/Carbon are 44.1 and 43.8 MJ/kg, respectively, which is lower than conventional diesel but higher than biodiesel. This indicates that higher heating value of renewable diesel makes it a better fuel having the fined and comprehensive properties including cetane index, flash point, and kinematic viscosity. From future aspects, a fuel with high cetane index and calorific value among all the organic liquid fractions can be good cetane index improver. From this context, renewable diesel offers similar properties and hence can be used as a cetane index improver. In contrast, density, flash point, and viscosity of green diesel are found to be the least. Green diesel has the highest calorific value because hydrogenation of glyceride lipids leads to deoxygenation and production of hydrocarbons. Biodiesel contains oxygenated functional groups while green diesel appears to have superior fuel properties. A further investigation in this area is needed, considering the wide variations in lipid compositions from algae and vegetable oils (edible or not edible), the cost of hydrogen and methanol, as well as the severity of the conditions employed in the catalytic hydrogenation reactors. Due to the possibility of wide variations in the chemical composition of algal oils and vegetable oils, it would be advisable to be cautious regarding these variations [29, 30]. There may be benefits to blending biodiesel with green diesel and using them in some appropriate ratios. The fuel characteristics of biodiesel and green diesel show great differences, with biodiesel provides lubricity, green diesel provides more energy and has a higher calorific value [31]. In cold climates, green diesel may not perform well due to its higher pour

Table 3 Detailed comparison of properties of Algal biodiesel, Karanja biodiesel, and HVO/green diesel

Parameter	EURO-IV Diesel specifications	Algal biodiesel	Karanja biodiesel	HVO (Green diesel)
Acid number (mg KOH/g)	NA	0.4	0.43	33.3
Distillation (°C)	NA	NA	394	265–320
Oxygen content (wt%)	NA	NA	12.8	0
Kinematic viscosity at 40 °C (mmb/s)	2.1–4.5	3.6–5.5	4.45	2.5–3.5
Flash point, min. (°C)	35–66	98	168	120–138
Density (kg/mc)	820–850	800	880	785
Cloud point (°C)	NA	− 15 to 2	12	− 5 to −30
Calorific value (MJ/kg)	NA	40	37.98	44
Pour point, maximum (°C)	3.0–15.0	−14	5	9
Cetane number, minimum	50	53	51	80–90
Sulfur content, maximum (mg/kg)	50	54	20	<Detection limit
Lubricity (HFRR), maximum (μm)	460	NA	NA	360

point. Biodiesel may emit fewer pollutants than conventional diesel hence offering better emission performance. The pour point of biodiesel is also good. It may also be necessary to perform a detailed life cycle analysis of both biodiesel and green diesel for better understanding. Additionally, there is a need to study the fuel characteristics of CVO-diesel since this involves milder ambient pressure conditions for the production of said fuel with no use of hydrogen and alcohols.

2 Conclusion

The trend of alternative fuels is increasing in the present scenario. As far as their sustainability is concerned, they replace petrol–diesel in transportation. Several important properties were also analyzed, which led to the following points:

(1)   In accordance with ASTM D 975, RD fuel has physicochemical properties.
(2)   The high calorific value of RD fuel results in low fuel consumption.

Table 4 Properties of different vegetable oils

Vegetable oil	Cetane number (°C)	Cloud point (°C)	Pour point (°C)	Heating value (MJ/kg) approx	Kinematic viscosity at 38 °C mm^2/s	Flash point (°C)	Density (kg/l)
Sesame	40	−3.9	−9.4	39	36	260	0.913
Peanut	42	12.8	−6.7	40	40	271	0.903
Sunflower	37	7.2	−15	40	34	274	0.916
Linseed	35	1.7	−15	39	27	241	0.924
Soybean	38	−3.9	−12.2	40	32	254	0.914
Cottonseed	42	1.7	−15	40	34	234	0.915
Palm	42	31.0	NA	NA	40	267	0.918
Rapeseed	38	−3.9	−31.7	40	37	246	0.912
Corn	38	−1.1	−40	40	35	277	0.909
Diesel	50	NA	−16	44	3.1	76	0.855

(3) Decarboxylation does not require hydrogen, but it keeps the catalyst in operation. The degree of unsaturation of the feedstock determines catalytic performance.

(4) In renewable diesel, longer H-C chains have a high cetane value and shorter or isomerized H-C chains have better cold flow properties. Up to 20% of blending of fossil fuel can improve the cold flow properties of renewable diesel.

(5) As a result of the oxygen present in its molecular structure, BD readily decomposes, while renewable diesel has greater stability due to the presence of no oxygen in the molecular structure and a saturated HC chain.

(6) Since RD has a greater viscosity than BD, it produces a better atomization and more power.

(7) RD has better combustion than BD as a result of its higher flash point, and thus has a shorter ignition delay between when fuel is injected and when combustion begins.

Lubricating properties of biodiesel are excellent. In comparison with regular diesel, green diesel is higher in calorific value and lower in oxygen content. According to the fuel characteristics, green diesel is a good fuel, except for its pour point, which may need additives. Using both biodiesel and green diesel may be the best option. Combining the two biofuels would complement their properties and help in the biosequestration of CO_2 and the renewable nature of both fuels. There is plenty of scope for further research in establishing biorefineries and establishing a bioeconomy in this area.

References

1. Baiju B, Naik MK, Das LM (2009) A comparative evaluation of compression ignition engine characteristics using methyl and ethyl esters of Karanja oil. Renew Energy 34:1616–1621
2. Biswas S, Mohanty P, Sharma DK (2014) Studies on co-cracking of jatropha oil with bagasse to obtain liquid, gaseous product and char. Renew Energy 63:308–316
3. Biswas S, Sharma DK (2013) Studies on cracking of Jatropha oil. J Anal Appl Pyrol 99:122–29; Biswas S, Sharma DK (2014) Effect of different catalysts on the cracking of Jatropha oil. J Anal Appl Pyrol 110:346–352
4. Deeba F, Kumar V, Gautam K, Saxena RK, Sharma DK (2012) Bioprocessing of jatropha curcas seed oil and deoiled seed hulls for the production of biodiesel and biogas. Biomass Bioenerg 40:13–18
5. Demirbas A, Dincer K (2008) Sustainable green diesel: a futuristic view. Energy Sources, Part A: Recovery, Utilization, and Environmental Effects 30:1233–1241
6. Dhar A, Agarwal AK (2014) Effect of karanja biodiesel blend on engine wear in a diesel engine. Fuel 134:81–89
7. Gautam K, Gupta NC, Sharma DK (2014) Physical characterization and comparison of biodiesel produced from edible and non-edible oils of Madhuca indica (mahua), Pongamia pinnata (karanja), and Sesamum indicum (til) plant oilseeds. Biomass Conversion and Biorefinery 4:193–200
8. Gautam K, Pareek A, Sharma DK (2015) Exploiting microalgae and macroalgae for production of biofuels and biosequestration of carbon dioxide—a review. Int J Green Energy 12:1122–1143
9. Kham-Or P, Suwannasom P, Ruangviriyachai C (2016) Effect of agglomerated NiMo HZSM-5 catalyst for the hydrocracking reaction of jatropha curcas oil. Energy Sources, Part A: Recovery, Utilization, and Environmental Effects 38:3694–3701
10. Kiss AA, Dimian AC, Rothenberg G (2006) Solid acid catalysts for biodiesel production—towards sustainable energy. Adv Synth Catal 348:75–81
11. Shirazi Y, Viamajala S, Varanasi S (2016) High-yield production of fuel- and oleochemical precursors from triacylglycerols in a novel continuous-flow pyrolysis reactor. Appl Energy 179:755–764
12. Yigezu ZD, Muthukumar K (2015) Biofuel production by catalytic cracking of sunflower oil using vanadium pentoxide. J Anal Appl Pyrol 112:341–347
13. Wisniewski A, Wosniak L, Scharf DR, Wiggers VR, Meier HF, Simionatto EL (2015) Upgrade of biofuels obtained from waste fish oil pyrolysis by reactive distillation. J Braz Chem Soc 26:224–232
14. Vu HX, Schneider M, Bentrup U, Dang TT, Phan BMQ, Nguyen DA et al (2015) Hierarchical ZSM-5 materials for an enhanced formation of gasoline-range hydrocarbons and light olefins in catalytic cracking of triglyceride-rich biomass. Ind Eng Chem Res 54:1773–1782
15. Zhao X, Wei L, Cheng S, Cao Y, Julson J, Gu Z (2015) Catalytic cracking of carinata oil for hydrocarbon biofuel over fresh and regenerated Zn/Na-ZSM-5. Appl Catal A 507:44–55
16. Yigezu ZD, Muthukumar K (2014) Catalytic cracking of vegetable oil with metal oxides for biofuel production. Energy Convers Manage 84:326–333
17. Abbasov V, Mammadova T, Andrushenko N, Hasankhanova N, Lvov Y, Abdullayev E (2014) Halloysite-Y-zeolite blends as novel mesoporous catalysts for the cracking of waste vegetable oils with vacuum gasoil. Fuel 117:552–555
18. Weber B, Stadlbauer EA, Stengl S, Hossain M, Frank A, Steffens D et al (2012) Production of hydrocarbons from fatty acids and animal fat in the presence of water and sodium carbonate: reactor performance and fuel properties. Fuel 94:262–269
19. Ramya G, Sudhakar R, Joice JAI, Ramakrishnan R, Sivakumar T (2012) Liquid hydrocarbon fuels from jatropha oil through catalytic cracking technology using AlMCM- 41/ZSM-5 composite catalysts. Appl. Catalysis a-General 433:170–178
20. Nimkarde MR, Vaidya PD (2016) Toward diesel production from karanja oil hydrotreating over CoMo and NiMo catalysts. Energy Fuels 30:3107–3112

21. Porwal J, Bangwal D, Garg MO, Kaul S (2012) Reactive-extraction of Pongamia seeds for biodiesel production. J Sci Ind Res 1:822–828
22. Prathima A, Karthikeyan S (2017) Characteristics of micro-algal biofuel from botryococcus braunii. Energy Sources, Part A: Recovery, Utilization, and Environmental Effects 39:206–212
23. Rajesh M, Sau M, Malhotra RK, Sharma DK (2015) Hydrotreating of Gas Oil, Jatropha oil, and their blends using a carbon supported cobalt-molybdenum catalyst. Pet Sci Technol 33:1653–1659
24. Rajesh M, Sau M, Malhotra RK, Sharma DK (2016) Synthesis and characterization of Ni-Mo catalyst using Jatropha curcas leaves as carbon support and its catalytic activity for hydrotreating of gas oil, Jatropha oil, and their blends. Pet Sci Technol 34:240–246
25. Rajesh M, Sau M, Malhotra RK, Sharma DK (2016) Synthesis and characterization of Ni-Mo catalyst using pea pod (Pisum sativum L) as carbon support and its hydrotreating potential for gas oil, Jatropha oil, and their blends. Pet Sci Technol 34:394–400
26. Sharma DK (2015) Emerging biomass conversion technologies for obtaining value-added chemicals and fuels from biomass. Proc Indian Natl Sci Acad 81:755–764
27. Sharma R, Sharma DK (2010) Emerging trends in the sequestration of CO2–Role of geomicro-biology, biosequestration, knowledge management and industrial approaches. J Appl Geochem 12:520–534
28. Vijayaraghavan K, Hemanathan K (2009) Biodiesel production from freshwater algae. Energy Fuels 23:5448–5453
29. Wang WC, Thapaliya N, Campos A, Stikeleather LF, Roberts WL (2012) Hydrocarbon fuels from vegetable oils via hydrolysis and thermo-catalytic decarboxylation. Fuel 95:622–629
30. Yenumala SR, Maity SK, Shee D (2016) Hydrodeoxygenation of karanja oil over supported nickel catalysts: Influence of support and nickel loading. Catal Sci Technol 6:3156–3165
31. Zarchin R, Rabaev M, Vidruk-Nehemya R, Landau MV, Herskowitz M (2015) Hydroprocessing of soybean oil on nickel-phosphide supported catalysts. Fuel 139:684–691

Chapter 9
Current Status of the Green Diesel Industry

Mohammad Aslam, Himansh Kumar, Anil K. Sarma, and Pramod Kumar

Abstract In the post-COVID-19 era, sustainability is a crucial aspect that includes natural resource management, environmental impact reduction, and socioeconomic issues. Due to the depletion of world petroleum reserves and the greenhouse gas emissions connected with their usage, it has become more evident that continuing reliance on fossil fuel energy resources is unsustainable. As a result, there are active research efforts underway to produce alternative renewable and carbon–neutral biofuels as alternative energy sources. Because of the easy access to feedstock, decreased emissions, and unusually high cetane number, green diesel is becoming increasingly popular in academics and research. This chapter focuses on the background of biodiesel and green diesel fuels and the current status of the industry, and provides insight into the future of green diesel in the world. This chapter mainly deals with biodiesel/green diesel production routes, catalysts employed for the production of green diesel, factors affecting the green diesel production process, fuel properties of different diesel fuels. It further addresses the potential production of biodiesel/green diesel in the world and examines the status and contribution of diesel fuels derived from vegetable oils and animal fats. It was observed that biofuels could progressively substitute a significant proportion of the fossil fuels required to meet the growing energy demands.

M. Aslam (✉)
Department of Chemistry, National Institute of Technology, Srinagar (J&K), India
e-mail: maslam@nitsri.ac.in

H. Kumar
Department of Mechanical Engineering, Teerthanker Mahaveer University, Moradabad, U.P., India
e-mail: himansh.rmu@gmail.com

H. Kumar · P. Kumar
Department of Mechanical Engineering, Dr. B R Ambedkar National Institute of Technology, Jalandhar, Punjab, India
e-mail: kushwahapramod@nitj.ac.in

H. Kumar · A. K. Sarma
Chemical Conversion Division, Sardar Swaran Singh-National Institute of Bio-Energy, Kapurthala, Punjab, India
e-mail: anil.sarma@nibe.res.in

Keywords Green diesel · Biodiesel · Hydroprocessing · Fuel properties · Industry scenario

1 Introduction

By 2050, the world's population is predicted to be over 10 billion people. Global population growth has resulted in a constant and ever-increasing demand for energy. The COVID-19 epidemic, on the other hand, has had a huge impact on worldwide energy production and consumption forecasts. Global liquid fuel production and consumption, as well as carbon emissions, are depicted in Figs. 1 and 2. The biggest source of concern is the carbon emissions produced by the usage of fossil fuels, as well as their negative influence on the environment. Due to COVID-19 pandemic, national and international transit restrictions, and government-imposed societal lockdowns, the lowest carbon emissions were recorded in 2020. As a result, biofuels derived from renewable resources are clearly a viable option for fulfilling future energy demands while reducing carbon emissions [1–5].

Biodiesel is the most widely used liquid biofuel in the transportation business (also known as fatty acid methyl esters or FAME). According to ASTM D6751, biodiesel is "a fuel comprised of mono-alkyl esters of long-chain fatty acids generated from vegetable oils or animal fats." Biodiesel is referred to as FAME (fatty acid methyl ester) or RME in Europe (rapeseed methyl ester). In Europe, RME is the most extensively used biofuel. Biodiesel is an alternative fuel made from a variety of feedstocks, including raw vegetable oil (sunflower, rapeseed, soy, jatropha, pongammia pinnata,

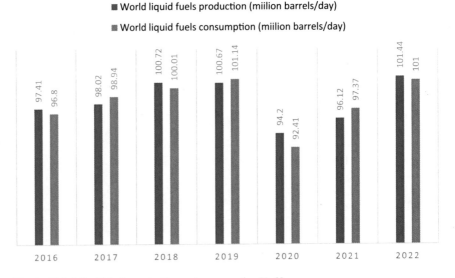

Fig. 1 Global liquid fuels production and consumption [1, 2]

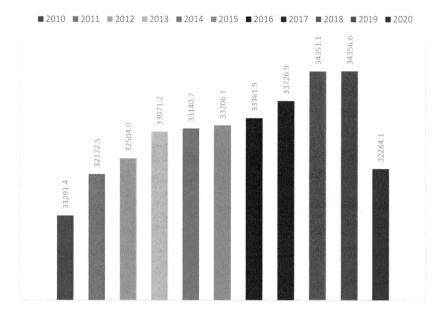

Fig. 2 Global CO_2 emissions (in million tonnes) [1, 2]

etc.), animal oils/fats (tallow, lard, yellow grease, chicken fat), and Omega-3 fatty acid by-products (fish oil, tallow oil, waste cooking oil, etc.). Biodiesel has several environmental benefits, including the ability to be a carbon–neutral fuel [6–10].

Green diesel is a transportation fuel made from biomass that may be used in diesel engines. It complies with the ASTM D975 petroleum specification in the United States and EN 590 in Europe. Hydrotreating or hydroprocessing, pyrolysis, gasification, and other biochemical and thermochemical methods can all be used to make green diesel. Hydroprocessing technology is the most widely used of them. It's important to understand that green diesel and biodiesel aren't the same thing. Green diesel is a mixture of straight-chain and branched paraffin with typical carbon numbers of C15–C18 and chemically identical to petroleum diesel. Biodiesel is a mono-alkyl ester produced through the transesterification process, whereas green diesel is a mixture of straight-chain and branched paraffin with typical carbon numbers of C15–C18 and chemically identical to petroleum diesel. Green diesel is regarded as the lowest carbon fuel in the planet. It is an alternative fuel that is chemically comparable to petroleum diesel and has performance characteristics that are virtually equivalent. Green diesel uses the same oil, fat, and grease feedstock as biodiesel, however unlike biodiesel blends, green diesel may be blended into petroleum diesel at greater blend percentages. Hydrotreating at a biorefinery or co-processing at a petroleum refinery are both common ways to make green fuel [11–15]. This chapter focuses on green diesel production routes, catalysts used in green diesel production, factors impacting the green diesel production process, fuel qualities of various diesel fuels, and the current global scenario for green diesel production.

2 Processing Technologies for Commercial Diesel Fuels

Petrodiesel is the most widely utilized traditional diesel fuel in CI engines. Various researchers have written on this fuel in the scientific literature. As a result, there is no need to go over this again. Diesel fuels originating from renewable/alternative sources are divided into two categories, i.e., Biodiesel and Green diesel. Only biodiesel and green diesel fuels, as well as their conversion process/technology from liquid biomass/vegetable oils, are discussed in this section.

Vegetable oils are a premium renewable resource that can be used as a promising feedstock for transportation fuel generation. In addition to poor atomization, polyunsaturated character, and lubricity, they have high viscosity, density, iodine value, and poor non-volatility. As a result, they cannot be utilized directly in CI engines and must be transformed into appropriate fuels through methods. Thermal cracking (pyrolysis), micro-emulsification, dilution, transesterification (alcoholysis), hydroprocessing, and other procedures are commonly used to handle triglyceride difficulties. The subject has been extensively covered in a variety of publications. Transesterification and hydroprocessing are commercially viable methods for producing transportation fuels from vegetable oils, as shown in Fig. 3 [14–17].

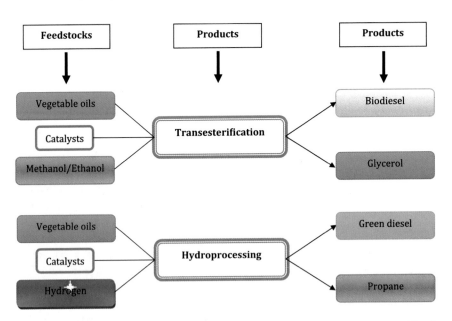

Fig. 3 Conversion routes for the production of biodiesel and green diesel fuels from vegetable oils

2.1 Transesterification Process for Biodiesel Production

Biodiesel is made by alcoholyzing vegetable oils with a simple alcohol such as methanol, ethanol, propanol, butanol, or amyl alcohol in the presence of a catalyst (typically NaOH or KOH, $NaOCH_3$) (Fig. 4). The main product is fatty acid methyl ester (FAME), known as biodiesel, and the by-product is glycerol (approximately 10%). To increase triglyceride conversion, excess alcohol is routinely employed. Biodiesel is a proven clean-burning renewable alternative fuel that is now being produced commercially as a potential replacement for petroleum diesel. For a multitude of reasons, biodiesel is now widely regarded as the best alternative to petroleum. It's manufactured from readily available vegetable oils or fats, has a commercially feasible synthesis process, is compatible with petrodiesel (Table 1), doesn't necessitate major engine changes, and can be distributed using existing infrastructure. [6–9].

Biodiesel is a diesel fuel substitute in diesel engines. The feedstock is derived from a range of sources, including waste oils and vegetable oils such as rapeseed and soybean oil. Biodiesel has several benefits, including biodegradability, reduced reliance on conventional fuels, and reduced greenhouse gas emissions. According to the US Energy Information Administration (2021), biodiesel has been commercially produced for 53 billion liters per year around the world. Europe (32%) is the world's largest producer of biodiesel, followed by the United States (19%), Indonesia (15%), and the rest of the world (34%). The global biodiesel market is expected to develop at a CAGR of 5.39% from US\$ 34.7 billion in 2020 to US\$ 50.1 billion in 2027. Without a doubt, transesterification is a globally recognized technology. New biodiesel plants, on the other hand, require a significant financial investment, and enormous quantities of by-product glycerol require a long-term market [1, 2, 18–20].

FAME has been used as a diesel fuel for many years, however, it has a number of operating difficulties when compared to petroleum-based diesel fuel. During

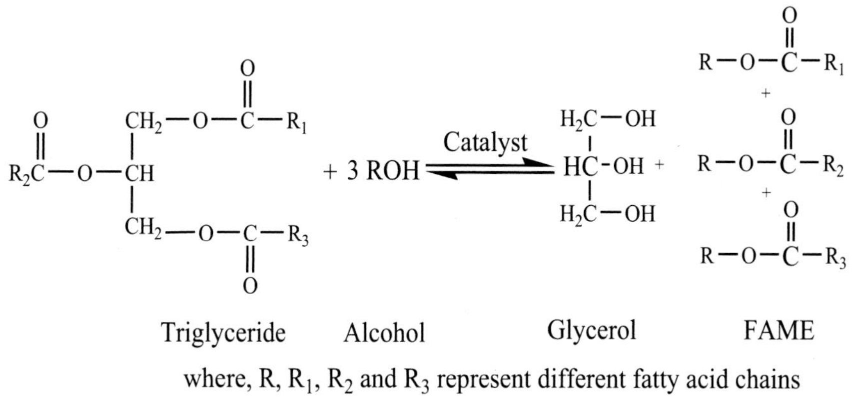

Triglyceride　　Alcohol　　　　　Glycerol　　　　FAME

where, R, R_1, R_2 and R_3 represent different fatty acid chains

Fig. 4 Schematic representation of biodiesel synthesis via tranesterification

Table 1 Comparative analysis of fuel properties of green diesel vs. other diesel fuels [22, 33–40]

Properties	Unit	Petrodiesel	Biodiesel (B100)	Green diesel	FT diesel
	Specifications	ASTM D975 & EN 590	ASTM D6751 & EN 14,214	ASTM D975 & EN 590	–
Chemical formula	–	C_8-C_{25}	$C_{12}-C_{22}$	$C_{15}-C_{18}$	–
Molecular weight	$gmol^{-1}$	200	292	–	–
Composition, carbon	(wt%)	87	77	85	–
Hydrogen		13	12	15	–
Oxygen		0	11	0	–
Density @ 60 °F	lb/gal	6.7–7.4	7.328	6.5	–
Sp. gravity, 60 °F/60 °F	–	0.85	0.88	0.78	0.77
Viscosity @ 60 °F	cP	2.6–4.1	1.9–2.6	–	–
Specific heat	Btu/lb °F	0.43	18,145	–	–
Lower heating value (LCV/NCV)	Btu/gal	128,450	119,550	117,059	123,670
Higher heating value (HCV/GCV)	Btu/gal	137,380	127,960	125,294	130,030
Flash point	°F	140–176	212–338	–	–
Auto-ignition temp	°F	600	300	–	–
Boiling temperature	°F	356–644	599–662	–	–
Lower flammability limits	Vol. %	1	–	–	–
Higher flammability limits	Vol. %	6	–	–	–
Freezing point	°F	−40 to −22	26–66	–	–
Reid vapor pressure	psi	0.2	<0.04	–	–
Cetane no	–	40–55	48–65	70–90	>75
Water solubility, fuel in water @70 °F	Vol. %	Negligible	–	–	–

(continued)

Table 1 (continued)

Properties	Unit	Petrodiesel	Biodiesel (B100)	Green diesel	FT diesel
	Specifications	ASTM D975 & EN 590	ASTM D6751 & EN 14,214	ASTM D975 & EN 590	–
Water solubility, water in fuel @70 °F	Vol. %	Negligible	0.05 (max)	–	–
Stoichiometric air/fuel ratio	Weight	14.7	13.8	–	–
Aromaticity	Vol. %	35	–	–	–
Cloud point	°F	23	23–59	−4 to 68	–
Sulfur content	ppm	<10	<1	<1	<1
Cold filter plugging point (CFPP)	°F	–	−0.4	–	–
Stability	–	Good	Marginal	Good	Good

long-term storage, FAME destroys some fuel system building materials (seals) and develops sludge and deposits. When there is a lot of FAME in the diesel fuel or when pure biodiesel is burnt, motor oil degrades more quickly. Carbon build-up is frequent when utilizing a biodiesel blend in a long-term engine. FAME's comparatively high water solubility, compared to mineral diesel fuel, may cause metal parts to rust more quickly. Due to unsaturated C = C double bonds and C = O bonds (which persist in FAME molecules), FAME has a limited anti-oxidation capacity, and it has a high flash point since it is less flammable than paraffins. Furthermore, new biodiesel plants necessitate a large upfront investment, and large quantities of by-product glycerol necessitate a market. Reduced cetane number, engine incompatibility, poor performance in cold weather, poor emission, and decreased oxidation stability are some of the other difficulties with bio-diesel [6–8, 21, 22]. The majority of FAME biodiesel's drawbacks have shifted focus to another method (hydroprocessing) for turning vegetable oils into renewable hydrocarbons like green diesel.

2.2 Hydroprocessing Technology for Green Diesel Production

Green diesel is a type of petrodiesel-like fuel derived from biological sources (plant or animal-based feedstocks) that is not an ester and so differs chemically from biodiesel. In terms of chemistry, green diesel is comparable to petrodiesel, except it is made from recently living biomass. The definition of green diesel is more complicated than that of biodiesel. The word "renewable diesel" has been defined by the Department of

Energy (DOE) in consultation with the Internal Revenue Service (IRS) and the Environmental Protection Agency (EPA) in the United States. All biomass-based diesel fuels that meet ASTM D975 standards and are not mono-alkyl esters are referred to as "renewable diesel". Green diesel, third-generation diesel, super-cetane diesel, renewable diesel, biocetano, hydrotreated/hydrogenated/hydroprocessed vegetable oil (HVO), hydrotreated biodiesel, paraffinic renewable diesel, non-esterified renewable diesel (NERD), bio-hydrogenated diesel (BHD), and hydrogenation derived renewable diesel (HDRD) are all terms that have been used as synonyms for "bio-based alkanes" derived from plant's vegetable oil or animal fat [23–28].

Apart from transesterification, hydroprocessing is another method of converting liquid biomass to biofuels. Hydroprocessing is a well-known and important process in the petroleum refinery/petrochemical complex for cracking bigger molecules and/or removing hetero-atoms (S, N, O, and metals) from petroleum-derived feedstock like heavy gas oil or vacuum gas oil. Hydroprocessing is a method that uses existing refinery equipment and infrastructure to convert non-edible seed oils, vegetable oils, waste cooking oil, and fat into usable liquid fuels (straight chain alkanes). Instead of the phrase "biodiesel", hydroprocessing is a modern technology for producing high-quality bio-based diesel fuels, which are frequently referred to as "renewable diesel" or "green diesel". The hydroprocessing of vegetable oils primarily produces green diesel (Fig. 5) with a high cetane number (Cetane No. 90) and hydrocarbons in the C15–C18 range, called "Super Cetane Diesel" [23, 28, 29].

2.2.1 Mechanism for Green Diesel Production

There are three methods to crack long chain of hydrocarbons (triglycerides) to short chain of hydrocarbons. The first method is called "thermal cracking", and it involves using heat to create a lighter product. The second method is "catalytic cracking", which uses less heat energy than thermal cracking and is carried out in the presence of an acid catalyst without the usage of hydrogen. The third process, dubbed "catalytic hydrocracking", happens when a bifunctional catalyst is combined with a high hydrogen pressure. The catalytic hydrocracking process uses less thermal energy, and the presence of hydrogen reduces coke production, which reduces pore blockage and hence catalyst deactivation. The simple transition of fatty acid triglycerides into hydrocarbons occurs during hydroprocessing of vegetable oils. During the hydroprocessing of vegetable oil, three major chemical events take place [17, 30].

(i)   Hydrogenation: It involves the hydrogenation of double bonds present in unsaturated chains of bonded fatty acids.

(ii) Hydrodeoxygenation (HDO): It involves the removal of oxygen atoms from carboxylic group in the form of water.

(iii) Hydrodecarboxylation (HDC): This reaction results in the release of carbon dioxide from the carboxylic group.

During the vegetable oil hydroprocessing, it was found that hydrogenation of double bonds precedes hydrodeoxygenation (HDO) and hydrodecarboxylation

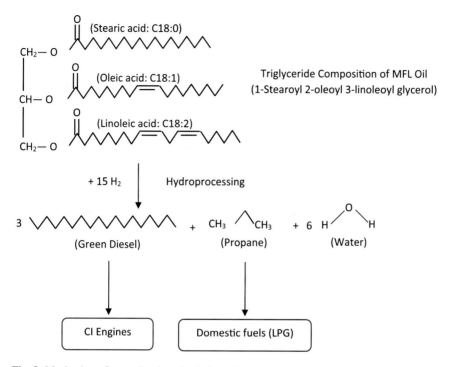

Fig. 5 Mechanism of green diesel synthesis from Mesua ferrea L triglyceride via hydroprocessing reaction [12]

(HDC) reactions. However, the reaction rates of HDO and HDC of saturated triglycerides varies depending on the reaction conditions and catalyst utilized. In general, the major reaction products of vegetable oil hydroprocessing are n-alkanes having the same carbon atom number as fatty acids present in the original vegetable oil (called HDO) and n-alkanes with the carbon atom number lower by one as compared to the carbon atom number of fatty acids in the original vegetable oil (called HDC) [29–31].

The hydrogenation of the double bonds of the side chains and the removal of oxygen on the metal sites of the catalysts are involved in the hydroprocessing of triglycerides. Most vegetable oils (e.g., rapeseed, sunflower, soybean, and palm oil, etc.) are hydroprocessed to produce C15–C18 hydrocarbon mixtures. This liquid mixture is usually referred to as "green diesel", or "renewable diesel" or "biohydrogenated diesel" because it resides within the boiling point range of diesel. It has the same chemical nature that of petroleum-derived diesel. Furthermore, the major output of vegetable oil hydroprocessing is straight-chain paraffins (n-alkanes), which can be hydroisomerized to produce high-quality diesel or hydrocracked to produce kerosene. In addition, propane, a gas by-product, is a good feedstock for the petrochemical industry and is also used as a motor fuel [31, 32].

2.2.2 Catalysts Employed for Green Diesel Production

Catalytic hydroprocessing of vegetable oils is a technique that is still in its early stages of development, with plenty of room for improvement in catalysis research. The use of typical commercial catalysts (refinery streams) for catalytic hydroprocessing of vegetable oils is well established. There are no commercially available catalysts that are particularly designed and developed for such applications. Cobalt and molybdenum (CoMo) or nickel and molybdenum (NiMo) supported on alumina substrate (Al2O3) with high surface area are the most well-known commercial catalysts. Hydroprocessing catalysts are a type of dual-action catalytic material that may be used to perform both hydrogenation and cracking/isomerization events. First, the active metals (such as Mo, Ni, Co, Pd, Pt, etc.) hydrogenate the feedstock molecules, making them more active in the event of cracking and heteroatom removal while minimizing coke formation on the catalyst. Hydrogenation also aids cracking by producing an active olefinic intermediate molecule during the dehydrogenation process. Finally, both cracking and isomerization reactions take place in acidic environment such as amorphous oxides (Al_2O_3-SiO_2) or crystalline zeolites (mainly Z-zeolites) or mixtures of zeolites with amorphous oxides [11–15, 19, 23, 29, 30].

Mono-metallic catalysts are quite expensive since they need the use of very precious metals (Noble metals) in their synthesis. The examples of monometallic catalysts used for the hydroprocessing of vegetable oils are Pd/C (5 wt% & 1 wt%), Pd/Al2O3, Pt/Al2O3, Pt/SAPO-11, Pt/SAPO-31 (1%Pt), Pt/Al2O3-SAPO-11, Pt/Al-SBA-15, Pt/β-zeolite, Pt/H-Y, Pt/H-ZSM-5, Ru/γ-Al2O3, Ru/SiO2, Ru/H-Y, Ru/Al-13-Mont, etc. Bimetallic catalysts were frequently utilized in vegetable oil hydroprocessing. Commercial hydrocracking/hydrotreating catalysts, such as sulfided Ni-Mo/-Al2O3 and Co-Mo/-Al2O3, as well as other industrial hydrotreating catalysts like FCC and HZSM-5, were widely utilized for vegetable oil hydroprocessing. Moreover, other sulfided catalysts including NiMo/H-ZSM-5, NiMo/SiO2-Al2O3, NiMo/γ-Al2O3-SiO2-TiO2, NiW/SiO2-Al2O3, NiMoP/Al2O3, NiMo/ZSM-5-Al2O3, etc., were also used as hydroprocessing catalysts [29–32].

As a result, selecting a good hydroprocessing catalyst is a vital step in determining the hydroprocessing product yield and quality, as well as the process's predicted run-length (operating parameters). However, choosing a hydrocracking catalyst for vegetable oil applications is especially important and difficult for two reasons; (i) Catalyst activity varies significantly, as commercial catalysts are designed for different feedstocks (i.e., feedstocks with high sulfur and oxygen contents, feedstocks containing large molecules, etc. (ii) Commercial hydroprocessing catalysts for lipid feedstocks are currently unavailable. Thus, commercial hydrotreating catalysts for vegetable oil hydroprocessing must be investigated and evaluated [23].

2.3 Process Parameters Affecting Green Diesel Production

Vegetable oil hydroprocessing is a complicated process that involves a number of sequential and parallel reactions. Temperature, pressure, sulfiding agents, feedstock nature and flow, and other process parameters have an impact on them. These variables were briefly explored in this section [29, 30].

2.3.1 Influence of Temperature

One of the most important parameters affecting product yields is the temperature of the hydrotreatment process. At lower reaction temperatures, oxygen removal is minimal, whereas at higher temperatures, oxygen removal from the final products is favored. However, as the temperature rises, the undesirable hydrocracking events compete with the good hydrotreatment reactions, lowering product selectivity in the diesel range. As a result, a lower reaction temperature is better for making diesel, whereas a higher reactor temperature is better for making gasoline. Furthermore, due to an increase in iso-alkanes concentration, cold flow characteristics (such as cloud point and pour point) were observed to improve with rising temperature. Sulfur and nitrogen are the most effective heteroatom removers in all circumstances (above 99.4%), while oxygen removal is favored by hydrotreating temperature.

2.3.2 Influence of Hydrogen Pressure and Flow

Hydrogen pressure and flow are critical in the hydrotreatment of vegetable oils because it not only acts as a reactant but also aids in the elimination of water from the catalyst, preventing deactivation. The relative activity of catalysts toward decarboxylation and hydrodeoxygenation are critical during triglyceride hydroprocessing because these processes determine hydrogen consumption, product yields, heat balance, and catalyst inhibition. Furthermore, it was shown that increasing the hydrogen pressure reduces decarboxylation and favors triglyceride hydrodeoxygenation. Furthermore, the increased hydrogen pressure reduces product cracking, resulting in increased product yield. It's worth noting that a larger H2/oil ratio favors hydrodeoxygenation over decarboxylation, as evidenced by the products' more evenly distributed hydrocarbons.

2.3.3 Influence of Sulfidation

For the hydroprocessing of vegetable oils, sulfided catalysts are the most active and commonly utilized catalysts. Due to the removal of sulfur from the catalysts, the water produced as a by-product of the deoxygenation processes, as well as the presence of oxygenates, limits the catalyst's activity. To keep the catalysts in active sulfided form, certain sulfiding agents must be added to the vegetable oil feedstock (low sulfur concentration). Furthermore, the addition of H_2S (or H_2S produced by a sulfiding agent) not only maintains the sulfided condition of the catalysts but also compensates for activity loss owing to water.

Furthermore, as the catalyst is continuously used and coke is deposited, the catalyst activity decreases and the catalyst becomes deactivated. The rate of catalyst deactivation, on the other hand, is mostly determined by temperature and hydrogen partial pressure. It has been observed that when temperatures rise, catalyst deactivation speeds up, whereas a high hydrogen partial pressure slows down catalyst deactivation. Most catalyst's catalytic activity can be regained through catalyst regeneration.

2.3.4 Influence of Feedstocks

With a few exceptions, such as coconut oil, vegetable oil includes fatty acids with carbon values ranging from C_{14} to C_{18}. It was observed that hydroprocessing vegetable oils produces alkanes with the same number of carbons as the source fatty acids, or one less. As a result, the fatty acid composition of the feed can be used to predict the composition of renewable diesel produced from vegetable oil hydroprocessing reactions. However, certain variances in product yields were observed, which were determined to be dependent on the chain length and degree of saturation of the vegetable oils.

Furthermore, it was observed that under low hydrogen pressure or a low hydrogen to oil ratio, cracking, cyclization, and aromatization reactions were favored, resulting in the synthesis of cycloalkane and aromatics. Moreover, cycloalkane selectivity increases as the degree of unsaturation of the feedstock increases, whereas alkane selectivity increases as the degree of saturation increases. However, because most unsaturation becomes saturated at process temperatures and pressures, and hence does not contribute to cyclization or aromatization, these compounds were not commonly observed in hydrotreating.

3 Fuel Properties Analysis of Green Diesel Verses Other Diesel Fuels

Green diesel has similar fuel qualities as high-quality sulfur-free petroleum diesel. Previously, the Fischer–Tropsch (FT) diesel was thought to be the greatest option for engines and emissions. Green diesel has many similarities to FT diesel in terms of composition, and it is also totally renewable. Green diesel is a safer fuel for a variety of uses due to its low sulfur level, metal-free nature, and lack of ash. It has a higher heating value than other biofuels, as well as a larger energy content than biodiesel. Green diesel also has outstanding cold qualities, making high bio-mandate blending ratios possible throughout the year, even in the harshest winters. Even at a low cloud point, its density remains practically constant. It is more ideal for compression ignition engines due to its high cetane number and low-density values. Furthermore, when compared to biodiesel, the flashpoint is higher, allowing for safer storage and transportation. Paraffinic hydrocarbons are present in green diesel. These paraffinic hydrocarbons have a higher calorific value and cetane index, as well as superior density, kinematic viscosity, flash point, and other fuel properties. As published by numerous researchers/industrial resources, Table 1 compares the fuel qualities of green diesel to biodiesel and petrodiesel [22, 24, 27, 33–40].

Green diesel is 100% renewable and can be made from a variety of renewable, bio-based feedstocks at existing oil refineries. It may be used in current diesel-powered engines and vehicles without requiring modifications in the existing CI engines. Because green diesel is chemically comparable to petroleum diesel, compatibility difficulties with current infrastructure and engines are minimized. It can be blended with petroleum diesel or biodiesel fuels at any varying ratio. It can also be used neat (100% Renewable Diesel). Green diesel has a higher cetane number, ranging from 70 to 95, which means it provides faster and more complete ignition, lowers NOx and PM emissions, and allows for a smoother start with less noise. When compared to traditional fuels, carbon dioxide absorbed by producing feedstocks reduces overall greenhouse gas emissions by balancing carbon dioxide generated from burning renewable hydrocarbon biofuels. It can be stored for a long period without losing quality or accumulating water. There are no aromatics or contaminants in it, therefore there is no odor (less harmful for operators/drivers and the environment) [35, 37].

Green diesel has several advantages over biodiesel, including compatibility with existing engines, feedstock flexibility (e.g., the content of free fatty acids in the vegetable oil is unimportant), higher cetane number (70–90), higher energy density, higher oxidation stability (i.e., zero O_2 content), and better cold weather performance. Furthermore, it has no effect on NOx emissions. It does not necessitate the use of water. There are no by-products that need to be treated further (e.g., glycerol). Renewable diesel distribution does not add to pollution because it may be delivered over existing pipes that are now used to distribute petrodiesel [35, 37].

4 Commercial Status of Green Diesel Production

Catalytic hydroprocessing of vegetable and waste oils has been used extensively on a commercial scale to produce renewable fuel/green diesel. Neste Oil Corporation (a European Company) is the world's leading producer of renewable diesel and sustainable aviation fuel produced from waste and residue raw materials. Neste is the world's largest producer of renewable diesel and jet fuel, with an annual capacity of 1.98 million metric tonnes (2 Mt/a), or 675 million gallons, distributed among four global facilities in Finland, Rotterdam, and Singapore. Neste's Singapore Refinery (started in 2010) has a production capacity of 1.3 million tonnes and serves customers in Europe and North America. Neste made a 1.4-billion-euro investment decision in 2018 to expand the Singapore refinery, which will increase Neste's renewable product production capacity by 1.3 million ton/a and bring total global renewable product capacity close to 4.5 million ton/a by 2023. Neste Oil converts biomass into fuel using the NExBTL® (Next Generation Bio-to-Liquid) process/technology. The renewable diesel is marketed under the brand name NExBTL®. NExBTL renewable diesel has a higher cetane number (75–90) than petroleum diesel and contains no aromatic compounds or sulfur. It's also a fungible, low-carbon, and low-emission paraffinic biofuel. Table 2 shows a list of renewable diesel producers, along with their product names, production capacities, and technology [30, 35, 37].

Furthermore, Diamond Green Diesel (USA), which started operations in 2013, is the world's second largest renewable diesel manufacturer. In 2021, the company intends to grow its sustainable diesel fuel production from 290 million gallons to 400 million gallons per year. In collaboration with Valero and Darling, Diamond has also approved a new 470 million gallons per year renewable diesel project at Valero's Port Arthur, Texas refinery. Diamond Green Diesel expects to begin operations in 2023, boosting its total annual renewable diesel production to about 1.2 billion gallons [38]. Oil and gas companies like ENI produce a considerable quantity of renewable diesel using technologies like Ecofining™ (an ENI patented technology), Vegan® (an Axens technology), etc. The companies employing HydroFlex™ hydrotreating process by Haldor Topsoe are UPM and Preem. The Ecofining™ process is also used by Diamond Green Diesel, a joint venture between Diamond Alternative

Table 2 World scenario of green/renewable diesel production [30, 35, 37–40]

Industry	Product	Production capacity (Million tons/year)	Process/technology	Industry location
Neste	Neste MY renewable diesel	2.6	NExBTL	Finland, Holland, Singapore
Diamond green diesel	Renewable diesel	0.4	Ecofining	USA
Eni S.p. A	Green diesel	0.5 to 1	Ecofining	Italy
Total S.A	Renewable diesel	0.5	Vegan technology by axens	France
Renewable energy group, Inc	Renewable diesel	0.2	Bio-synfining	USA
UPM	BioVemco renewable diesel	0.1	Hydroflex	Finland
Preem	Preem evolution diesel	Unknown	Hydroflex	Sweden
Cetane energy	Renewable diesel	Unknown	Hydrotreating	USA
AltAir paramount	Renewable jet	0.15	UOP renewable jet fuel Process	USA
Petrixo oil & gas	Renewable jet	0.5	UOP renewable jet fuel Process	UAE

Energy LLC (a subsidiary of Valero Energy Corporation) and Darling Ingredients Inc., which is the largest producer in the United States. Apart from Neste Oil in Finland and Diamond in the United States, other oil companies are establishing commercial units/pilot plants, including Universal Oil Products (UOP)-Eni (UK, Italy), Nippon Oil (Japan), Petrobras/H-Bio (Brazil), British Petroleum (Australia), ConocoPhillips/Tyson (USA, Ireland), Dynamic Fuels/Syntroleum/Tyson (USA), SK Energy/HBD (Korea), etc. [30, 35]. Table 2 depicts the global picture for green/renewable diesel production, including plant capacity and location, as well as the process technology used.

4.1 Green Diesel Policies and Regulations

The United States of America has established a number of federal and state-level environmental rules that boost the market for renewable diesel. The Renewable Fuel Standard (RFS) was established by the US Environmental Protection Agency (EPA) to encourage biodiesel and renewable diesel refiners and blenders through a tradable credit system known as "Renewable Identification Numbers" (RINs). Renewable fuels (biodiesel or renewable diesel) must be blended with gasoline or diesel fuel under the Renewable Fuel Standard (RFS) program [42]. California has adopted state-level rules, such as the Low-Carbon Fuel Standard (LCFS) programme, to gradually reduce the carbon intensity of its transportation fuel. California intends to reduce the carbon intensity of fuels by 10% in 2020 and 30% in 2030 under this regulation. Under the LCFS program, biomass-based diesel also generates credits, i.e., low-carbon fuels with a carbon intensity less than the annual standard earn credits, whereas fuels with a carbon intensity greater than the annual standard generate deficits. The net supply of renewable diesel in California has increased since the LCFS program was implemented. In 2018, the net supply of renewable diesel hit 100 million gallons. In 2019, most of the US biomass-based diesel imports came from Singapore since 2015. In 2019, imports from Singapore increased by 49% (i.e., up to 17,000 barrels per day), while overall imports of US biomass-based diesel (which includes biodiesel/renewable diesel) increased by 26% (i.e., 27,000 barrels per day). In addition, other states such as Oregon and Washington in the United States, as well as British Columbia in Canada, are following California's lead in reducing their carbon footprints from transportation fuels [42].

The European Commission announced a new rule in 2018, the renewable energy directive-II (RED-II), requiring fuel suppliers to supply a minimum of 14% renewable energy in road and rail transport by 2030. A 3.5% sub-goal for advanced biofuels made from biomass and waste base feedstock is included in the 14% target. The 14% target cannot be achieved by biodiesel alone due to several reasons including chemical composition of biodiesel, blending limitation (7% max. v/v to avoid engine damage), and technology readiness. The available technology/solutions for producing biofuels from most advanced feedstocks are more suited to the manufacture of petroleum biofuels than FAME. As a result, renewable diesel might be used by refineries to bridge the renewable energy gap for FAME and meet RED-II targets [42].

5   Future Prospects of Green Diesel

Commercial green diesel plants have been built all over the world. Globally, more than 5.5 billion liters of renewable diesel are produced each year, with that number expected to rise to 13 billion liters by 2024. It is a commercial fuel produced in USA and imported from Asia. There are five renewable diesel facilities in operation in the United States, with a total capacity of around 590 million gallons per year. The production is expected to expand in the near future, with a capacity of 2 billion gallons under development at six plants and expansion at three current operations. The development of renewable diesel production is deeply influenced by worldwide legislation and policies. The regulation-driven demand in the US and EU is expected to spur a USD 5 billion investment in new renewable diesel plants over the next 5 years [41–44].

Neste is currently the world's leading producer of renewable diesel fuel. It is the world's largest producer, with four facilities: two in Finland, one in Rotterdam, and one in Singapore. In addition, various proposed renewable diesel production capacity will be available to the global market in the next 5 years (Fig. 6). For instance, the plants such as NEXT Renewable Fuel's Oregon project, Diamond Green Diesel's expansion project in Louisiana, and World Energy in California are planned in the United States. Similarly, Neste's growth project in Singapore and ENI's Venice project extension in Italy are both planned for the coming years. As a result, the proposed renewable diesel production capacity expansions have the potential to grow the global market for renewable diesel [41–44].

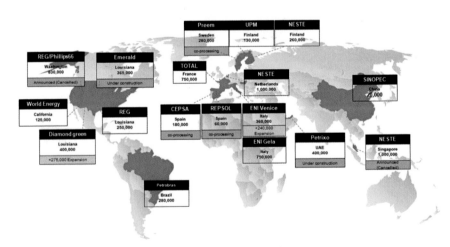

Fig. 6 Worldwide green diesel plants (installed and planned). *Source* [42, 44]

6   Conclusion

Green diesel production capacity is approaching 5.5 billion liters per year, with a projected increase to 13 billion liters by 2024. Hydroprocessing of vegetable oils and fats has been studied as the most widely utilized technology for producing green diesel all over the world. As discussed, the appropriate operating conditions and an effective catalyst may produce large yields of C15–C18 saturated hydrocarbons with the necessary isomer/normal composition ratios, which dictate the green diesel's cetane number and cold flow properties. They also showed much decreased engine emissions of CO, HC, smoke, solid particles, and NOx emissions, which are known to increase when using biodiesel. CO_2 emissions have been shown to be comparable to petroleum diesel, but one must also consider the green diesel's renewable nature, which means it does not contribute to net CO_2 accumulation in the atmosphere. The catalyst is the driving power of all chemical reactions involved in the vegetable oil hydroprocessing, which is the heart of the entire process. Despite the fact that the technology is mature, there is still room for new catalysts to be developed for emerging bio-feedstocks such as lignocellulosic biomass/microalgae. Green diesel is expected to play a significant role in the transportation industry in the near future, replacing fossil-derived diesel fuel from the global markets.

References

1. International Energy Agency (IEA). https://www.iea.org/, accessed 2022
2. US Energy Information Administration of DOE. https://www.eia.gov/, accessed 2022
3. Mandari V, Devarai SK (2021) Biodiesel production using homogeneous, heterogeneous, and enzyme catalysts via transesterifcation and esterifcation reactions: a Critical Review. BioEnergy Res. https://doi.org/10.1007/s12155-021-10333-w
4. Jinfang T, Longguang Yu, Rui X, Shan Z, Yuli S (2022), Global low-carbon energy transition in the post-COVID-19 era, Applied Energy, 307, 118205
5. Rita E, Chizoo E, Cyril US (2021) Sustaining COVID-19 pandemic lockdown era air pollution impact through utilization of more renewable energy resources. Heliyon 7:7
6. Mohammad A, Saxena P, Sarma AK (2014) Green technology for biodiesel production from Mesua ferrea L. seed oil, Energy and Environment Research, 4 (2), 11–21
7. Kumar P, Aslam M, Singh N, Mittal S, Bansal A, Jha MK, Sarma AK (2015) Characterization, activity and process optimization with a biomass-based thermal power plant's fly ash as a potential catalyst for biodiesel production, RSC. Advances 5:9946–9954
8. Sarma AK, Kumar P, Aslam M, Chouhan APS (2014) Preparation and characterization of musa balbisiana colla underground stem nano-material for biodiesel production under elevated conditions. Catal Lett 144(7):1344–1353
9. Singh Chouhan AP, Sarma AK (2011) Modern heterogeneous catalysts for biodiesel production: a comprehensive review. Renew Sustain Energy Rev 15:4378–4399
10. Brar KK, Sarma AK, Mohammad Aslam BS, Chadha, (2017) Potential of oleaginous yeast Trichosporon sp., for conversion of sugarcane bagasse hydrolysate into biodiesel. Biores Technol 242:161–168
11. Aslam M, Konwar LJ, Sarma AK, Kothiyal NC (2015) An investigation of catalytic hydrocracking of high FFA vegetable oils to liquid hydrocarbons using biomass derived heterogeneous catalysts. J Anal Appl Pyrol 115:401–409

12. Mohammad A, Kothiyal NC, Sarma AK (2015) True boiling point distillation and product quality assessment of biocrude obtained from mesua ferrea L. seed oil via hydroprocessing. Clean Technology and Environment Policy 17(1):175–185

13. Kumar H, Konwar LJ, Aslam M, Sarma AK (2016) Performance, combustion and emission characteristics of a direct injection VCR C.I. engine using Jatropha curcas oil microemulsion: a comparative assessment with JCO B100, JCO B20 and Petrodiesel. RSC Adv 6:37646–37655

14. Deepak Singh SS, Sandhu AK, Sarma (2019) A comprehensive experimental investigation of green diesel as a fuel for CI engines. Int J Green Energy 16(14):1152–1164

15. Himansh Kumar AK, Sarma PK (2021) Experimental investigation of 2-EHN effects upon CI engine attributes fuelled with used cooking oil-based hybrid microemulsion biofuel. Int J Environ Sci Technol. https://doi.org/10.1007/s13762-021-03751-y

16. No S (2014) Application of hydrotreated vegetable oil from triglyceride-based biomass to CI engines—a review. Fuel 115:88–96

17. Sotelo-Boyas R, Trejo-Zarraga F, Hernandez-Loyo FDJ (2012). Hydroconversion of triglycerides into green liquid fuels. https://doi.org/10.5772/48710

18. Srinivas D, Satyarthi JK (2012) Challenges and opportunities in biofuels production. Indian J Chem. (51A):174–185

19. Sinha AK, Anand M, Rana BS, Kumar R, Farooqui SA, Sibi MG, Kumar R, Joshi RK (2013) Development of hydroprocessing route to transportation fuels from non-edible plant-oils. Catal Surv Asia 17:1–13

20. Crocker M, Santillan-Jimenez E, Morgan T, Lacny J, Mohapatra S (2013) Catalytic deoxygenation of triglycerides and fatty acids to hydrocarbons over carbon-supported nickel. Fuel 103:1010–1017

21. Liu Y, Sotelo-Boyas R, Murata K, Minowa T, Sakanishi K (2012) Production of biohydrogenated diesel by hydrotreatment of high-acid-value waste cooking oil over ruthenium catalyst supported on Al-polyoxocation-pillared montmorillonite. Catalysts 2:171–190. https://doi.org/10.3390/catal2010171

22. Kubicka D, Simacek P, Sebor G, Pospisil M (2009) Hydroprocessed rapeseed oil as a source of hydrocarbon-based biodiesel. Fuel 88:456–460

23. Bezergianni S, Dimitriadis A (2013) Comparison between different types of renewable diesel. Renew Sust Energ Rev. 21:110–116

24. Yoon JJ (2011) Advanced biofuels USA. http://advancedbiofuelsusa.info/wp-content/uploads/2011/03/11-0307-Biodiesel-vs-Renewable_Final-3-JJY-formatting-FINAL.pdf

25. Herskowitz M, Zarchin R, Rabaev M, Vidruk-Nehemya R, Landau MV (2015) Hydroprocessing of soybean oil on nickel-phosphide supported catalysts. Fuel 139:684–691

26. Perego C, Ricci M (2012) Diesel fuel from biomass. Catal Sci Technol 2:1776–1786

27. Knothe G (2010) Biodiesel and renewable diesel: a comparison. Prog Energy Combust Sci 36:364–373

28. Dahiya A (2015). Cutting-edge biofuel conversion technologies to integrate into petroleum-based infrastructure and integrated biorefineries. https://doi.org/10.1016/B978-0-12-407909-0.00028-6

29. Chiranjeevi T, Satyarthi JK, Gokak DT, Viswanathan PS (2013) An overview of catalytic conversion of vegetable oils/fats into middle distillates. Catal Sci Technol 3:70–80

30. Bo Zhang, Wu, Jinsheng, Yang C, Qi, Qiu, Yan Q, Li R, Wang B, Wu, Jinlong, Ding Y (2018) Recent developments in commercial processes for refining bio-feedstocks to renewable diesel. BioEnergy Res 11:689–702

31. Chen S (2012) Green oil production by hydroprocessing. Int J Clean Coal Energy 1:43–55

32. Pinto F, Varela FT, Gonçalves M, Andre RN, Costa P, Mendes B (2014) Production of biohydrocarbons by hydrotreating of pomace oil. Fuel 116:84–93

33. Kim DS, Hanifzadeh M, Kumar A (2018) Trend of biodiesel feedstock and its impact on biodiesel emission characteristics. Environ Prog Sustain 37:7–19

34. Hoekman SK, Broch A, Robbins C, Ceniceros E, Natarajan M (2012) Review of biodiesel composition, properties, and specifications. Renew Sustain Energy Rev 16:143–169

35. Douvartzides SL, Charisiou ND, Papageridis KN, Goula MA (2019) Green diesel: biomass feedstocks. Production Technologies, Catalytic Research, Fuel Properties and Performance in Compression Ignition Internal Combustion Engines. Energies 12(809):1–42
36. Singh D, Subramanian KA, Garg MO (2018) Comprehensive review of combustion, performance and emissions characteristics of a compression ignition engine fueled with hydroprocessed renewable diesel. Renew Sustain Energy Rev 81:2947–2954
37. Neste Oil. http://www.nesteoil.com, assessed in 2022
38. https://www.diamondgreendiesel.com/, accessed in 2022
39. https://uop.honeywell.com/en/industry-solutions/renewable-fuels/green-diesel, accessed in 2022
40. https://petrobras.com.br/en/news/renewable-diesel-brings-more-quality-competition-and-sustainability-to-the-biofuels-segment-in-brazil.htm, accessed in 2022
41. Alternative Fuels Data Center. https://afdc.energy.gov/fuels/, accessed in 2022
42. Renewable Diesel: The Fuel of the Future. https://www.futurebridge.com/, accessed in 2022
43. https://www.iea.org/reports/renewables-2019/transport, accessed in 2022
44. Building up the future, technology status and reliability of the value chains, Sub group on advanced biofuels: sustainable transport forum, https://op.europa.eu/en/publication-detail/-/publication/f1c977d1-67a4-11e8-ab9c-01aa75ed71a1.

Chapter 10
Biodiesel, Green Diesel and Petrodiesel: A Comparison

Mohd Razali Shamsuddin, Wan Nor Adira Wan Khalit, Surahim Mahmud, M. Safa-Gamal, Tresylia Ipah Anak Ujai, Azizul Hakim Lahuri, and Tengku Sharifah Marliza

Abstract This chapter emphasized a comparison between biodiesel, green diesel and petrol diesel. In general, diesel is referred to any liquid fuel specifically designed for the automotive engine as a fuel. Generally, biodiesel is an alternative fuel that has similar properties to conventional diesel fuel. Biodiesel can be produced from vegetable oils, animal fats or waste cooking oil via catalysed transesterification process. Biodiesel is renewable energy, nontoxic and biodegradable and may reduce net carbon monoxide emission by 78% compared to petrol diesel. Green diesel is one of the alternative fuels which has a similar molecule structure as petrol diesel and yet it provides better diesel properties in terms of higher heating value, energy density, very high cetane numbers and outstanding cold flow properties compared to conventional biodiesel. Green diesel is produced by hydrotreating triglycerides in vegetable oils with hydrogen. The most common type of diesel fuel is petrol diesel. It is produced from the fractional distillation of crude petroleum between the temperature of 200 and 350 °C at atmospheric pressure resulting in a mixture of carbon chains that typically contain between 8 and 21 carbon atoms per molecule.

Keywords Biodiesel · Green diesel · Petrodiesel · Comparative analysis

M. R. Shamsuddin
Preparatory Centre for Science and Technology, Universiti Malaysia Sabah, 88400 Kota Kinabalu, Sabah, Malaysia

W. N. A. W. Khalit · S. Mahmud · M. Safa-Gamal
Department of Chemistry, Faculty of Science, Universiti Putra Malaysia, Selangor 43400 UPM Serdang, Malaysia

T. I. A. Ujai
Department of Chemistry, Faculty of Resource Science and Technology, Universiti Malaysia Sarawak, 94300 Kota Samarahan, Sarawak, Malaysia

A. H. Lahuri · T. S. Marliza (✉)
Department of Science and Technology, Universiti Putra Malaysia Bintulu Campus, Sarawak 97008 Bintulu, Malaysia
e-mail: t_marliza@upm.edu.my

A. H. Lahuri
e-mail: azizulhakim@upm.edu.my

M. Aslam et al. (eds.), *Green Diesel: An Alternative to Biodiesel and Petrodiesel*,
Advances in Sustainability Science and Technology,
https://doi.org/10.1007/978-981-19-2235-0_10

1   Introduction

As the world experienced rapid growth of industrial activity and a drastic increase in the human population in the preceding century, the demand for alternate energy sources has become vital. The widespread use of fossil resources, especially petroleum and its derivatives as energy resources or raw materials for other purposes, has resulted in serious environmental concerns primarily related to the production of certain harmful gases such as SOx, NOx and COx. In this situation, biomass has become an important alternative for a renewable source of energy.

1.1   Biodiesel

Biodiesel fuel is a sustainable source of energy that can be manufactured from biological feedstock that rapidly grows demands in recent years, especially in Asia [5]. It is also known as the mono-alkyl esters of vegetable oils or animal fats which is obtained by transesterification of oil or treating a fat with alcohol, usually methanol or ethanol. These types of esters have better properties than triglycerides due to their smaller size [77].

Transesterification is a process where glycerine is isolated from fat or vegetable oil which offers the production of biodiesel via four basic routes from oils and fats which are base catalysed transesterification, direct acid catalysed transesterification, conversion of oil to fatty acid followed by biodiesel, and lastly non-catalytic transesterification of oils and fats [74]. This process also decreases the viscosity of highly viscous vegetable oils making it is a suitable process to produce biodiesel. Researchers reported that high viscosity and low volatility will cause operational problems in engines such as poor or incomplete combustion of fuel and form deposits [16, 41]. Other ways that are used to produce biofuels from vegetable oil are cracking, supercritical methanol, enzyme hydrolysis and hydroprocessing [41]. In addition to the process, parameters that affected the process are temperature, molar ratio of the alcohol to oil, type and quantity of catalyst, type of procedure and the composition of the reactants mixture [77]. According to [16], the utilization of vegetable oils as fuel is liquid nature portability, ready availability, renewability, have higher heat content, lesser sulphur content, reduce aromatic content and ability to biodegradable. In addition, biodiesel also produces no sulphur, no net carbon dioxide, less carbon monoxide, particulate matter, smoke and hydrocarbon emissions and they produce more oxygen which leads to complete combustion and reduced emissions [9].

1.2   Green Diesel

Green diesel has been proposed as an alternative for a renewable diesel that is derived from the same feedstocks as biodiesel but undergoes different processes such as catalytic hydrodeoxygenation (HDO) of triglyceride and fatty acid molecules [77]. Production of n-alkanes requires the removal of oxygen without the fragment of triglyceride side chains which is called selective deoxygenation (SDO) comes in three different pathways, for instance, decarboxylation (deCO$_2$) of the intermediate fatty acids, decarbonylation (deH$_2$O–deCO) of the intermediate fatty aldehydes and dehydration (deH$_2$O) of the intermediate fatty alcohols then hydrogenation which emerges in the formation of alkenes [40]. One of the interesting facts about green diesel is that the combustion made by green diesel has the ability to decrease carbon dioxide emissions by 46.7%, particulate matter by 66.7% and unburned hydrocarbon by 45.2%. It is also reported that methyl esters of vegetable oils are better than neat vegetable oils when it gives reduced smoke levels and increased thermal efficiencies [50]. This fact gives green diesel to have a better quality than biodiesel from the aspect of oxygen efficiency, specific gravity, sulphur content (ppm), heating value, cloud point, cetane number and stability. Moreover, green diesel has a high cetane number and lower density which enable refiners to enhance a significant amount of refinery streams that can be blended into the diesel pool while fulfilling all the diesel requirements. Green diesel requires more energy to manufacture than petrol diesel but it is renewable and saves more fossil energy per tonne if being compared to biodiesel and also contributes to cleaner diesel fuel [29].

1.3   Petrodiesel

Aside from renewable diesel such as biodiesel and green diesel, petrodiesel or petroleum diesel is a fossil diesel that is obtained from the process of fractional distillation in the temperatures ranging from 200 to 350 °C at atmospheric pressure to achieve a mixture of carbon chains that consists of 8 to 21 carbon atoms per molecule [28]. It is also non-renewable and raises environmental concerns from the burning of petroleum products and crude oil which lead to the destruction of the natural landscape and wildlife habitats [11, 56]. Generally, crude oil that is used for the process is sent to a distillation column and heated to extract different fuels, for instance, lighter fuels propane and butane that are reduced from the top and denser fuels such as gasoline and petrodiesel from the bottom of the column. Subsequently, many other processes will occur to extract, purify and optimize the extracted oil to meet demanded requirements [11]. Petrodiesel has a few disadvantages that include poor lubricity that could lead to the failure of fuel injectors and pumps in the engine parts as lubricity improves with the chain length and the presence of double bonds. Some researchers also reported that using biodiesel in contrast with petrodiesel has enhanced the brake specific consumption but shows a decrease in pollutants such as particulate matter (PM) and carbon monoxide (CO) [28]. According to Ramírez-Verduzco and Hernández-Sánchez [56], the quality and cetane number of renewable diesel such as biodiesel and green diesel is higher when petrodiesel is used as the additive that the cetane index in renewable diesel is 77.25 which is 29.6% higher than in petrodiesel.

2 Biodiesel

2.1 Production Method

The production of biodiesel has become an essential commodity because of its properties as a clean energy source and as an environmental-friendly substance. Thus, biodiesel manufacturing requires a chemical process such as direct mixing, pyrolysis, micro-emulsion, esterification or transesterification.

Direct Blending/Use

Edible oil and animal fats can be used in direct injection (DI) engines due to their properties such as good heating value. The advantages of using renewable fuel as direct use in engines are a cleaner resource as it produces a lower amount of greenhouse gases, high mileage than using petrodiesel and is safe to use [76]. However, few problems were also identified such as it cannot be used without modification in DI engine, adverse effect on engine performance and tend to solidify at cold temperature [54]. With half of their nominal output, the engine's chamber temperature stays below 200 °C. Edible oils have poor volatility, higher flash point and viscosity than conventional fossil fuels. Direct use of edible oil resulted in tar deposit in the injector nose which will affect the spraying characteristics.

To circumvent these issues, renewable fuels are blended directly with petrodiesel, improving fuel quality and lowering fossil fuel consumption. For the best results, Prabu et al. [53] recommended a 20% palm oil blend (PO20); meanwhile, to improve engine efficiency and reduce CO_2 emissions, they added antioxidants like butylated hydroxytoluene (BHT) and n-butanol [76]. In order to reduce the pour and flash point of B10 biodiesel, [26] added magnesium to the chicken fat and petrodiesel blend and Ali et al. [4] also claimed that a 30% of biodiesel blend reduced the pour point from 14 °C to 0 °C (palm oil and diesel fuel).

Pyrolysis

Plant oil pyrolysis was the first attempt to make fuel from plant oil. During the process, the heat created increased molecular vibration, causing the oil or animal fat to stretch and break down into tiny particles [44]. As a result, it produces fuel with reduced viscosity, less sulphur and greater cetane numbers, resulting in faster ignition [7]. The pyrolysis process occurs at 250–350 °C and the vaporized oil or animal fat is piped to the condenser. Then, the condenser cools the vaporized liquid biodiesel [54]. The black residuals with high viscosity from the bottom of the reactor is called as bio-mast. Biodiesel that is produced via pyrolysis using biomass has an advantage with less than half the total amount of greenhouse gas emissions than fossil fuel. Zhao et al. [80] used a ZSM-5 catalyst with 20% Zn to pyrolyze camelina oil to get excellent biodiesel yield.

Micro-emulsion

There are two types of micro-emulsions: oil in water and water in oil. These colloidal dispersions contain isotopic fluids ranging in size from 0.01 to 0.05 m [76]. The micro-emulsion with alcohol reduces the viscosity of edible oil, smoke formation, improves fuel atomization and spray properties [31, 45]. Sanchez-Cantú et al. [60] reported 97.1% biodiesel yield using micro-emulsion of methanol and soybean oil in shear mixing equipment. However, using micro-emulsified fuel in diesel engines can produce issues like incomplete combustion, carbon build-up and nozzle failure [54].

Esterification and Transesterification

Esterification and transesterification are commonly used in the production of biodiesel. In the esterification process, high FFA feedstock is preferred to yield methyl esters and water as by-products [83]. Both reactions require alcohol, but esterification is generally faster due to one-step conversion reaction [24]. Yet, the esterification reaction required higher energy to remove OH from the carboxylic acid. The loss of hydrogen from alcohol and OH from carboxylic acid causes water production.

Lipid feedstocks like oil and animal fats contain triacylglycerols (TAGs) that are chemically transesterified into FAME in the presence of alcohol using an acid or base catalyst [31]. This process converts triglycerides into mono-, di- and glycerol (by-product) and reduces vegetable oil viscosity. This reaction, also known as alcoholysis, changes an ester by exchanging alkyl groups [10]. Because the reaction is reversible, excess alcohol can be employed to change the equilibrium. Despite the uses of ethanol, methanol and butanol, as alcohols; however, methanol is often utilized in transesterification because of its low cost and high reactiveness.

Employment of esterification and transesterification on the production of biodiesel recently reported nowadays. In the esterification reaction of PFAD, Lokman et al. [38] achieved 94.5% FFA conversion and 90.4% FAME yield utilizing sulfonated-starch solid acid as catalyst. Meanwhile, Mansir et al. [42] reported transesterification of WCO using CaO-assisted W-Mo catalyst yielded 96.2% FAME.

Simultaneous Esterification and Transesterification

Generally, high free fatty acids (FFA) and triglyceride feedstock may be transformed into fatty acid methyl esters (FAME) in two steps: first, the FFA are esterified using acid catalyst, then the unreacted triglycerides are transesterified using a base catalyst [34]. This technique has several drawbacks such as high purification costs, difficulty extracting catalyst from a liquid mixture and contamination of catalyst in FAME. This two-step method raises the cost of biodiesel manufacturing.

Recently, a heterogeneous acid catalyst demonstrated excellent yield in converting FFAs and triglycerides into FAME through simultaneous esterification and transesterification. This acid catalyst may also be used as a one-step process with a better transesterification rate than the base catalyst [34]. Utilization of acid catalyst was reported by Lien et al. [34] with more than 90% of FAME yield via solid acid carbon catalyst. Meanwhile, Nb_2O_5/SO_4 produced 98% of FAME yield using *Chlorella minutissima* microbial oil [39], and 92% FFA conversion of soybean oil using $SnSO_4$ for 10 reusability cycle [23].

2.2 Technologies in Biodiesel Production

In the recent past, a few biodiesel manufacturing methods such as traditional reflux, microwave-aided reaction, ultra-sound-assisted reaction and supercritical methanol have all been developed. Each of these innovations has benefits and drawbacks of its own. Table 1 compares the benefits and drawbacks of the previous study of several biodiesel manufacturing methods.

Table 1 The advantages and disadvantages of technologies for biodiesel production [2, 33, 35, 69, 76]

Technologies	Advantages	Disadvantages	Previous study
Conventional reflux	Minimal alcohol required Reduce manufacturing costs Methanol regeneration high All-purpose feedstock	Scalability issues Not for industry High-temperature and high-pressure incompatible	[21, 43, 70, 81]
Autoclave reactor system	The feedstock and alcohol reaction temperature, pressure, stirring rate and flow rate may be regulated	Temperature of reaction may be greater than alcohol boiling point Intensive energy use	[30, 71, 72]
Microwave-assisted transesterification	Rapid reaction rate Reduced reaction time Heat loss is minimal Heat transfer efficiency	Catalyst performance is important Not for solid feedstock	[20, 25, 35, 76]
Ultrasound-assisted transesterification	Less energy uses A faster reaction times Affordability	Catalyst quantity required high Large quantity of wastewater generated	[12, 14, 33]
Supercritical methanol	No catalyst needed Faster reaction time	Costly production Extensive conditions High energy use	[37, 46, 68, 76, 78]

2.3 Biodiesel Quality and Properties

Biodiesel is a well-known diesel substitute. Biodiesel quality may be affected by a variety of variables, including feedstock type, manufacturing method and catalyst activity. Before employment in a diesel engine, the physicochemical characteristics of the generated biodiesel should meet the standard specifications given by the US biodiesel standard (ASTM D6751) and the European biodiesel standard (EN 14,214). There are a variety of fuel characteristics to consider such as density, cetane number, acid value, kinematic viscosity, pour point, cloud point and sulphur contents. The cetane number of biodiesels is nearly identical to that of petroleum diesel. In comparison to conventional diesel, biodiesel has a higher flash point, making it safer to carry and store. As shown in Table 2, the test procedure compares the biodiesel and diesel standard requirements [38, 42, 70].

3 Green Diesel

Green diesel is a new generation biofuel also known as "renewable diesel", "second-generation diesel", "Hydrogenated Esters and Fatty Acids (HEFA)", "Bio-Hydrogenated Diesel (BHD)", "Hydrogenation Derived Renewable Diesel (HDRD)", "Hydrotreated Vegetable Oil" or "Hydrogenated Vegetable Oil". The last two names share the same acronym HVO and have been used during the last decade since vegetable oils were the most usual biomass feedstock for the production of this biofuel. Today, green diesel is also produced from other biomass sources such as animal or fish fats but the term HVO is still in use in the industry as well as in the fuel standards and the European regulation [18]. Green diesel is a mixture of straight chain and branched saturated hydrocarbons free oxygen which typically

Table 2 Comparison standards of biodiesel

Fuel property	Test method	Biodiesel		Diesel
		ASTM D6751	EN 14,214	ATM D975
Cetane number	ASTM D613	47 min	51 min	48 min
Acid value mg KOH g^{-1}	AOAS Cd 3d-63	0.80 max	0.50 max	–
Kinematic viscosity (mm^2 s^{-1}; 40 °C)	ASTM D445	1.9–6.0	3.5–5.0	2.6
Density (15 °C), kgL^{-1}	ASTM D5002	0.87–0.9	0.86–0.9	0.85
Flash point (°C)	ASTM D93	93 min	120 min	59 min
Cloud point (°C)	ASTM D2500	−3 to 12	–	
Pour point (°C)	ASTM D97	−15 to 16	–	3 to 9
Sulphated ash content (mg kg^{-1})	ASTM D482	0.02 wt% max	–	–
Sulphur content (mg kg^{-1})	ASTM D2494	15 max	10 max	50 max

contains 15 to 18 carbon atoms per molecule (C15 to C18). Due to its chemical affinity and compatibility with fossil petroleum diesel, green diesel can be used with existing diesel engines [56] in pure form (called R100) or as a blend with any desired blending ratio, for example, a blend of 20% renewable diesel and 80% petroleum diesel is called R20, and a blend of 5% renewable diesel and 95% of petroleum diesel is called R5, offering several advantages over biodiesel, including lower greenhouse gas emissions.

3.1 Green Diesel Production Method

Green diesel has a molecular structure that is similar to petroleum diesel, but with some chemical and physical properties of better quality, for instance, high cetane number, good chemical–thermal stability and low sulphur content. Recently, the hydroprocessing of triglycerides/fatty acid/fatty acid methyl ester under high hydrogen pressure and temperature in the presence of a catalyst is the most preferable for biofuel production. Hydroprocessing is a general term used for the catalytic reactions that use hydrogen to eliminate the heteroatoms such as sulphur, nitrogen, oxygen and metals, and also to saturate the olefins and aromatics. In this process, unsaturated fatty acids are firstly converted into saturated fatty acids via hydrogenation followed by the hydrocracking of fatty acids into shorter chain hydrocarbon through simultaneous reactions of decarboxylation/decarboxylation/hydrodeoxygenation [66].

Hydrogenation is the reaction where the addition of hydrogen occurs without the cleavage of bonds. The process is commonly used to saturate or reduce organic compounds under a hydrogen atmosphere [48]. The purpose of hydrogenating renewable feedstocks on an industrial scale is to produce hydrocarbon molecules boiling in the diesel range, which are directly compatible with existing fossil-based diesel and meet all current specifications for ultra-low sulphur diesel (ULSD), as specified in ASTM D973. With the introduction of feedstocks stemming from renewable sources, new types of molecules with a significant content (10–15 wt%) of oxygen are present and must be properly treated by both the hydrotreating process and catalysts. High oxygenated compounds give undesirable properties to bio-oil since they contribute to the high corrosiveness, low thermal stability, low volatility and low energy content [47]. It is an essential process in the biofuel industry as biomass-derived compounds need to be hydrogenated or hydrodeoxygenated to produce biofuels and chemicals.

Additional catalysts play a significant role in the hydroprocessing process. Sotelo-Boyás et al. [63] carried out hydrocracking process in a batch reactor over a temperature range of 300 to 400 °C, initial hydrogen pressure of 5 to 11 MPa and reaction times 1 to 6 h using NiMo/γ-Al2O3 and Pt-zeolitic-based catalysts found that the metal function of the catalyst and a high hydrogen pressure contribute to the saturation of the side chains of the triglycerides while the acid function of the catalyst contributes to the cracking of the C–O bond and to the isomerization of the n-olefins formed, which are then transformed in iso-paraffins. A high enough temperature is important to increase the cracking activity. However, temperatures higher than 380 °C will further increase the hydrocarbon cracking to lighter products hence decreasing the diesel yield. There are many catalysts used for hydrocracking reactions such as nickel (Ni), niobium phosphate (NbOPO4), zeolite, rhodium (Rh), platinum (Pt) and palladium (Pd). Among catalysts, noble metal catalysts (Pd and Pt) are the most favourable for hydrocracking reactions due to their high hydrogenation ability [66].

As above mentioned, green diesel has been mainly produced by the hydrotreating of triglycerides in vegetable oils in the presence of some catalysts. Free fatty acids are formed by the scission of propane from the glycerol backbone of the triglyceride molecules in presence of hydrogen. The three moles of linoleic acid, palmitic acid and oleic acid are formed. In the second step, the unsaturated fatty acids are converted to saturated fatty acids by hydrogenation. Then, hydrodeoxygenation (HDO), decarbonylation (deCO) and decarboxylation (deCO$_2$) are occurred to eliminate oxygen known as the deoxygenation process (DO). Deoxygenation involves the removal of oxygenated compounds from a molecule in the form of water, CO$_2$ or CO. Deoxygenation can be accomplished with or without the presence of hydrogen during the reaction. Three main mechanisms for deoxygenation include (1) hydrodeoxygenation that involves C–O bond cleavage under hydrogen and yields hydrocarbons and water, and for fatty acids, it involves reducing the oxidation state of the carbon atom of the carboxylic groups using H$_2$; (2) decarboxylation that removes carboxyl groups to yield paraffins and CO$_2$ and (3) decarbonylation that yields olefins, CO and water; [48]. The hydrodeoxygenation (HDO) keeps the same carbon atoms as in the original fatty acids whereas the decarbonylation (deCO) and decarboxylation (deCO$_2$) form n-alkane that have one carbon atom less than the original fatty acids [32].

Hydrodeoxygenation

The primary objectives of HDO are to reduce the O/C ratio and at the same time to increase the H/C ratio [52]. Hydrodeoxygenation, which consumes the highest amount of H$_2$ is an exothermic reaction process that removes oxygen as water at higher temperature and H$_2$ pressures. It is H$_2$-intensive process and hence requires continuous H$_2$ supply to achieve the desired result. Hydrodeoxygenated vegetable oil possesses lower viscosity, superior atomization, lubricity and high cetane number [73]. The absence of oxygen in the diesel also reduces its reactivity and makes it more stable [1] and hence suitable as a commercial replacement to diesel.

Figure 1 shows the reaction mechanism for HDO for tristearin and stearic acid as a model compound of triglycerides and fatty acids into straight chain hydrocarbons. In the HDO process, the triglycerides present in vegetable oils are converted into n-alkanes along with water and propane gas as by-products. Separation of oxygen in HDO reaction takes place via C–O bond rupture, C = O bond hydrogenation and C–C bond cleavage. The larger hydrocarbons are selectively formed in the presence of a catalyst which enables the C = O bond hydrogenation and C–O bond cleavage, thereby preventing the C–C bond breaking [52]. The route of C–C and C–O bond cleavage and the degree of cleavage are known to be the primary factors that determine the selectivity of the HDO reaction and most likely influenced by the role of catalysts.

However, therein lies an issue regarding the HDO reaction. The formation of water as the major by-product is a drawback that may reduce the catalyst activity [75]. To

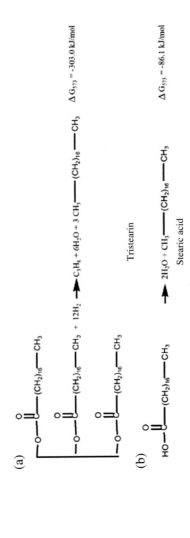

Fig. 1 Reaction mechanism of **a** tristearin and **b** stearic acid in HDO pathway [1]

overcome the issue, many researchers use the supported noble metal catalyst such as Pd/SiO, Pd/C and Pt/C [36] in HDO reaction but due to high cost, it becomes less favourable. Thus, modifications have been proposed in recent years with the introduction of a new catalyst such as supported bi-metal catalyst [8]. The consumption of high H_2 gas which is typically a major requirement for the hydrogenation reaction to occur makes HDO less environmental and economical friendly [27]. The reason for this is because of the costs associated with the use of H_2 and the fact that the majority of the world's H_2 production comes from fossil fuel reforming. Ideally, biofuels, which are supposed to be considered sustainable and renewable, should not be heavily dependent on non-renewable sources [58].

Decarboxylation and Decarbonylation

Decarbonylation (deCO) removes carbonyl group (C = O) as both water and CO to produce alkanes/alkenes with one less carbon atom than the precursor fatty acid with the consumption of 0–1 mol H_2 per mol of fatty acid, depending on the reaction mechanism. Rogers and Zheng [58] decarbonylation removes carbonyl groups from the product to improve its stability and reduce its heating value [1]. Decarboxylation (deCO$_2$) is a process of the removal carboxyl group (-COOH) from the molecule as CO_2 and produces alkanes with one less carbon unit making it particularly suitable for producing diesel with lower acidity. deCO$_2$ effectively consumes no H_2 which seems attractive from the economic standpoint. Unlike the HDO reaction pathway, the decarboxylation reaction pathways are overall endothermic reactions and generally favour increasing reaction temperatures. Figure 2 shows the reaction mechanism for tristearin and stearic acid as model compounds of triglycerides and fatty acids into straight chain hydrocarbons via decarboxylation and decarbonylation process.

The deCO and deCO$_2$ reaction routes are claimed to be the more appealing reaction routes, as they require fewer activated hydrogen sites and thus less H_2. As deoxygenation reactions are related to cracking reactions, they have been shown to

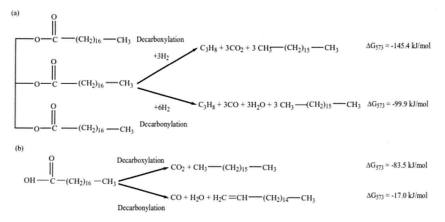

Fig. 2 Reaction mechanism of **a** tristearin and **b** stearic acid in decarboxylation and decarbonylation pathway [52]

be the most prevalent at elevated temperatures (typically > 300 °C) owing to their endothermicity [58].

3.2 Green Diesel Properties

Renewable diesel is a biomass-derived transportation fuel suitable for use in diesel engines. It meets the ASTM D975 specification for petroleum in the United States and EN 590 in Europe as mentioned in Sects. 10.2 and 10.4. Table 3 shows that the typical green diesel produced also meets the specifications of "EN15940: Automotive

Table 3 Comparison standard of green diesel [18]

	Green Diesel European Standard		Typical green diesel
	EN15940:2016/A1:2018		
	Min	Max	
Carbon (wt%)	–	–	84.9
Hydrogen (wt%)	–	–	15.1
Oxygen (wt%)	–	–	0
Cetane number (CN)	70/51	–	70–90
Lower heating value (LHV) MJ/kg	–	–	43.7–44.5
Density at 15 °C (kg/m^3)	765	800	770–790
Polycyclic aromatic hydrocarbons (wt%)	–	–	<0.1
Aromaticity	–	1.1 wt%	0 max
Sulphur content (mg/kg)	–	5	<5
Flash point (°C)	55	–	>59
Cloud point (°C)	Down to −34		
Ash content (wt%)	–	0.01	<0.001
Water content (mg/kg)	–	200	<200
Carbon residue on 10% distillation (wt%)	–	0.3	<0.1
Total contamination (mg/kg)	–	24	<10

(continued)

Table 3 (continued)

| | Green Diesel European Standard | | Typical green diesel |
| | EN15940:2016/A1:2018 | | |
	Min	Max	
Water and sediment (vol%)	–	0.02	≤ 0.02
Fatty acid methyl esters (FAME) (vol%)	–	7	0
Viscosity at 40 °C (mm^2/s)	2	4.5	2–4

fuels-Paraffinic diesel fuel from synthesis or hydrotreatment-Requirements and Test Methods" which specifically applies to non-oxygenated green diesel fuels made by hydroprocessing or Fischer–Tropsch synthesis.

4   Petrodiesel

Petroleum diesel, alternatively known as petrodiesel or diesel, is the most widely used in the transportation sector, such as cars, buses, trucks, trains, boats and ships. Petrodiesel has a wide variety of performance, efficiency and safety characteristics as a transportation fuel. It also has a higher energy density than other liquid fuels; thus, it delivers more usable energy per unit of volume. Petrodiesel is also defined as a low sulphur diesel fuel derived from crude oil. U.S. Environment Protection Agency (US EPA) has regulated the standard for sulphur content limits for diesel fuel to facilitate significant reductions in hazardous SOx emissions from diesel engines, contributing significantly to acid rain and air pollution. Hence, to comply with US EPA regulations, the petroleum sector is developing Ultra-Low Sulphur Diesel (ULSD) fuel, cleaner-burning diesel with a sulphur level of no more than 15 parts per million (ppm) [17]. Currently, crude oil is a significant source of energy and fuel. It is originated from fossil fuel that is discovered underground in subterranean reservoirs. Density, sulphur content and acidity are used to classify crude oil, and these characteristics influence how it is refined [62]. Each crude oil source has a distinct combination of molecules that determine its physical and chemical features, such as colour and viscosity. Thus, the petrodiesel produced from the crude oil is generated using a process known as fractional distillation, in which crude oil is heated to 350 °C under atmospheric pressure, and the constituents of the oil are separated into fractions [79]. Saturated hydrocarbon derivatives (75% v/v), mainly paraffin hydrocarbons, including n-paraffins, iso-paraffins and cycloparaffins and aromatic hydrocarbon derivatives (25% v/v), including alkylbenzenes and naphthalene derivatives, are found in crude oil, along with low amounts of sulphur, nitrogen, oxygen and metals [59, 61]. Diesel fuels are mainly composed of C_{10} to C_{19} hydrocarbons [65]. It is worth noting that the petrodiesel molecules are hydrocarbons, consisting

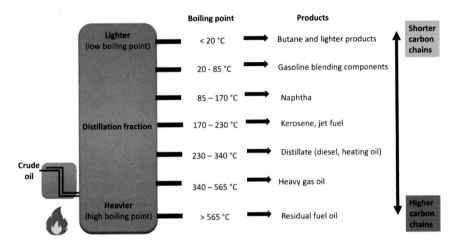

Fig. 3 Fractional distillation process that separates crude oil into fractions. *Source* U.S. Energy Information Administration [49]

of hydrogen and carbon atoms in addition to minor portions of heteroatoms. Thus, in ideal conditions, with enough oxygen, combustion of diesel product produces CO_2 and H_2O and traces of pollutants.

Figure 3 illustrates the fractional distillation of crude oil into its constituent fractions, including diesel. Nevertheless, all the fractions are further processed in other refinery units. Typically, there are three steps in petroleum refineries: separation, conversion and treatment [15]. In modern separation, crude oil is heated through piped furnaces to nearly 350 °C. The liquids and vapours from the furnace are discharged into the distillation column. All refineries utilize atmospheric distillation units, while more advanced refineries employ vacuum distillation units. The distillation units divide the liquids and vapours into petroleum fractions depending on their different boiling points. At the bottom are the heavy fractions, while at the top are the light fractions. The lightest fractions, such as gasoline and liquefied petroleum gases, evaporate and ascend to the highest part of the distillation fraction column. Meanwhile, medium-weight fractions, such as kerosene and distillates (diesel), stay in the centre of the distillation fraction column. Heavier fractions (gas oils) dissociate farther down in the distillation fraction column, as the heaviest fractions have the highest boiling points.

However, the heavy and low-value distillation fractions will be further treated to produce diesel and gasoline. Cracking is the most frequently used in conversion method, since it employs heat, pressure, catalysts and hydrogen to convert high molecular weight hydrocarbon into lighter molecules such as gasoline and diesel fuel. Thermal cracking, hydrocracking and catalytic cracking are three distinct techniques that may break down high molecular weight hydrocarbon. Nonetheless, the cracking technique is not the sole process.

4.1 Production Method

Fluid Catalytic Cracking (FCC)

Oil refineries utilize FCC to transform high molecular weight hydrocarbon fractions into more lucrative gasoline, olefinic gases and other by-products. The feedstock for FCC is typically heavy gas oil or vacuum gas oil (HVGO), which is a fraction of crude oil having an average boiling point at 340 °C or above atmospheric pressure. The feedstock is heated to an elevated temperature and mild pressure before being introduced to the catalyst in the FCC process (Fig. 4). The catalyst converts the heavy molecular weight of hydrocarbon into shorter chain molecules [67].

Hydrocracking

Hydrocracking is a flexible catalytic refining process that is frequently used to improve the heavy molecular weight of hydrocarbon fractions produced after crude oil distillation and residue. The process involves the addition of H_2, which eliminates contaminants such as sulphur to create a product that satisfies environmental requirements. It also consists of converting heavy molecular weight molecules to smaller molecular weight molecules via the breaking of C–C bonds and the addition of H_2. The significant products possess lower boiling temperatures, are highly saturated and typically range from heavy diesel to light naphtha [67].

Hydrodesulphurization

Hydrodesulphurization (HDS) is a catalytic reaction and it is widely used in petroleum distillation. It is used to remove sulphur from petroleum products and natural gas through the hydrogenolysis process. Typically, HDS is carried out by co-feeding oil and hydrogen into a fixed-bed reactor loaded with effective inorganic catalysts (e.g. $CoMo/Al_2O_3$, $NiMo/Al_2O_3$), as shown in Fig. 5 [22]. The sulphur in

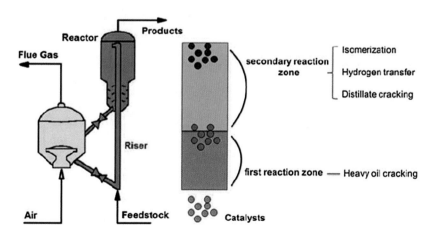

Fig. 4 A simplified diagram representation of a petroleum refinery for fluid catalytic cracking unit [22]

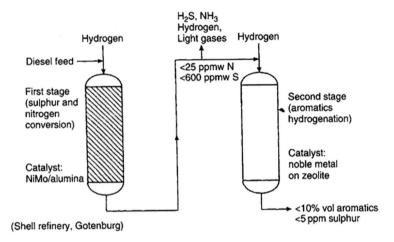

Fig. 5 Line-up of Gothenburg two-stage HDS unit [57]

organo-sulphur compounds is transformed to H₂S via HDS under high pressure (1–18 MPa) and elevated temperature (200–450 °C), with the exact parameters varying based on the degree of desulphurization needed and the type of the sulphur molecules in the feed [6].

Hydrotreatment

The diesel hydrotreating (DHT) unit as shown in Fig. 6 mainly used to decrease unfavourable species from diesel fractions via preferentially reacting unfavourable species with H₂ in a reactor at high temperatures and mild pressures. On the other hand, this process is also used to enhance the colour, odour, thermal, oxidative

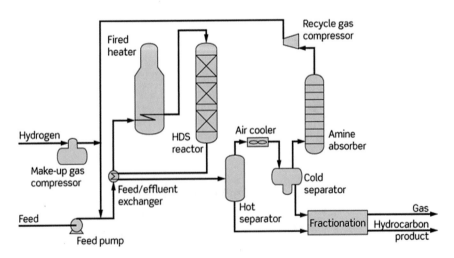

Fig. 6 Schematic diagram of hydrotreatment unit

stability and demulsibility of basic lubricant stocks [65]. On the DHT catalyst surface, several processes take place simultaneously, involving hydrodesulphurization (HDS), hydrodenitrogenation (HDN) and aromatic saturation/hydrogenation (HDA) [82].

4.2 Diesel Grades

In a sophisticated refinery, the petroleum diesel fuel pool is comprised of all accessible streams, including (i) straight-run product, (ii) light cycle oil from a fluid catalytic cracking unit and (iii) hydrocracked gas oil. Furthermore, straight-run diesel may be used or may need modest enhancements before being used in off-road diesel fuel. Thus, the refinery must combine all accessible streams to ensure that the product fulfills all performance, regulatory, economic and inventory criteria. In addition, petroleum diesel fuels are commercially available in many grades, classified into different standards according to the application purposes.

Generally, the standard usually specifies the qualities of the diesel fuel, such as the flashpoint, cetane number, sulphur content and density. Thus, several diesel fuel specifications are available, including EN 590 (European Union), ASTM D975 (United States), GOST R 52,368 (Russia; similar to EN590) and NATO F 54 (equivalent to EN 590). For instance, in the United States, ASTM D975 refers to grades 1-D and 2-D distillate fuels, frequently utilized in high-speed mobile engines, medium-speed stationary engines and rail locomotive engines. On the other hand, Grade 4-D refers to the more viscous distillates and mixes of these distillates with residual fuel oils. No. 4-D fuels are suitable for low- and medium-speed engines that need a steady load and a consistent speed [65].

4.3 Diesel Fuel Properties and Uses

Diesel oil is defined by several characteristics that determine its structure and existence, including its colour, ignition quality, viscosity, gravity, volatility and stability. The colour of diesel oil is a helpful indicator of contaminants by residual components, water or small solid particles. Thus, visual inspection is essential to ensuring the delivery of clean fuel. The appearance of diesel oil, predominantly the colour, should be determined since it is used for manufacturing control purposes. Likewise, acknowledgement is significant as to smell, and it is typically necessitated that diesel fuel is sensibly free of contaminants like mercaptans (RSH, thiol derivatives), which bestow unsavoury scents to the fuel. On the other hand, the primary feature of diesel fuel quality is the cetane number, which measures a diesel fuel's ignition delay and is based on the combustion properties of two hydrocarbons: n-hexadecane (cetane) and 2, 3, 4, 5, 6, 7, 8-heptamethylnonane. Cetane has a short delay time upon combustion and a cetane number of 100; heptamethylnonane has a longer delay time and a cetane number of 15.

Similarly, to how the octane number is important for automotive fuels, the cetane number is used to determine the ignition quality of diesel fuels. It is similar to the percentage by volume of cetane in the blend with heptamethylnonane that matches the test fuel's ignition quality (ASTM D613). Fuels with cetane values between 40 and 55 may be used in diesel engines [13].

Furthermore, cloud and pour points are used to determine the extent to which wax has coagulated in the oil. They do not evaluate the actual wax composition of the oil. The oil temperature is decreased until the appearance of haziness; this defied as the cloud point. Further decrease in the temperature causes the oil to cease flow, known as the pour point [3].

While the viscosity of oil indicates its resistance to internal flow and, therefore, its lubricating properties, it is standard practice in the oil sector to provide viscosities in centistokes (the unit for kinematic viscosity), Saybolt universal seconds, or Redwood seconds. Viscosity data are often obtained in the laboratory at 100 °F, 130 °F or 210 °F [64].

4.4   Latest Research

Diesel innovation continues to evolve to meet today's and the future's requirements for effective, perfect and solid force in a variety of industries around the world. Despite increasing efforts to GHGs emissions via the development of emerging technologies like batteries and fuel cells, which may eventually replace diesel in specific applications, diesel will remain the dominant technology for the time being. The future of diesel will be defined by four main strategies: emission closer to zero, improved energy performance, increased use of low-carbon renewable biofuels and hybridization [19].

Emissions closer to zero

Diesel is still the most popular long-haul transportation technology, supplying 97% of trucks in the US. Greener diesel fuel improved engines and efficient pollution management work together to achieve near-zero pollution of fine particles and smog-forming chemicals such as nitrogen oxides (NOx). California approved new emission regulations in 2020, requiring lower levels of NOx and particulate matter beginning in 2027 [55].

Improve energy performance

The probable efficacy of the diesel engine is an intriguing topic since the US EPA and the National Highway Traffic Safety Administration (NHTSA) is investigating the potential severity of new productivity requirements for diesel engines. According to certified engines from 2013 to 2014, modern compression-ignition diesel engines lead the commercial trucking sector, converting approximately 43–44% of fuel energy into engine output. Tractor engines will likely decrease their fuel consumption and CO_2 emissions by 6% between 2010 and 2017, or approximately 1% per year, to satisfy current efficiency and carbon requirements [51].

Expanded use of renewable fuels

Diesel, biodiesel and renewable diesel are all important measures in the fight against climate change. Excellent biofuels are affordable, accessible and proven because they are suitable for both new and current cars. They provide significant near-term reductions in GHGs emissions across broad sectors of the economy that depend heavily on diesel engines now and in the future. Currently, most diesel engines across the world run on ULSD fuel. The majority of heavy-duty diesel engines can run on biodiesel mixes up to 20% (B20). Renewable diesel fuel is designed to follow the same design guidelines as petroleum diesel fuel and may be used as a complete replacement for petroleum. Even though biodiesel and renewable diesel fuel are made from waste agricultural feedstocks, the US EPA classifies them as advanced biofuels since they reduce GHG emissions by at least 50% [17]. Consumption of these advanced biofuels is relatively low compared to petroleum diesel fuel, but it is growing. Biodiesel and renewable diesel were consumed in 3 billion gallons in 2020. During the same period, 68 billion gallons of ULSD petroleum diesel fuel were used.

The United States Renewable Fuel Standard (RFS) requires the use of specific biofuels mixed with petroleum-based diesel and gasoline with the ultimate aim of developing locally produced fuel with the potential to decrease GHGs emissions. The RFS mandates the use of biofuels such as biodiesel and renewable diesel fuel throughout the country.

Hybridization

Another way to enhance diesel engines and save even more fuel while reducing GHGs emissions from internal combustion engines (ICEs) is to use hybrid innovation in broader applications. Hybridization is a technique that captures wasted energy, stores it and then uses it to do beneficial work. It will become more prevalent in the future in specific applications. Furthermore, hybrid system components such as electric motors, controllers and energy storage systems work with ICEs and conventional transmissions to use energy from both sources to move the vehicle or machine with less overall energy consumption than an engine alone.

5   Conclusion

The current chapter has discussed the difference between biodiesel, green diesel and petrol diesel. As conclusion, biodiesel and green diesel have enhanced quality than petrodiesel since they can be produced from biomass of low and high acidity. The fuels are also eco-friendly, reduce sulphur content and have low particulate matter, THC and carbon monoxide emissions. Green diesel also does not contain oxygen which makes it very stable progressively.

References

1. Abdulkareem-Alsultan G, Asikin-Mijan N, Lee HV, Rashid U, Islam A, Taufiq-Yap YH (2019) A review on thermal conversion of plant oil (edible and inedible) into green fuel using carbon-based nanocatalyst. Catal 9(4):350
2. Aboelazayem O, Gadalla M, Saha B (2017) Biodiesel production from waste cooking oil via supercritical methanol: optimisation and reactor simulation. Renew Energy 124:144–154
3. Akhil AG, MAKPK, Akhilesh S, MACA, Khan S, Kanna R (2017) Determination of cloud and pour point of various petroleum products. Int Ref J Eng Sci 6(9):1–4
4. Ali OM, Mamat R, Abdullah NR, Adam A (2016) Analysis of blended fuel properties and engine performance with palm biodiesel-diesel blended fuel. Renew Energy 86:59–67
5. Altaie MAH (2020) Performance and emission analysis of enriched biodiesel and its combined blends with petrodiesel. Biofuels 11(3):251–260. https://doi.org/10.1080/17597269.2017.135 8949
6. Alves L, Paixão SM, Pacheco R, Ferreira AF, Silva CM (2015) Biodesulphurization of fossil fuels: energy, emissions and cost analysis. RSC Adv 5(43):34047–34057
7. Ambat I, Srivastava V, Sillanpää M (2018) Recent advancement in biodiesel production methodologies using various feedstock: a review. Renew. Sust Energ Rev 90:356–369
8. Asikin-Mijan N, Lee HV, Abdulkareem-Alsultan G, Afandi A, Taufiq-Yap YH (2017) Production of green diesel via cleaner catalytic deoxygenation of Jatropha curcas oil. J Clean Prod 167:1048–1059
9. Atabani AE, Silitonga AS, Badruddin IA, Mahlia TMI, Masjuki HH, Mekhilef S (2012) A comprehensive review on biodiesel as an alternative energy resource and its characteristics. Renew Sust Energ Rev 16(4):2070–2093
10. Avhad MR, Marchetti JM (2015) A review on recent advancement in catalytic materials for biodiesel production. Renew Sust Energ Rev 50:696–718
11. Cai, G (2016) Performance of petrodiesel and biodiesel fuelled engines: a fundamental study of physical and chemicals effects. Dissertation, The University of Adelaide
12. Cebrián-garcía, S, Balu AM, Luque R (2018) Ultrasound-assisted esterification of valeric acid to alkyl valerates promoted by biosilicified lipases. Front Chem 6(197): 1–7
13. Chang PRR, Hsu S, Springer Handbook of Petroleum Technology (2017) (Springer International Publishing.pdf). Springer, 2017
14. Costa-felix RPB, Figueiredo MKK, Alvarenga AV (2018) An ultrasonic method to appraise diesel and biodiesel blends. Fuel 227:150–153
15. Demirbas A (2010) Biorefineries. Springer-Verlag, London
16. Demirbas A, Dincer K (2008) Sustainable green diesel: a futuristic view. Energy Sources, Part A: recovery, utilization and environmental effects 30(13):1233–1241
17. Diesel Fuel Standards and Rulemakings (2021) United States Environmental Protection Agency, Washington. https://www.epa.gov/diesel-fuel-standards/diesel-fuel-standards-and-rulemakings, accessed 23 Nov 2021
18. Douvartzides SL, Charisiou ND, Papageridis KN, Goula MA (2019) Green diesel: biomass feedstocks, production technologies, catalytic research, fuel properties and performance in compression ignition internal combustion engines. Energies. 12(5): 1–41 (809)
19. Driving the Future (2021) Diesel Technology Forum, Maryland. https://www.dieselforum.org/policy/driving-the-future, accessed 23 Nov 2021
20. Duz MZ, Saydut A, Ozturk G (2011) Alkali catalyzed transesterification of sunflower seed oil assisted by microwave irradiation. Fuel Process Technol 92:308–313
21. Farooq M, Ramli A (2015) Biodiesel production from low FFA waste cooking oil using heterogeneous catalyst derived from chicken bones. Renew Energy 76:362–368
22. Feng R, Qiao K, Wang Y, Yan ZF (2013) Perspective on FCC catalyst in China. Appl Petrochemical Res. 3:63–70
23. Fonseca JM, Teleken JG, Almeida VDC, Silva C (2019) Biodiesel from waste frying oils: methods of production and purification. Energy Convers Manag 184:205–218

24. Gonçalves M, Mantovani M, Carvalho WA, Rodrigues R, Mandelli D, Silvestre Albero J (2014) Biodiesel wastes: an abundant and promising source for the preparation of acidic catalysts for utilization in etherification reaction. Chem Eng J 256:468–474

25. Gude VG, Patil P, Martinez-guerra E, Deng S (2013) Microwave energy potential for biodiesel production. Sustain Chem Process 1: 1–31(5)

26. Guru M, Koca A, Can O, Can Ç, Fatih S (2010) Biodiesel production from waste chicken fat-based sources and evaluation with Mg based additive in a diesel engine. Renew Energy 35(3):637–643

27. Huber GW, O'Connor P, Corma A (2007) Processing biomass in conventional oil refineries: production of high-quality diesel by hydrotreating vegetable oils in heavy vacuum oil mixtures Appl Catal A-Gen 329:120–129

28. Innocent SD, Sylvester PO, Yahaya FM, Nwadike IN, Okoro L (2013) Comparative analysis of biodiesel and petroleum diesel. Int J Educ Res 1(8):1–8

29. Kalnes TN, Marker T, Shonnard DR, Koers KP (2008) Green diesel production by hydrorefining renewable feedstocks. Biofuels Technology 4:7–11

30. Karnjanakom S, Kongparakul S, Chaiya C, Reubroycharoen P, Guan G, Samart C (2016) Biodiesel production from Hevea brasiliensis oil using SO₃H-MCM-41 catalyst. J Environ Chem Eng 4(1):47–55

31. Kirubakaran M, Arul Mozhi Selvan V (2018) A comprehensive review of low-cost biodiesel production from waste chicken fat. Renew Sustain Energy Rev 82:390–401

32. Kittisupakorn P, Sae-ueng S, Suwatthikul A (2016) Optimization of energy consumption in a hydrotreating process for green diesel production from palm oil. Computer Aided Chemical Engineering 38:751–756

33. Korkut I, Bayramoglu M (2017) Selection of catalyst and reaction conditions for ultrasound assisted biodiesel production from canola oil. Renew Energy 116:543–551

34. Lien Y, Hsieh L, Wu JCS (2010) Biodiesel synthesis by simultaneous esterification and transesterification using oleophilic acid catalyst. Ind Eng Chem Res 49(5):2118–2121

35. Lin, J., Chen, Y. (2017) Production of biodiesel by transesterification of Jatropha oil with microwave heating. J. Taiwan Institute of Chemical Engineers. 75(c):43–50 0, 1–8

36. Liu D, Chen EYX (2013) Diesel and alkane fuels from biomass by organocatalysis and metal-acid tandem catalysis. Chemsuschem 6(12):2236–2239

37. Lokman IM, Goto M, Rashid U, Taufiq-Yap YH (2016) Sub- and supercritical esterification of palm fatty acid distillate with carbohydrate-derived solid acid catalyst. Chem Eng J 284:872–878

38. Lokman IM, Rashid U, Hin Y, Yunus R (2015) Methyl ester production from palm fatty acid distillate using sulfonated glucose-derived acid catalyst 81:347–354

39. Loures CCA, Amaral MS, Da PCM, Zorn SMFE, De Castro HF, Silva MB (2018) Simultaneous esterification and transesterification of microbial oil from Chlorella minutissima by acid catalysis route: a comparison between homogeneous and heterogeneous catalysts. Fuel 211:261–268

40. Lycourghiotis S, Kordouli E, Kordulis C, Bourikas K (2021) Transformation of residual fatty raw materials into third generation green diesel over a nickel catalyst supported on mineral palygorskite. Renew Energy 180:773–786

41. Manchanda T, Tyagi R, Sharma DK (2018) Comparison of fuel characteristics of green (renewable) diesel with biodiesel obtainable from algal oil and vegetable oil. Energy Sources, Part A: recovery, utilization and environmental effects. 40(1):54–59

42. Mansir N, Hwa S, Ibrahim ML, Taufiq-Yap Hin Y (2017) Synthesis and application of waste egg shell derived CaO supported W-Mo mixed oxide catalysts for FAME production from waste cooking oil: effect of stoichiometry. Energy Convers Manage 151:216–226

43. Mardhiah HH, Chyuan H, Masjuki HH, Lim S, Ling Y (2017) Investigation of carbon-based solid acid catalyst from Jatropha curcas biomass in biodiesel production. Energy Convers Manage 144:10–17

44. Mehta A, Mehta N, Mehta A (2019) Optimisation of performance parameters for biodiesel production with slow pyrolysis using response surface methodology. Int J Ambient Energy 42(16):1823–1829

45. Melo-espinosa EA, Piloto-rodríguez R, Goyos-pérez L, Sierens R, Verhelst S (2015) Emulsification of animal fats and vegetable oils for their use as a diesel engine fuel: an overview. Renew Sust Energ Rev 47:623–633
46. Modi, DN (2010) Biodiesel production using supercritical methanol. Dissertation, Missouri University of Science And Technology
47. Nie L, Resasco DE (2014) Kinetics and mechanism of m-cresol hydrodeoxygenation on a Pt/SiO$_2$ catalyst. J Catal 317:22–29
48. Oi LE, Choo MY, Lee HV, Ong HC, Hamid SBA, Juan JC (2016) Recent advances of titanium dioxide (TiO$_2$) for green organic synthesis. RSC Adv 6(110):108741–108754
49. Oil and petroleum products explained (2020) U. S. Energy Information Administration, Washington. https://www.eia.gov/energyexplained/oil-and-petroleum-products/refining-crude-oil-the-refining-process.php, accessed 23 Nov 2021
50. Onukwuli OD, Esonye C, Ofoefule AU (2021) Combustion exhaust release impact, diesel engine performance, and optimization studies of green diesel-petrodiesel blend in a high compression ratio direct-injection compression-ignition engine. Adv Mech Eng 13(5):1–25
51. Oscar D, Nic L (2015) Advanced tractor-trailer efficiency technology potential in the 2020–2030 timeframe. The International Council on Clean Transportation. https://theicct.org/public ations/advanced-tractor-trailer-efficiency-technology-potential-2020%E2%80%932030-tim eframe. Accessed 23 Nov 2021
52. Pattanaik BP, Misra RD (2017) Effect of reaction pathway and operating parameters on the deoxygenation of vegetable oils to produce diesel range hydrocarbon fuels: a review. Renew Sust Energ Rev 73:545–557
53. Prabu SS, Asokan MA, Prathiba S, Ahmed S, Puthean G (2018) Effect of additives on performance, combustion and emission behavior of preheated palm oil/diesel blends in DI diesel engine. Renew Energy 122:196–205
54. Rajalingam A, Jani SP, Kumar AS, Khan MA (2016) Review article production methods of biodiesel. J Chem Pharm Res 8(3):170–173
55. Raju ASK, Wallerstein BR, Johnson KC (2021) Achieving NOx and greenhouse gas emissions goals in California's heavy-duty transportation sector. Transp Res Part D Transp Environ 97:102881
56. Ramírez-Verduzco LF, Hernández-Sánchez MJ (2021) Blends of green diesel (synthetized from palm oil) and petroleum diesel: a study on the density and viscosity. Bioenergy Research 14(3):1002–1013
57. Rigutto MS, vatto Veen R, Huve L (2007) Chapter 24—Zeolites in hydrocarbon processing. Stud Surf Sci Catal 168: 855–913
58. Rogers KA, Zheng Y (2016) Selective deoxygenation of biomass-derived bio-oils within hydrogen-modest environments: a review and new insights. Chemsuschem 9(14):1750–1772
59. Saleh TA (2020) Characterization, determination and elimination technologies for sulfur from petroleum: toward cleaner fuel and a safe environment. Trends Environ Anal Chem 25: e00080
60. Sanchez-Cantu M, Tellez MM, Lydia MP, Zeferino-díaz R, Hilario-martínez JC, Sandoval-ramírez J (2019) Biodiesel production under mild reaction conditions assisted by high shear mixing. Renew Energy 130:174–181
61. Sharma M, Sharma P, Kim JN (2013) Solvent extraction of aromatic components from petroleum derived fuels: a perspective review. RSC Adv 3:10103–10126
62. Solomon LH, Carruthers, JE, Waddams AL (2021) Petroleum refining. Encyclopedia Britannica. https://www.britannica.com/technology/petroleum-refining, accessed 23 Nov 2021
63. Sotelo-Boyás R, Liu Y, Minowa T (2008) Production of green diesel by hydrocracking of canola oil on Ni-Mo/γ- Al$_2$O$_3$ and Pt-zeolitic based catalysts. Proceeding of AIChE Annual Meeting, Philadelphia
64. Speight JG (2017) Handbook of petroleum refining. CRC Press, United Kingdom
65. Speight JG (2019) Handbook of industrial hydrocarbon processes. Elsevier, United Kingdom
66. Srihanun N, Dujjanutat P, Muanruksa P, Kaewkannetra P (2020) Biofuels of green diesel–kerosene–gasoline production from palm oil: Effect of palladium cooperated with second metal on hydrocracking reaction. Catalysts 10(2):1–13

67. Sullivan D, Metro S, Pujadó PR (eds) (2015) Handbook of petroleum processing. Springer, Switzerland
68. Syazwani ON, Ibrahim ML, Kanda H, Goto M (2017) Esterification of high free fatty acids in supercritical methanol using sulfated angel wing shells as catalyst. J Supercrit Fluids 124:1–9
69. Taherkhani M, Sadrameli SM (2017) An improvement and optimization study of biodiesel production from linseed via in-situ transesterification using a co-solvent. Renew Energy 119:787–794
70. Thushari I, Babel S (2018) Sustainable utilization of waste palm oil and sulfonated carbon catalyst derived from coconut meal residue for biodiesel production. Biores Technol 248:199–203
71. Thushari I, Babel S (2020) Biodiesel production from waste palm cooking oil using solid acid catalyst derived from coconut meal residue. Waste and Biomass Valorization 11(9):4941–4956
72. Thushari I, Babel S, Samart C (2019) Biodiesel production in an autoclave reactor using waste palm oil and coconut coir husk derived catalyst. Renew Energy 134:125–134
73. Tiwari R, Rana BS, Kumar R, Verma D, Kumar R, Joshi RK, Garg MO, Sinha AK (2011) Hydrotreating and hydrocracking catalysts for processing of waste soya-oil and refinery-oil mixtures. Catal Commun 12(6):559–562
74. Venkateswarulu TC, Raviteja CV, Prabhaker KV, Babu DJ, Ranganadha Reddy A, Indira M, Venkatanarayana A (2014) Review on methods of transesterification of oils and fats in bio-diesel formation. Int J Chemtech Res 6(4):2568–2576
75. Veriansyah B, Han JY, Kim SK, Hong SA, Kim YJ, Lim JS, Shu YW, Oh SG, Kim J (2012) Production of renewable diesel by hydroprocessing of soybean oil: effect of catalysts. Fuel 94:578–585
76. Vern R, Hua Y, Mubarak NM, Khalid M, Abdullah EC, Nolasco-hipolito C (2019) An overview of biodiesel production using recyclable biomass and non- biomass derived magnetic catalysts. J Environ Chem Eng 7(4): 103219
77. Vonortas A, Papayannakos N (2014) Comparative analysis of biodiesel versus green diesel. WIREs Energy Environ 3:3–23
78. Wei C, Huang T, Chen H (2013) Biodiesel production using supercritical methanol with carbon dioxide and acetic acid. J Chem 7895942:1–6
79. Yoon JJ (2017) What's the difference between biodiesel and renewable (green) diesel. Advance Biofuels USA. https://advancedbiofuelsusa.info/wp-content/uploads/2011/03/11-0307-Biodie sel-vs-Renewable_Final-_3_-JJY-formatting-FINAL.pdf, accessed 24 Nov 2021
80. Zhao X, Wei L, Cheng S, Huang Y, Yu Y, Julson J (2015) Catalytic cracking of camelina oil for hydrocarbon biofuel over ZSM-5-Zn catalyst. Fuel Process Technol 139:117–126
81. Zhou Y, Niu S, Li J (2016) Activity of the carbon-based heterogeneous acid catalyst derived from bamboo in esterification of oleic acid with ethanol. Energy Convers Manag 114:188–196
82. Zhu FXX, Hoehn R, Thakkar V, Yuh E (2017) Hydroprocessing for clean energy: design, operation, and optimization.Wiley, United State
83. Zillillah TG, Li Z (2012) Highly active, stable, and recyclable magnetic nano-size solid acid catalysts: efficient esterification of free fatty acid in grease to produce biodiesel. Green Chem 14(11):3077–3086

Chapter 11
Biodiesel and Green Diesel Fuels: A Techno-Economic Analysis

J. Aburto and M. A. Amezcua-Allieri

Abstract Climate change represents a major challenge for our world's equilibrium and society. Therefore, we need to implement an adaptive energy matrix where fuels like fossil diesel are used wisely in coming years, but where sustainable and renewable biofuels increase its contribution to the transport sector or in electric power generation. Among those renewable biofuels, biodiesel and green diesel may play significant roles through circular economy of biomass, with local and regional production into a biorefinery with minima emissions and waste streams. The assessment of a particular scientific development may be approached by the techno-economic analysis (TEA) of the production process, which may serve to identify scientific and technological limitations and challenges that impact economic and financial parameters. In this work, we present and discuss the advantages and challenges of blending biodiesel or green diesel with fossil diesel, the technological, economic and financial issues that must be approached by TEA and its constrains, and the comparison and discussion of several TEA works dealing with biodiesel and green diesel production. Even if green diesel is just recently available with respect to biodiesel, it represents an important opportunity in co-processing plant oils into a petroleum refinery, access to high volume capacity plants and distribution, which may help to decarbonize the petroleum fuel sector.

Keywords Energy transition · Green diesel · Biodiesel · Techno-economic analysis · Biofuel processes

J. Aburto (✉) · M. A. Amezcua-Allieri
Gerencia de Transformación de Biomasa, Instituto Mexicano del Petróleo, Eje Central Lázaro Cárdenas Norte 152, Col. San Bartolo Atepehuacan, Alcaldía Gustavo A. Madero, 07730 Mexico City, Mexico
e-mail: jaburto@imp.mx

309

1 Introduction

Currently, the main source of energy is still coming from hydrocarbons that are extracted from the subsoil, which are not renewable and whose use generates greenhouse gases that are directly related to climate change (CC). It is well known that CC affects the planet; therefore, this phenomenon is the main topic of international agreements for countries to take the necessary measures for CC mitigation. Renewable energy represents an option for clean transport fuels. Therefore, some countries are facing the development of a new policy related to sustainable energy transition. However, a paradigm shift to renewable energy is not only essential for environment but also calls for economy changes. The approach of linear economy must migrate to a circular economy in which wastes are reduced, reused and recycled, and energy intensity is decoupled from population growing.

Biomass, as a raw material, may have a major impact on the environment and the transition into a circular economy. A smarter use of raw materials can lower CO_2 emissions and mitigate some environmental impacts, due to conversion of lignocellulosic and oleaginous biomass into a wide range of value-added products (such as biofuels, bioproducts, and industrially important chemicals) that may contribute to decarbonization. The most representative liquid biofuels obtained from oleaginous biomass are biodiesel and green diesel. Their production is highly correlated to conversion technology improvements and the right types of biomass for a successful process. The assessment of a particular scientific development may be approached by the techno-economic analysis (TEA) of the production process, which may serve to identify scientific and technological limitations and challenges that impact economic and financial parameters.

This chapter deals with biodiesel and green diesel quality in terms of the chemical composition, properties and specifications, in addition to their blending with fossil diesel. The production process for both biofuels is also described focus on TEA of the production process. Particular emphasis is put on technology advancement for more energy efficient and cost-effective energy, taking the advantage of biomass, as a primary source of renewable energy used as a sustainable alternative to fossil fuels.

2 Biodiesel and Green Diesel: Properties, Specifications and Blending

Renewable fuels for compression-ignition (CI) engines comprise biodiesel and green/renewable diesel and may be used blended with fossil diesel. Hence, biodiesel is a mixture of long-chain Fatty Acid Methyl Esters (FAME) derived from the transesterification of triacylglycerides from renewable lipid sources, such as vegetable oil, animal fat, tallow and waste cooking oil (WCO) with methanol but ethanol can also be used [19]. The largest possible source of suitable oils comes from oil crops such as rapeseed, palm or soybean but alternatives are Jatropha, castor oil, camelina, among others. Hence, biodiesel, known as B100 or neat biodiesel, is a liquid biofuel in its pure and unblended form. Like fossil diesel, biodiesel is used in fuel compression-ignition engines with little or no modifications. For biodiesel (100% Biodiesel, B100), specifications are proposed to standardize the quality of this biofuel with the main international standards. In this way, the performance and durability of the cars will not be affected. The proposed specifications are those in force in the USA [6] and in Europe EN 14214 [21]. The sulfur content must be less than or equal to 15 ppm (mg/kg).

Moreover, green diesel is a second-generation biofuel, which is generally a mixture of lineal paraffins similar to those present in fossil diesel and may enhance some diesel blend properties as cetane, cold flow, time storage and particulate matter in exhaust. The green diesel may be produced by several technologies that affect such final composition.

On the other hand, fossil diesel is a light middle distillate fuel for use in a diesel engine with several grades depending mainly on the sulfur content from 15 to 5000 ppm [5].

Fuel specifications for fossil diesel and biodiesel fuels are issued by the USA [5–7], Europe [23], and national regulations as in Mexico [17, 22].

Renewable fuels as biodiesel and green diesel may be added to fossil diesel as an additive to enhance some properties (less than 0.5–1.0% v/v) or in a larger amount in order to contribute to volume pool but more importantly, to decarbonize the blended fuel and contribute to less net CO_2 emissions. The use of biodiesel or green diesel is practically restricted as a blend with fossil diesel. As the renewable fuel may come from different oleaginous sources and therefore be composed by different fatty chains or paraffins, it is necessary to characterize its quality as neat or blended fuel by already mentioned standards. As far as it is reported, the properties of diesel blend with biodiesel are not altered in mixtures with concentrations less than or equal to 5% v/v according to experience in its use in various countries of the world with the exception of the enhancement on lubricity with a least 0.5% v/v biodiesel into fossil diesel [68].

Since green diesel composition is like fossil diesel, the blends must meet the standards ASTM D975 and EN 590. Now, the properties of such fuels reveal some differences that limit blending (Table 1). Indeed, blending policy considers a maximum 20% v/v biodiesel (B20) in the USA, followed by B10 in countries like China, Brazil and Argentina, while Canada considers only B2. We observed that the net calorific value (41–43 MJ/kg) and specific weight (0.78–0.86) of neat biodiesel and green diesel are quite similar among the fuels and fewer incidences are expected at high biodiesel or green diesel contents. Nevertheless, the ignition and clouding temperatures as well as kinematic viscosity are higher for biodiesel and green diesel that may affect engine performance. The high paraffin content of green diesel increases the cetane number and it also contributes to the high clouding temperature (18 °C). Other properties like S, N and aromatic hydrocarbon content are not relevant since such compounds [4] are not present in biodiesel or in small quantities in green diesel. Another alternative diesel fuel is gas to liquids (GTL) diesel and is obtained through the Fischer–Tropsch process of natural gas. Its properties are similar to other fuels,

Table 1 Property comparison of fossil diesel, biodiesel, green diesel, GTL diesel and blends

Property	Unit	Standard	Diesel[1]	Biodiesel[2]	Green diesel[3]	GTL diesel[4]	HVO30[5]	B20[6]
Net calorific value	MJ/kg	ASTM D240	42.34	41.3	43.47	34.08	43.1	38.9
Ignition temperature	°C	ASTM D93	104	174	138	–	–	111
Clouding temperature	°C	ASTM D2500	3	16	18	−7	−7	–
Kinematic viscosity @ 40 °C	mm^2/s	ASTM D445	3.81	4.5	3.94	4.03	2.83	4.12
Specific weight	–	ASTM D1298	0.841	0.855	0.778	0.774	0.826	0.858
S	mg/kg	ASTM D5453	303	–	3.2	'	6.6	–
N	mg/kg	ASTM D5291	62	–	<0.3	–	–	–
Aromatic hydrocarbons	% vol	ASTM D1319	22.4	–	0.6	'	2.8% w/w	–
Saturated hydrocarbons	% vol	ASTM D1319	68	–	99.1	'	–	–

[1] Diesel obtained from Mexican Tula Refinery. Experimental data obtained from Mexican Petroleum Institute; [2] Nagi et al. [48], [3] Green diesel obtained from palm oil at the Mexican Petroleum Institute; [4] Armas et al. [4], [5] ASTM [10], [6] Ravi Teja et al. [58]

but its calorific value is minor with a low clouding temperature [4]. This neat fuel was tested on CI engine without modifications and showed reduction on hydrocarbons and CO emissions, but NOx exhaust remains unchanged to those of fossil diesel.

Blending of renewable fuels with fossil diesel is usually done at the refinery and must comply already mentioned standards. Hence, a blend composed by 30% hydrotreated vegetable oil (HVO) and fossil diesel (HVO30) showed a relevant −7 °C clouding temperature, which allows it to be used in cold weather (Table 1). Other relevant properties are like neat fossil and renewable diesels. The CO_2 exhaust emission of such blend showed a reduction with 569 g/kWh when compared with fossil diesel (580 g/kWh). The HVO30 blend also showed a significant drop on CO, hydrocarbons, particulate matter and NOx exhaust emissions [10]. This combustion behavior was attributed to the lower distillation range, the higher cetane number and the absence of aromatic hydrocarbons in HVO.

Concerning biodiesel blends with fossil diesel, we observed that relevant properties of a B20 blend using cashew nutshell biodiesel are very close to those of fossil diesel although kinematic viscosity outranges the standard of No. 1 diesel but complies with No. 2 diesel (Table 1). The brake thermal efficiency (BTE), which refers to the efficiency of heat conversion to work, is slightly minor for the B20 blend when compared to fossil diesel and attributed to the higher viscosity and specific weight and smaller calorific value of B20 blend [58]. Moreover, it has been observed that the blends of WCO or karanja biodiesel with fossil fuel larger to 40% (B40) does not meet the specification of filter blocking tendency and indicating that such blends may clog a fuel filter [70]. The presence of salt particles, polymers from oxidative biodiesel degradation or the formation of clouds at low temperature may be the cause of operational problems of the engine. Also, a minor deposition of carbon was detected on the cylinder head, the injector tip and the piston crown of a diesel engine using a B20 blend with rice bran biodiesel. Such improved operation was attributed to a better fuel combustion, less soot formation and better lubricity [33]. As in the case of HVO blends, the use of B20 blend in a diesel engine resulted in lower exhaust emissions like CO_2, CO and hydrocarbons, but NOx incremented. The authors concluded that the performance of the CI engine does not differ significantly if a B20 blend or fossil diesels are used [69].

Biodiesel performance in cold weather depends on the blend of biodiesel, the feedstock and the fossil diesel characteristics. In general, blends with smaller percentages of biodiesel perform better in cold temperatures. Typically, 5% biodiesel (B5) performs about the same that fossil diesel in cold weather. Both biodiesel and No. 2 diesel have some compounds that crystallize in very cold temperatures. In winter weather, fuel blenders and suppliers consider the use of a cold flow improver to avoid crystallization. For the best cold weather performance, users should work with their fuel provider to ensure that the blend is appropriate [42].

In some cases, additional fuel additives added to diesel fuels would perform some specific functions such as: (1) reduction of exhaust emissions as NOx, COx and particulate matter, (2) increase the oxygen concentration in the engine and in the particulate filter to improve combustion, (3) improve the fluid stability over wide-ranging conditions, (4) increase the viscosity index, (5) reduction in ignition delay time, and flash point [59] and (6) increase the cetane number, lubricity and conductivity. Indeed, the use of nano-additives enhances the quality of diesel–biodiesel fuel blends [64]. The nano-additives are used to achieve specific fuel properties and to improve the performance characteristics and to attain a good emission control of the compression ignition engine without any modification. Khond et al. [40] reported that nano-Al particles are more conducive to the creation of micro-explosions during combustion, leading to the air–fuel mixing and resulting in a complete combustion whilst manganese-based nanoparticle additives reduce emissions such as polycyclic aromatic hydrocarbons (PAHs; [26]). The use of antioxidants improves oxidation stability of a B20 blend while exhaust emissions like NOx diminishes about 6–9% but hydrocarbons are augmented between 8 and 11% and attributed to a reduction of oxidative free radicals [57].

3   Production Process for Biodiesel and Green Diesel

Low carbon fuels like biodiesel and green diesel may be produced by a certain number of technologies implying first the kind of biomass feedstocks like edible and non-edible oils and fats from plants, animals and microorganisms and their wastes [1, 3, 12, 13, 20, 52, 55, 66, 73, 78]. Nevertheless, such kind of feedstocks is greater when corresponds to the production of green diesel, since microorganisms, as well as forest, agricultural, agro-industrial and urban solid residues may be used. Secondly, the technology choice corresponds to the definition of each biofuel, i.e. biodiesel is the blend of methyl or ethyl esters of fatty acids while green diesel is composed mostly by large paraffins (C_{12}–C_{18}) and some cyclic alkanes and alkenes as stated above.

Hence, biodiesel is commonly produced via trans-esterification between an oil/fat and methanol/ethanol to obtain the corresponding ester blend and glycerol as coproduct and usually with a homogeneous alkali catalyst but an acid one or enzymes may also be employed [11, 30, 39, 45, 51, 54]. Nevertheless, the use of heterogeneous catalysts or free/immobilized enzymes has gained some interest due to the reduction of processing issues like ester/water separation, wastewater treatment and corrosion [3, 77]. Processing of fatty feedstock in the presence of alcohol raises some mass and heat transfer problems that limit reaction and biodiesel yield. Typical batch reactors suffer from these issues, but they are the more frequently used. Heterogeneous catalysts allow the use of continuous reactors while the application of microchannel, ultrasound and microwave reactors can reduce transfer problems, but ultrasound needs special attention to avoid emulsification, while the two firsts must be scale-up. They are also efforts to produce biodiesel without any catalyst, and this approach requires high temperatures and pressures as well as high volume of alcohol like in methanol supercritical conditions [3, 11, 63]. Also, separation of biodiesel and glycerol/water/alcohol phase is enhanced by methanol supercritical approach, but sedimentation and centrifugation are the most popular separation methods when using a batch reactor [30].

In the case of green diesel, it might be obtained by the hydrotreatment of oil/fat feedstocks in the presence of hydrogen and a heterogeneous catalyst, which are also named as hydrotreated vegetable oils, HVO, like the NexBTL® and ShiftoGreen® processes [30, 45, 46, 56]. Moreover, there are some studies concerning different kinds of processes like pyrolysis, hydrothermal liquefaction and gas/Fischer–Tropsch (called gas to liquids, GTL, when natural gas is the feedstock) that produce a bio-oil that requires further refining to diesel and gasoline-like fractions [4, 14]. All these approaches require new or dedicated installations that increment capital investment (CAPEX), but an interesting one is the co-processing of oil/fat with a middle distillate through hydrotreatment, which allows the use of existing hydrotreatment units inside a refinery and the in-built production of decarbonized green diesel/diesel blend [30] (Table 2).

Table 2 Main technologies for the conversion of oily biomass to biodiesel and green diesel

Technology	Catalyst	Reaction conditions	References
Transesterification	Homogeneous Heterogeneous	Oil/MeOH ratio 1:20 30–100 °C	Farobie and Hartulistiyoso [25]
In-situ transesterification	Homogeneous Heterogeneous	Simultaneous oil extraction, esterification and transesterification	Thoppil and Hein [71]
Hydrotreatment	Ni NiMoS$_x$ Ni-Mo/SiO$_2$-Al$_2$O$_3$	30–92 bars 180–450 °C	Long et al. [45]
Pyrolysis/hydroprocessing	Sulfided Ni-Mo/γAl$_2$O$_3$ Pt/Al$_2$O$_3$	50–200 bars 300–400 °C	Patel and Kumar [55]
Hydrothermal liquefaction	Catalytic Uncatalytic	100–250 bars 280–370 °C	Chiaramonti et al. [13]
Gasification/Fischer–Tropsch	Co	7–12 bar 210–260 °C	Cruz-Neves et al. [18]

4   Techno-Economic Analysis of Biodiesel and Green Diesel Production

Techno-economic analysis (TEA) may be applied to early studies of basic and applied research to get insights about the scientific endeavors and technical parameters (1<TRL<5) like reaction stoichiometry, kinetics, yield, purity, experimental conditions, and others that impact financial parameters, which may guide near-to-market technology projects, 5<TRL<9 [37, 49]. Biodiesel and its blends with fossil diesel have been in the market in several countries for many years as seen before, while green diesel is still an embryonic technology with a limited market. Indeed, biodiesel and green diesel production accounts for 150,000 and 23 million liters per year, respectively [35, 36]. Therefore, biodiesel technology and market may be considered as mature when compare to green diesel. Nevertheless, TEA is a significant tool that might guide the deployment of mature technologies in new regions and countries considering cases as feedstocks, market conditions, policies, among others. In case of early-stage products like green diesel, TEA may be used to prioritize research and development and to build scenarios of deployment. The process simulation requires the definition of a conceptual process engineering, identification of all entries and outs (feedstocks, chemicals, products, coproducts, emissions, etc.), definition of plant capacity to obtain the economic metrics like internal rate of return (IRR), payback time (PBt) net present value (NPV), among others. There are some proprietary programs like Superpro Designer, Aspen Plus, Pro/II, IMP-Bio2Energy and CHEMCAD as well as open-source programs like SIM42, OMChemSim and

DWSIM and BioSTEAM [16, 47]; but not all were designed for biomass processing simulation.

The study of biofuels production by TEA involves different feedstocks depending on regional and national availabilities and policies, different technologies, catalysts, process flow configurations, separation units, upgrading processes, capital expenses (CAPEX) and unit production costs (Tables 3 and 4). Sometimes the co-processing of oil/fat blends with/without fossil fractions is considered [8, 31, 75]. This latter approach may accelerate the decarbonization of fossil fuels by using existing petroleum refineries and lowering CAPEX. Also, the valorization of whole biomass results in production of energy, fuels and bio-based chemicals that may be implemented with carbon capture, use and storage (CCUS), which leads to a circular bioeconomy that will be discussed further [15, 43].

Hence, biodiesel production involves the use of oily-rich feedstocks such as microalgae, oleaginous plants, animal fats and their wastes (Table 3). The origin of the feedstock impacts the conversion process to biodiesel and therefore the techno-economic parameters of every technology. Most processes require the initial oil/fat extraction and condition/refining of the oily fraction in order to be able to put through certain process conditions and reactions. Then, the oily fraction composed by triacyl glycerides/free fatty acids is converted to methyl or ethyl fatty esters mainly by esterification and transesterification reactions. Such reactions usually involve the use of catalysts that may be homogeneous (NaOH, KOH, HCl, H_2SO_4, free enzymes), heterogeneous (various active phases supported on inert materials, supported enzymes). In supercritical conditions, un-catalytic reactions may be accomplished but higher temperature, pressure and oil/alcohol ratio are required [3, 11, 63]. Esterification and transesterification reactions are usually accomplished at low to medium temperatures (30–100 °C, see Table 3) and ambient pressure when a homogeneous catalyst or free enzymes are employed. But the use of heterogeneous catalysts frequently requires higher temperatures and pressures. This is attributed to a more constrained accessibility to the active sites on supported or solid catalyst that also requires a higher oil/alcohol ratio when compared to the homogeneous catalytic approach. Then, product obtention implies the separation of (1) the oily phase which contains the biodiesel and unreacted oily fraction and (2) the water phase composed by glycerol, unreacted alcohol and mono- and diacyl glycerides. All these will affect techno-economic parameters such as CAPEX and unit production cost (Table 3). Indeed, in situ transesterification of whole microalgae cells reduces CAPEX due to the minor requirement of processing units and then less kind of equipment and installations, i.e. 2.1, 3.9 and 4.2 million $US for in situ, catalytic and enzymatic transesterification, respectively. Nevertheless, the unit production cost of transesterification alternatives is higher than catalytic approach due to higher enzyme's cost, higher demand of solvent, alcohol, or utilities like cooling water, high-pressure steam and fired heat [32]. Considering more reliable feedstocks as oleaginous plants, the biodiesel's unit production cost drops considerably (0.7–1 $US/kg) due to supported extensive agriculture and existing conventional markets that make available the oil to the biodiesel's producers at a more competitive price compared to alternative sources as Jatropha curcas, Salicornia spp., Ricinus communis (castor bean), Camelina sativa

Table 3 Comparison of techno-economic analysis of biodiesel production

Feedstock	Main product	Technology	CAPEX (million $US)	Unit production cost ($US/kg)	References
Microalgae	Biodiesel	Catalytic transesterification	3.9	4.9–6.5	Heo et al. [32]
Microalgae	Biodiesel	Enzymatic transesterification	4.2	12.5–17.2	Heo et al. [32]
Microalgae	Biodiesel	In situ transesterification	2.1	7.6–22.2	Heo et al. [32]
Safflower	Biodiesel Bioethanol	Transesterification Saccharification-fermentation	93.5–95.8	0.7* 0.43	Khounani et al. [41]
Lignocellulosic waste	Biodiesel	Oleaginous yeast/transesterification	n/a	6.0	Sae-ngae et al. [61]
Syrup waste	Biodiesel	Oleaginous yeast/transesterification	n/a	9.0	Sae-ngae et al. [61]
African palm	Biodiesel Hydrogen	Transesterification Palm residues/Gasification Glycerol/steam reforming	2.2	0.89 2.7	Niño-Villalobos et al. [50]
Waste cooking oil	Biodiesel	Homogeneous transesterification	4.4	1.1	Al-Sakkari et al. [2]
Soybean oil	Biodiesel	Heterogeneous transesterification	7.2	0.78	Al-Sakkari et al. [2]
Palm oil	Biodiesel	Transesterification	5.2	1	Oke et al. [53]
Acidic oil	Biodiesel	Ionic liquid catalyzed transesterification	6.3	0.97	Gebremariam et al. [29]

* Unit price

Table 4 Comparison of techno-economic analysis of green diesel production and co-products

Feedstock	Main product	Technology	CAPEX (million $US)	Minimum selling price ($US/L)	References
Carinata oil	Jet fuel Gasoline Diesel	Catalytic hydrothermolysis	191	1.32* 1.22* 1.33*	Eswaran et al. [24]
Forest residues	Upgraded bio-oil	Catalytic pyrolysis Non-catalytic pyrolysis	122.3 104.4	0.8 0.3	van Schalkwyk et al. [72]
Pine woodchips	Bio-oil	Pyrolysis/co-processing with petroleum	91	0.5**	Talmadge et al. [67]
Sorghum biomass	Green diesel	Hydrolysis/Fermentation to fatty acids/Fischer–Tropsch	354	1.6**	Larnaudie et al. [44]
Palm oil	Green diesel Propane	Hydrotreating	13.8	0.8	Martínez-Hernández et al. [46]
Microalgae oil	Green diesel	Hydrothermal liquefaction	150–250	0.5–1	Roles et al. [60]
Forest residues	Green diesel	Catalytic fast pyrolysis/hydrocracking	213	2	Spatari et al. [65]
Forest residues	Green diesel	Fast pyrolysis/hydrocracking	427	1.68	Spatari et al. [65]
Waste vegetable oil	Green diesel Propane	Hydrogenation co-processing	27.5	0.81 $US/kg*	Glisic et al. [28]
Waste vegetable oil	Green diesel Propane	Hydrogenation unit stand-alone	56.8	1 $US/kg*	Glisic et al. [28]
Corn stover	Green naphtha and diesel	Fast pyrolysis and bio-oil upgrading	200–287	0.56–0.82**	Wright et al. [74]

* Unit price; ** Unit production cost

L., microalgae, among others. However, oils from African palm, soybean, sunflower, safflower, canola, rapeseed are essential for human food [78] and their use for biofuel applications have raised concerns [27]. CAPEX is similar among these oleaginous plants since stand-alone units are considered with similar process scheme and production capacity. Another attractive source of biomass are wastes as brewer's spent grains (lignocellulosic), expired soft drinks (syrup) and waste cooking oil (WCO). The two first may be used to grow oleaginous yeast that produces an oil able to convert it to biodiesel, but their unit production cost is still high (6–9 $US/kg). On the other hand, the valorization of WCO avoids the pollution of soil and water and is considered as the more reliable source for biodiesel production, but their availability is not big enough to account to the market. In addition, the total utilization of biomass into an integrated biorefinery that produces energy (thermal, power), biofuels (ethanol, biodiesel, biogas) and bio-based chemicals has attracted attention since is a unique renewable approach that shows such capacity to mitigate climate change. In that sense, the use of lignocellulosic residues such as safflower straw for ethanol and the oil for biodiesel may account for the economic biorefinery feasibility [41]. Also, the valorization of palm residues from palm oil extraction as well as glycerol co-produced with biodiesel might be used to obtain syngas through gasification or hydrogen by steam reforming, respectively [50]. This versatility supports the water–food–energy nexus that accounts for the sustainability of a biorefinery that considers the water footprint, food and energy production.

In the case of green diesel, there are different kinds of technologies that depend strongly on the chemical nature of the feedstock, since the biomass oil fraction to more complex biomass like wood residues, microalgae, sorghum biomass among others (Table 4). Such oil fraction may be treated by hydrotreating (hydrogenation), hydrogenation co-processing, pyrolysis or hydrothermolysis (hydrothermal liquefaction) to obtain mainly paraffins in the two first processes or a bio-oil in the third one that requires further processing and refining to obtain a diesel-like fraction. CAPEX strongly depends on process scheme and units as well as installed capacity, but we observed that green diesel requires a higher investment when compared to biodiesel (Tables 3 and 4). This is associated to the very different kinds of involved reactions for every biofuel and the need of conditioning and refining. Indeed, the hydrogenation of oils/fats converts the triacyl glycerides into linear paraffins, propane and water in which separation is easy and further refining is not necessary. Such paraffins may be blended with fossil diesel as long as their properties comply with diesel specifications [5], On the other side, green diesel synthesis requires the cracking and partial deoxygenation of biomass through hydrothermolysis or pyrolysis that yields a complex bio-oil composed by a myriad of molecules [55, 72] that required further processing and refining in order to obtain a diesel-like fraction.

Now, these processes require high temperature and pressure in contrast to biodiesel synthesis as well as hydrogen and sometimes the use of more expensive heterogeneous catalyst for conditioning and refining. All these affect OPEX and then the unit production cost and minimum selling price are higher (Table 4). The co-processing of bio-oil with petroleum results in lower unit production cost.

The environmental impact of renewable diesel production using life cycle assessment has been reported [79]. Authors showed that the production of renewable diesel by Hydro-processed Esters and Fatty Acids (HEFA) is more environmentally friendly than fossil diesel production. In particular, CO_2 emission decreases around 110% (i.e. mitigation occurred) compared with conventional diesel production. However, renewable diesel production has a relevant environmental impact on human toxicity because of agrochemical consumption during biomass cultivation.

5 Conclusions

Renewable fuels like biodiesel and green diesel may contribute to lesser CO2 net emissions when they are blended with fossil diesel. Although the composition of biodiesel and green diesel is quite different, fatty acid methyl esters vs C12-C18 paraffins, they modified some properties of blends like cetane number, lubricity, clouding temperature, brake thermal efficiency, and resulted in lower emissions of COx, particulate matter and hydrocarbons but generally NOx incremented. As these renewable diesels have a reduced or absent sulfur content, blends with fossil biodiesel may contribute to the obtention of ultra-low sulfur diesel as well as a decarbonized fuel due to neutral CO_2 emissions. The properties of blends with fossil diesel are not altered in mixtures with concentrations less than or equal to 5%, with the exception of lubricity that's it is enhanced by biodiesel in a 0.5–1.0% v/v content.

There are different technologies that lead to the obtention of biodiesel and green biodiesel. Technology choice depends upon the availability of lignocellulosic or oleaginous biomass, infrastructure, chemicals, catalysts, the matureness of technology and the CAPEX and OPEX associated costs. Techno-economic analysis is fundamental to get insights about the scientific endeavors and technical parameters that must be considered in order to enhance economic and financial parameters. Such an approach must also consider the specific case under study, the market conditions as well as government and industrial policies.

Acknowledgements The authors acknowledge the financial support from British Council, Newton Fund Impact Scheme through project NFIS-540821111 as well as Instituto Mexicano del Petróleo co-financing through project Y.62001 "A decision support platform for bioenergy technology deployment and policymaking in Mexico".

References

1. Aggarwal M, Remya N (2021) The State-of-the-Art production of biofuel from microalgae with simultaneous wastewater treatment: influence of process variables on biofuel yield and production cost. BioEnergy Res. https://doi.org/10.1007/s12155-021-10277-1
2. Al-Sakkari EG, Mohammed MG, Elozeiri AA, Abdeldayem OM, Habashy M, Ong ES, Rene ER, Ismail I, Ashour I (2020) Comparative technoeconomic analysis of using waste and virgin cooking oils for biodiesel production. Front Energy Res 583357
3. Alsultan AG, Asikin-Mijan N, Ibrahim Z, Yunus R, Razali SZ, Mansir N, Seenivasagam S, Taufiq-Yap YH (2021) A short review on catalyst, feedstock, modernised process, current state and challenges on biodiesel production. Catalysts 1261
4. Armas O, García-Contreras R, Ramos A, López AF (2015) Impact of animal fat biodiesel, GTL, and HVO fuels on combustion, performance, and pollutant emissions of a light-duty diesel vehicle tested under the NEDC. J Energy Eng C4014009
5. ASTM (2017) D975–17 Standard specifications for diesel fuel oils
6. ASTM (2015) Standard specification for biodiesel fuel blend stock (B100) for middle distillate fuels. D6751–15ce1
7. ASTM (2020) Standard specificications for fiesel fuel oils. D975

8. Bezergianni S, Dimitraidis A, Kikhtyanin O, Kubicka D (2018) Refinery co-processing of renewable feeds. Prog Energy Combust Sci 29–64
9. BS (2017) EN 590:2013+A1:2017 Automotive fuels -diesel requirements and test methods
10. Bortel I, Vávra J, Takáts M (2019) Effect of HVO fuel mixtures on emissions and performance of a passenger car size diesel engine. Renew Energy 140:680–691
11. Chanthon N, Ngaosuwan K, Kiatkittipong W, Wongsawaeng D, Appamana W, Assabumrungrat S (2021) A review of catalyst and multifunctional reactor development for sustainable biodiesel production. ScienceAsia 531–541
12. Chiaramonti D, Buffi M, Rizzo AM, Prussi M, Martelli F (2015) Bio-hydrocarbons through catalytic pyrolysis of used cooking oils: towards sustainable jet and road fuels. Energy Procedia 343–349
13. Chiaramonti D, Prussi M, Buffi M, Casini D, Rizzo AM (2015) Thermochemical conversion of microalgae: challenges and opportunities. Energy Procedia 819–826
14. Chiaramonti D, Buffi M, Rizzo AM, Prussi M, Martelli F (2021) Bio-hydrocarbons through catalytic pyrolysis of used cooking oils: towards sustainable jet and road fuels. Energy Procedia 343–349
15. Conteratto C, Artuzo FD, Santos OIB, Talamini E (2021) Biorefinery: a comprehensive concept for the sociotechnical transition toward bioeconomy. Renew Sustain Energy Rev 111527
16. Cortes-Peña Y, Kumar D, Singh V, Guest JS (2020) BioSTEAM: a fast and flexible platform for the design, simulation, and techno-economic analysis of biorefineries under uncertainty. ACS Sustain Chem & Eng 3302–3310
17. CRE (2016) Lineamientos por lo que se establecen las especificaciones de calidad y características. NOM-016-CRE
18. Cruz-Neves R, Cooling-Klien B, da Silva RJ, Alves-Ferreira-Rezende MC, Funke A, Olivarez-Gómez E, Bonomi A, Maciel-Filho R (2020) A vision on biomass-to-liquids (BTL) thermo-chemical routes in integrated sugarcane biorefineries for biojet fuel production. Renew Sustain Energy Rev 109607
19. Demirbas A (2009) Progress and recent trends in biodiesel fuels. Energy Convers Environ 50:14–34
20. Dimian AC, Kiss AA (2019) Eco-efficient processes for biodiesel production from waste lipids. J Clean Prod 118073
21. DIN (2019) EN 14214 Liquid petroleum products - fatty acid methyl esters (FAME) for use in diesel engines and heating applications -requirements and test methods.
22. DOF (2016) NOM-016-CRE "Especificaciones de calidad de los petrolíferos
23. EN (2017) Automotive fuels- diesel requierments and test methods. 590:2013+A1 2017
24. Eswaran S, Subramaniam S, Geleynse S (2021) Dataset for techno-economic analysis of catalytic hydrothermolysis pathway for jet fuel production. Data in Brief 107514
25. Farobie O, Hartulistiyoso E (2021) Palm oil biodiesel as a renewable energy resource in Indonesia: current status and challenges. BioEnergy Res. https://doi.org/10.1007/s12155-021-10344-7
26. Fazliakmetov R, Shpiro G (1997) Selection and manufacture technology of antismoke additives for diesel fuel and boiler fuels oils. Izdetal Stvo Neft I Gaz 4:4355
27. Filip O, Janda K, Kristoufek L, Zilberman D (2019) Food versus fuel: an updated and expanded evidence. Energy Econ 82:152–166
28. Glisic SB, Pajnik JM, Orlovic AM (2016) Process and techno-economic analysis of green diesel production from waste vegetable oil and the comparison with ester type biodiesel production. Appl Energy 176–185
29. Gebremariam SN, Hvoslef-Eide T, Terfa MT, Marchetti JM (2019) Techno-economic performance of different technological based bio-refineries for biofuel production. Energies 3916
30. Hájek M, Vávra A, Carmona HdP, Kocik J (2021) The catalysed transformation of vegetable oils or animal fats to biofuels and bio-lubricants: a review. Catalysts 1118
31. Han X, Wang H, Zeng Y, Liu J (2021) Advancing the application of bio-oils by co-processing with petroleum intermediates: a review. Energy Convers Manag: X 100069

32. Heo H-Y, Heo S, Lee JH (2019) Comparative techno-economic analysis of transesterification technologies for microalgal biodiesel production. Ind & Eng Chem Res 18772–18779

33. Hoang AT, Tabatabaei M, Aghbashlo M, Carlucci AP, Ölcer AI, Le AT, Ghassemi A (2021) Rice bran oil-based biodiesel as a promising renewable fuel alternative to petrodiesel: a review. Renew Sustain Energy Rev 135:110204

34. IMP (2016) Report of project D.61013 "biojet fuel"

35. International Energy Agency (IEA) (2021) Renewable energy market update. outlook for 2021 and 2022, May 2021. Último acceso: November de 2021. https://www.iea.org/reports/renewable-energy-market-update-2021

36. International Energy Agency (IEA) (2020) Renewables 2020. Analysis and forecast to 2025, November 2025. Último acceso: November de 2021. https://www.iea.org/reports/renewables-2020

37. Kargbo H, Harris JS, Phan AN (2021) Drop-in fuel production from biomass: critical review on techno-economic feasibility and sustainability. Renew Sustain Energy Rev 110168

38. Keskin A, Gürü M, Altiparmak D (2007) Biodiesel production from tall oil with synthesized Mn and Ni based additives: effects of the additives on fuel consumption and emissions. Fuel 11391143

39. Khan Z, Javed F, Shamair Z, Hafeez A, Fazal T, Aslam A, Zimmerman WB, Rehman F (2021) Current developments in esterification reaction: A review on process and parameters. J Ind Eng Chem 80–101

40. Khond VW, Kriplani V (2016) Effect on nanofluid additives on performances and emissions of emulsified diesel and biodiesel fueled stationary CI engine: a comprehensive review. Renew Sustain Energy Rev 1338:1348

41. Khounani Z, Nazemi F, Shafiei M, Aghbashlo M, Tabatabaei M (2019) Techno-economic aspects of a safflower-based biorefinery plant coproducing bioethanol and biodiesel. Energy Convers Manag 112184

42. Knothe G, van Gerpen J, Krahl J (2016) The biodiesel handbook. AOCS Book Series

43. Koytsoumpa EI, Magiri-Skouloudi D, Karellas S, Kakaras E (2021) Bioenergy with carbon capture and utilization: A review on the potential deployment towards a European circular bioeconomy. Renew Sustain Energy Rev 111641

44. Larnaudie V, Bule M, San KY, Vadlani PV, Mosby J, Elangovan S, Karanjikar M, Spatari S (2020) Life cycle environmental and cost evaluation of renewable diesel production. Fuel 118429

45. Long F, Liu W, Jiang X, Zhai Q, Cao X, Jiang J, Xu J (2021) State-of-the-art technologies for biofuel production from triglycerides: a review. Renew Sustain Energy Rev 111269

46. Martínez-Hernández E, Ramírez-Verduzco LF, Amezcua-Allieri MA, Aburto J (2019) Process simulation and techno-economic analysis of bio-jet fuel and green diesel production. Minimum selling prices. Chem Eng Res Des 60–70

47. Martínez-Hernández E, Amezcua-Allieri MA, Aburto J (2021) Assesing the cost of biomass and bioenergy production in agroindustrial processes. Energies 4181

48. Nagi J, Ahmed SK, Nagi F (2008) Palm biodiesel an alternative green renewable energy for the energy demands of the future. ICCBT 79–94

49. NASA (2012) Technology Readines Level (TRL). Último acceso: November de 2021. https://www.nasa.gov/directorates/heo/scan/engineering/technology/technology_readiness_level

50. Niño-Villalobos A, Puello-Yarce J, González-Delgado AD, Ojeda KA, Sánchez-Tuirán E (2020) Biodiesel and hydrogen production in a combined palm and Jatropha biomass biorefinery: simulation, techno-economic, and environmental evaluation. ACS Omega 7074–7084

51. Nisar S, Hanif MA, Rashid U, Hanif A, Akhtar MN, Ngamcharussrivichai C (2021) Trends in widely used catalysts for Fatty Acid Methyl Esters (FAME) production: a review. Catalysts 1085

52. Niu FX, Liu Q, Bu YF, Liu JZ (2017) Metabolic engineering for the microbial production of isoprenoids: carotenoids and isoprenoid-based biofuels. Synth Syst Biotechnol 167–175

53. Oke EO, Okolo BI, Adeyi O, Adeyi JA, Ude CJ, Osoh K, Otolorin J, Nzeribe I, Darlinton N, Oladunni S (2021) Process design, techno-economic modelling, and uncertainty analysis

of biodiesel production from palm kernel oil. BioEnergy Res. https://doi.org/10.1007/s12155-021-10315-y

54. Pasha MK, Dai L, Liu D, Du W, Guo M (2021) Biodiesel production with enzymatic technology: progress and perspectives. Biofuels Bioprod & Biorefining 1526–1548

55. Patel M, Kumar A (2016) Production of renewable diesel through the hydroprocessing of lignocellulosic biomass-derived bio-oil: a review. Renew Sustain Energy Rev 1293–1307

56. Ramírez-Verduzco LF, Aburto-Anell JA, Amezcua-Allieri MA, Luna-Ramírez MRS, Díaz-García L, Medellín-Rivera L, Rodríguez-Rodríguez JE (2020) Hydrodeoxigenation process of vegetable oils for obtaining green diesel. USA Patente 10858594, 8 December 2020

57. Rashed MM, Kalam MA, Masjuki HH, Habibullah M (2016) Improving oxidation stability and NOx reduction of biodiesel blends using aromatic and synthetic antioxidant in a light duty diesel engine. Ind Crops Prod 89:273–284

58. Ravi Teja KMV, Issac Prasad P, Kumar Reddy K, Banapurmath NR, Soudagar MEM, Hossain N, Afzal A, Ahamed Saleel C (2021) Comparative analysis of performance, emission, and combustion characteristics of a common rail direct injection diesel engine powered with three different biodiesel blends. Energies 14:5597

59. Ribeiro NM, Pinto AC, Quintella CM, da Rocha GO, Teixeira LSG, Guarieiro LN, do Carmo-Rangel M, Veloso MCC, Rezende MJC, da Cruz RS, de Oliveira AM, Torres EA, de Andrade JB (2007) The role of additives for diesel and diesel blended (ethnaol and biodiesel) fuels: a review. Energy & Fuels 4:2433–2445

60. Roles J, Yarnold J, Hussey K, Hankamer B (2021) Techno-economic evaluation of microalgae high-density liquid fuel production at 12 international locations. Biotechnol Biofuels 133

61. Sae-ngae S, Cheirsilp B, Louhasakul Y, Suksaroj TT, Intharapat P (2020) Techno-economic analysis and environmental impact of biovalorization of agro-industrial wastes for biodiesel feedstocks by oleaginous yeasts. Sustain Environ Res 11

62. Saxena V, Kumar N, Saxena VK (2017) Comprehensive review on combustion and stability aspects of metal nanoparticles and its additive effect on diesel and biodiesel fuelled CI engine. Renew Sustain Energy Rev 70–56

63. Singh CS, Kumar N, Gautam R (2021) Supercritical transesterification route for biodiesel production: Effect of parameters on yield and future perspectives. Environ Prog & Sustain Energy 1–13

64. Soudagar MEM, Nik-Ghazali NN, Abul-Kalam IA, Badruddin NR, Banapurmath N, Akram N (2018) The effect of nano-additives in diesel-biodiesel fuel blends: a comprehensive review on stability, engine performance and emission characteristics. Energy Convers Manag 146–177

65. Spatari S, Larnaudie V, Mannoh I, Wheeler MC, Macken NA, Mullen CA, Boateng AA (2020) Environmental, exergetic and economic tradeoffs of catalytic- and fast pyrolysis-to-renewable diesel. Renew Energy 371–380

66. Stamenkovic OS, Gautam K, Singla-Pareek SL, Dhankher OP, Djalovic IG, Kostic MD, Mitrovic PM, Pareek A, Veljkovic VB (2021) Biodiesel production from camelina oil: present status and future perspectives. Food Energy Secur. https://doi.org/10.1002/fes3.340

67. Talmadge M, Kinchin C, Chum HL, de Rezande-Pinho A, Biddy M, de Almeida MBB, Casavechia LC (2021) Techno-economic analysis for co-processing fast pyrolysis liquid with vacuum gasoil in FCC units for second-generation biofuel production. Fuel 119960

68. Tat ME, Celik ON, Er U, Gasan H, Ulutan M (2022) Lubricity assessment of ultra-low sulfur diesel fuel (ULSD), biodiesel, and their blends, in conjunction with pure hydrocarbons and biodiesel based compounds. Int J Engine Res 23:214–231

69. Tesfa BC, Mishra R, Aliyu AM (2021) Effect of biodiesel blends on the transient performance of compression ignition engines. Energies 14:5416

70. Thangamani S, Sundaresan SN, Kannappan S, Barawkar VT, Jeyaseelan T (2021) Impacto of biodiesel and diesel blends on the fuel filter: a combined experimental and simulation study. Energy 227:120526

71. Thoppil Y, Zein SH (2021) Techno-economic analysis and feasibility of industrial-scale biodiesel production from spent coffee grounds. J Clean Prod 127113

72. van Schalkwyk DL, Mandegari M, Farzad S, Görgens JF (2020) Techno-economic and environmental analysis of bio-oil production from forest residues via non-catalytic and catalytic pyrolysis processes. Energy Convers Manag 112815
73. Walls LE, Rios-Solis L (2020) Sustainable production of microbial isoprenoid derived advanced biojet fuels using different generation feedstocks: a review. Front Bioeng Biotechnol 599560
74. Wright MM, Daugaard DE, Satrio JA, Brown RC (2010) Techno-economic analysis of biomass fast pyrolysis to transportation fuels. Fuel S2-S10
75. Wu L, Yang Y, Yan T, Wang Y, Zheng L, Qian K, Hong F (2020) Sustainable design and optimization of co-processing of bio-oil and vacuum gas oil in an existing refinery. Renew Sustain Energy Rev 109952
76. Zhang X, Yang R, Anburajan P, Van-Le Q, Alsehli M, Xia Ch, Brindhadevi K (2022) Assessment of hydrogen and nanoparticles blended biodiesel on the diesel engine performance and emission characteristics. Fuel 307:121780
77. Zambare V, Patankar R, Bhusare B, Christopher L (2021) Recent advances in feedstock and lipase research and development towards commercialization of enzymatic biodiesel. Processes 1743
78. Zhou Y, Zhao W, Lai Y, Zhang B, Zhang D (2020) Edible plant oil: global status, health issues, and perspectives. Front Plant Sci 1315
79. Arguelles A, Amezcua-Allieri MA, Ramírez-Verduzco LF (2021) Life cycle assessment of green diesel production by hydrodeoxygenation of palm oil. Front Energ Res 9:690725. https://doi.org/10.3389/fenrg.2019.00025.

Chapter 12
Performance, Combustion, and Emission Analysis of Green Diesel Derived from Mesua ferrea L. Oil on a CI Engine: An Experimental Investigation

Himansh Kumar, Mohammad Aslam, Anil K. Sarma, and Pramod Kumar

Abstract Renewable diesel or green diesel is a positive approach to overcome the related issues associated with other biofuel formulation techniques such as gum formation (biodiesel), poor fuel properties, and the issue of phase separation (pyrolysis oil, other advanced biofuel, etc.) and the long-term operation effects. Green diesel is a straight chain paraffinic hydrocarbon which is free of aromatics, oxygen, and sulfur, and also contains high cetane number. In this experimental investigation, formulation and application of Mesua Ferrea L. vegetable oil-based green diesel have been discussed. Comparative assessment of green diesel and petrodiesel had depicted that the performance, combustion, and emission of CI engine were superior to green diesel. Improved fuel properties of green diesel enhanced the combustion characteristics of CI engine which reflects its superiority w.r.t petrodiesel. The effects of green diesel were also measured at a low CI engine compression ratio (15:1), to measure its better suitability.

Keywords Green diesel · Combustion analysis · Cetane number · CI engine · Compression ratio

H. Kumar (✉)
Mechanical Engineering Department, Teerthanker Mahaveer University, Moradabad, UP, India
e-mail: himansh.rmu@gmail.com

H. Kumar · P. Kumar
Mechanical Engineering Department, Dr. B R Ambedkar National Institute of Technology, Jalandhar, Punjab, India
e-mail: kushwahapramod@nitj.ac.in

H. Kumar · A. K. Sarma
Chemical Conversion Division, Sardar Swaran Singh-National Institute of Bio-Energy, Kapurthala, Punjab, India
e-mail: anil.sarma@nibe.res.in

M. Aslam
Department of Chemistry, National Institute of Technology, Srinagar (J&K), India
e-mail: maslam@nitsri.ac.in

© The Author(s), under exclusive license to Springer Nature Singapore Pte Ltd. 2022   325
M. Aslam et al. (eds.), *Green Diesel: An Alternative to Biodiesel and Petrodiesel*,
Advances in Sustainability Science and Technology,
https://doi.org/10.1007/978-981-19-2235-0_12

1 Introduction

Ultra-low sulfur diesel (ULSD) free of heavy metals and ash content is the present demand of society to decrease environmental issues [1, 2]. In continuation, this may also increase the working life of CI engines in terms of low maintenance and decreased lubricating oil frequency with less burden on emission control systems, compared with the existing scenario [3, 4]. In addition, the rapid decrement of crude oil reserves is also a barrier to its availability in the future [5, 6]. Liquid fuel from biological sources is a positive approach to overcome this situation. But its higher viscosity creates poor fuel atomization during injection in the engines, injector coking, cold start, and unregulated emissions due to incomplete combustion [7, 8]. These downsides of bio-based fuels retard their use as fuel in CI engines. This can be overcome from various routes, and hydrogenation of vegetable oil is one of them [9, 10]. Hydrogenation of vegetable oils is an advanced technique to formulate high-end bio-based renewable diesel comprising of superior fuel properties w.r.t petrodiesel and even superior combustion behavior during combustion in CI engine [10]. This fuel is also named as "renewable diesel/green diesel" due to its origin from petro-crop and is also green in color [11, 12].

Chemically, green diesel is a mixture of paraffinic hydrocarbons free of sulfur and aromatics. The cold flow and other fuel properties of green diesel can be regulated as per the requirement by adjusting the hydrogenation reaction parameters [13, 14]. The cetane number of green diesel is reported very high as compared to petrodiesel, which meant that it shows superior ignitibility to petrodiesel [15, 16].

Ogunkoya et al. (2015) had reported 10% higher brake thermal efficiency of CI engine for renewable diesel in comparison with petrodiesel. It was mainly due to the long carbon chain which resulted in high energy content, hence superior results. Also, the decrement in carbon monoxide (CO), unburned hydrocarbon (UHC), particulate matter (PM), and smoke emission were observed for renewable diesel in comparison with petrodiesel. However, the NO_x emission was reported higher due to high combustion temperature [11]. Aslam et al. (2017) had reported that the green diesel, obtained from the hydrogenation, followed by fractional distillation of its biocrude has higher cetane number and gross calorific value in comparison with petrodiesel [1, 2].

It has been revealed from the literature that renewable diesel/green diesel is a promising fuel in all regards to CI engine application. With this inspiration, the current work is focused on the formulation of renewable diesel/green diesel from *Masua Ferrea L.* vegetable oil, its characterization and application on CI engine. The fuel properties of green diesel were also compared with petrodiesel as per ASTM/EN standards. CI engine performance attributes have been taken into consideration for comparative analysis in-between petrodiesel and green diesel. The effects of green diesel at a low compression ratio (CR 15) were also done to examine its effects at this operating condition.

2 Material and Methods

2.1 Materials

Mesua ferrea L. (MFL) vegetable oil seed was first crushed and converted into powdered form and the oil was extracted through the soxhlet solvent extraction method. After that, the MFL oil was separated from the solvent with the help of rota-vac apparatus. The separated MFL oil was centrifuged to remove the solid and suspended particles. The collected MFL oil was heated above 100 °C to remove water and other sediments. Other chemicals were procured from commercial sources and used in their original condition. These chemicals are listed in Table 1.

2.2 Catalysts

A bio-based renewable catalyst had been used and prepared from Musa balbisiana Colla underground stem (MBCUS) ash, which was collected from the local area of Darrang, Assam (India). This catalyst was prepared by the simple thermal treatment route, in which the MBCUS was cut into thin parts (1.5 cm × 15 cm) and dried in the presence of sunlight for 15 days. Later on, flamed at ambient conditions to convert it into ash. The collected ash was activated through heating at 550 °C in the muffle furnace for 2 h.

Table 1 Chemicals used for the formulation of green diesel from MFL oil

Chemical name	Purity (%)	Company name
N, O-Bis(trimethylsilyl)trifluoroacetamide (BSTFA)	99.5	Sigma-Aldrich
Na_2CO_3	99.8	Sigma-Aldrich
Oleic acid	ultra-pure	Sigma-Aldrich
Stearic acid	99	Sigma-Aldrich
Pyridine	99.9	Sigma-Aldrich
Pentane	99.9	SRL
Hexane	99.9	Sigma-Aldrich
Hepatane	99.9	Sigma-Aldrich
Octane	99.9	Sigma-Aldrich
Nonane	99.9	Sigma-Aldrich
Undecane	99.9	Sigma-Aldrich
Tridecane	99.9	Sigma-Aldrich
Tetradecane	99.9	Sigma-Aldrich
Pentadecane	99.9	Sigma-Aldrich
Hexadecane	99.9	Sigma-Aldrich
Heptadecane	99.9	Sigma-Aldrich

2.3 Formulation of Green Diesel from MFL Oil

The green diesel had formulated through the conversion of MFL vegetable oil into biocrude followed by its hydro-cracking [2]. The biocrude was prepared with the help of high-pressure high temperature (HPHT) batch reactor with reaction temp 400 °C, with insertion of hydrogen at 05 MPa pressure, catalyst loading of 3% by weight at the reaction mixing speed of 100 rpm for 1 h duration. Almost, 93% (vol.) of yield was found after this reaction. Afterward, the collected biocrude was refined into various components as per the petroleum refinery specifications (ASTM D 2892 and ASTM D 5236 procedures) using True Boiling Point (TBP) distillation unit (B/R Instruments Corporation, USA). Six different types of distilled components at different temperature ranges, such as 35–140 °C (petrol range), 140–180 °C (kerosene range), 180–300 °C, 300–325 °C, 300–370 °C (petrodiesel range), and 370–482 °C (wax range), were fractioned from the MFL oil biocrude with the help of TBP distillation unit [12, 13]. Usually, the obtained hydrocarbon products within the range of 180–370 °C are observed as 'Green diesel' or 'Renewable diesel' [1, 2]. In continuation, the obtained green diesel was fueled in a CI engine test rig to examine its performance attributes.

2.4 Characterization of MFL Based Green Diesel

The fuel properties such as density, kinematic viscosity, flash point temperature, gross calorific value, cetane number, and cloud point temperature of green diesel, biodiesel from MFL oil, and petrodiesel were examined as per the standard methods provided by ASTM/EN.

2.5 CI Engine Experimental Test Rig. Set-Up

A one-cylinder, 4-stroke, constant speed CI engine test rig. with rated power of 5.2 kW @ 1500 rpm was considered to conduct experiments. A variac-based eddy current dynamometer, coupled with the crankshaft was used to vary the load on the CI engine test rig. Two piezometric sensors were placed inside the engine cylinder to screen the in-cylinder pressure and injection pressure. One optical crank-angle sensor was placed near the crankshaft to record each degree of rotation. The all-recorded signals from the various sensors were interpreted with the help of a data acquisition device and conveyed to 'Enginesoft LV Version 9.0' (Apex Innovation Pvt Ltd) software for performance and combustion attributes [8, 9].

AVL DiGas 444 Analyzer (India) was used to measure the level of harmful exhaust gas emissions. Harmful emissions such as carbon monoxide (CO), unburned hydrocarbon (UHC), and oxides of nitrogen (NO_x) were recorded as per the testing method, approved by the Ministry of Road Transport and Highways, Government of India, and specified in MoRTH/ CMVR/TAP-115/116, Issue No. 3, Part-VIII for AVL DI Gas 444 analyzer [16, 17].

3   Results and Discussion

3.1   Fuel Properties of MFL Oil-Based Green Diesel

The fuel properties of green diesel from MFL, B100, biodiesel blend (B20) from MFL oil and petrodiesel have been shown in Table 2. The density of green diesel was found comparable with petrodiesel and lower than that of biodiesel. But, the kinematic viscosity of green diesel was found lower than that of petrodiesel and B20. Also, a higher calorific value was observed for green diesel along with higher cetane number in comparison with petrodiesel and B20, resulting due to long chain of straight paraffinic hydrocarbons. The flash point temperature of green diesel was higher than that of petrodiesel and B20. Overall, green diesel has shown superior values w.r.t B20 and petrodiesel.

3.2   CI Engine Performance Attributes

In this segment, performance attributes such as brake thermal efficiency (BTE) and brake specific fuel consumption (BSFC) of CI engine for MFL oil-based green diesel in comparison with B20, green diesel at CR 15, and petrodiesel have been discussed.

3.2.1   Brake Thermal Efficiency

Figure 1 has depicted the brake thermal efficiency (BTE) of green diesel in comparison with B20, green diesel at CR 15, and petrodiesel. The BTE of CI engine has

Table 2 Fuel properties of green diesel, B20, 100, and petrodiesel

Test Fuels	Density @15 °C gcm^{-3}	Kinematic Viscosity @40 °C mm^{2}s^{-1}	GCV MJkg^{-1}	Cetane number/index	Flash point (°C)	Cloud point (°C)	Carbon Residue (wt %)
B20	0.846	3.5	41.99	52	76	−5	0
B100	0.869	4.5	39.52	47	121	3	0
Green diesel	0.833	2.6	45.58	90	75	−9	0
Petrodiesel	0.823	2.85	42.86	44	66.5	−8	0.100

Fig. 1 Brake thermal
efficiency of CI engine for
test fuels w.r.t load variation

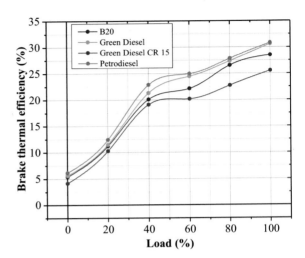

shown an increased trend w.r.t engine load. The high calorific value and cetane number of green diesel to that of B20 and petrodiesel had accelerated the combustion process which resulted in higher release of accumulated energy [8, 9]. The same has also been observed with combustion attributes (cylinder pressure, net heat release rate, and rate of pressure rate) (in the next section) which have shown higher value adjacent to top dead center (TDC). From Fig. 1, it has been observed that green diesel has shown comparable results w.r.to petrodiesel. However, the BTE for green diesel at compression ratio (CR) 15 is the lowest among other tested fuels which is justifiable as at lower CR the fuel is insufficient to burn properly.

3.2.2 Brake Specific Fuel Consumption

Figure 2 has depicted the brake specific fuel consumption (BSFC) of green diesel in comparison with B20, green diesel at CR 15, and petrodiesel. It has been detected from Fig. 2 that green diesel has comparable and lower BSFC in comparison with petrodiesel and B20, respectively. This was mainly because of the higher calorific value and cetane number of green diesel [11, 12]. However, the BSFC in case of CR 15 is higher than all test fuels which is resulted due to incomplete combustion [16].

3.3 *CI Engine Combustion Attributes*

The combustion attributes such as cylinder pressure (CP), net heat release rate (NHRR), and rate of pressure rise (ROPR) of CI engine for green diesel in comparison with B20, green diesel at CR 15, and petrodiesel were analyzed. The combustion data was recorded at full load condition because constant speed CI engine works

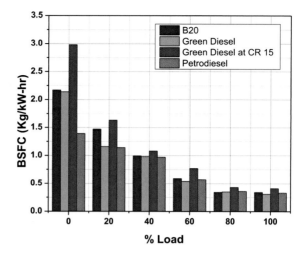

Fig. 2 Brake specific fuel consumption for test fuels w.r.t load variation

efficiently only at full load condition and an average of 100 cycles was taken into consideration.

3.3.1   Cylinder Pressure

Figure 3 has depicted the cylinder pressure (CP) of green diesel in comparison with B20, green diesel at CR 15, and petroleum diesel under full load condition. The maximum in-cylinder pressure for green diesel, B20, green diesel at CR 15, and petrodiesel was 71.26 bar @ 11° aTDC, 63.81 bar @ 10° aTDC, 57.56 bar @ 13° aTDC and 64.48 bar @ 11° aTDC, respectively. It has been observed from Fig. 3 that green diesel shows higher cylinder pressure than other test fuels. Also, the green diesel at CR 15 has shown the lowest combustion pressure with the highest delay among other test fuels. The highest cylinder pressure of green diesel among other test blends was resulted due to the higher cetane number and lower kinematic viscosity/density of green diesel which resulted in earlier accumulation of green diesel [11, 12]. Also, the combustion delay for green diesel at CR 15 was mainly due to insufficient combustion temperature resulting due to low compression ratio hence poor combustion [8, 16].

3.3.2   Net Heat Release Rate

Figure 4 has depicted the net heat release rate (NHRR) of green diesel in comparison with B20, green diesel at CR 15, and petrodiesel. Figure 4 shows about the maximum NHRR (NHRRmax) for green diesel (36.25 (J/Deg) @ 1° bTDC), B20 (27.22 (J/Deg) @ 1° aTDC), green diesel at CR 15 (55.05 (J/Deg) @ 5° aTDC) and petrodiesel (32.64@ 1° aTDC), respectively. It has been detected from Fig. 4 that the green

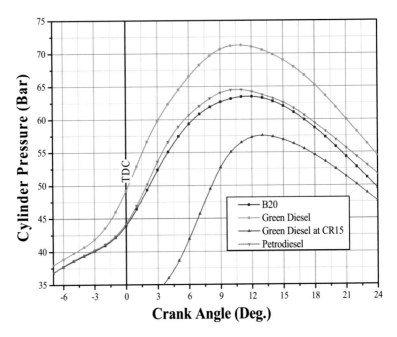

Fig. 3 Effect of test fuels on cylinder pressure (CP) at different crank angles under full load condition

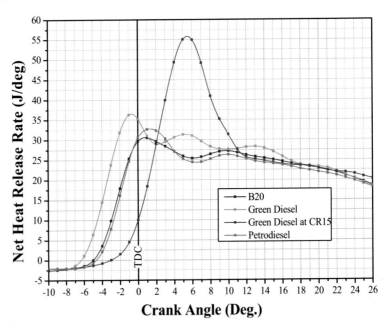

Fig. 4 Effect of test fuel on net heat release rate (NHRR) at different crank angles under full load condition

diesel shows maximum NHRR among other test fuels except green diesel at CR 15. Green diesel has also shown 1° earlier NHRRmax as compared with other test fuels. This was mainly because of the high cetane number and high calorific value along with low kinematic viscosity of green diesel which had created earlier atomization and this leads to shorter ignition delay, hence efficient combustion [12, 13]. Whereas for green diesel at CR 15, the NHRRmax is highest but 5° aTDC which has no positive contribution in CI engine application. This was due to more accumulation of green diesel with insufficient burning temperature [16, 17].

3.3.3 Rate of Pressure Rise

Figure 5 has depicted the rate of pressure rise (ROPR) of green diesel in comparison with B20, green diesel at CR 15, and petrodiesel. Figure 5 illustrates about the ROPR and maximum rate of pressure rise (ROPRmax) for green diesel (3.8 bar @ 1° bTDC), B20 (3 bar @ TDC), green diesel at CR 15 (3.86 bar @ 5° aTDC), and petrodiesel (3.16 bar @ 1° aTDC). The maximum ROPR signifies uncontrolled combustion in CI engine. It has been detected from Fig. 5 that as the load increases, the ROPR for green diesel shifts toward TDC which shows higher pressure force near TDC [8–10]. This shows higher conversion of chemical energy into pressure energy during

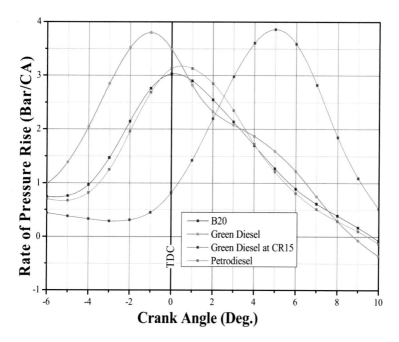

Fig. 5 Effect of test fuels on the rate of pressure rise (ROPR) at different crank angles under full load condition

expansion stroke, hence more power generation w.r.t other test fuels. ROPR of green diesel at CR 15 is highest but retard to TDC which has no significant role for CI engine application.

3.4 CI Engine Emission Attributes

The emission attributes such as carbon monoxide (CO), unburnt hydrocarbon (UHC), and oxides of nitrogen (NO_x) of green diesel in comparison with B20, green diesel at CR15, and petrodiesel were analyzed in CI engine emission analysis.

3.4.1 Carbon Monoxide Emission (CO)

Figure 6 has depicted the carbon monoxide (CO) emissions of green diesel in comparison with B20, green diesel at CR15, and petrodiesel. CO emissions are resulted due to incomplete combustion in CI engine [5–7]. It is clearly shown from Fig. 6 that green diesel has produced the lowest emission among other tested fuels in the entire range of load conditions. The highest CO emission reduction for green diesel is mainly because of its higher cetane number that accelerates the combustion process

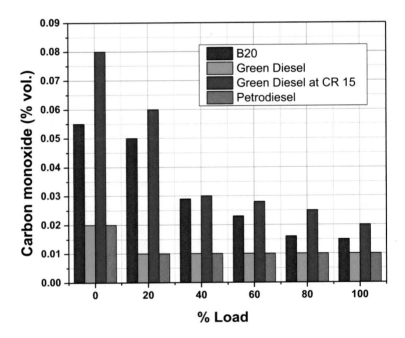

Fig. 6 Carbon monoxide emission at different load conditions on CI engine for test fuels

[14–16]. The CPmax, ROPRmax, and NHRRmax are also very near to TDC which also reflects more conversion of chemical energy into heat energy, hence proper mixing of air–fuel in comparison with other tested fuels [15–17].

3.4.2   Unburnt Hydrocarbon Emission (UHC)

Figure 7 has depicted the unburnt hydrocarbon (UHC) emissions of green diesel in comparison with B20, green diesel, and petrodiesel. Higher UHC emission results due to incomplete combustion of hydrocarbon molecules of doses fuel. It has clearly shown in Fig. 7 that the green diesel had shown the lowest UHC emission in comparison with other tested fuels in the entire range of load conditions. The highest UHC emission reduction is mainly because of the higher amount of cetane number that accelerates the mixing of fuel and air which resulted from complete combustion in the combustion chamber [11–13]. The NHRRmax also reflects the effective utilization of heat energy of green diesel in comparison with other test fuels [14–16].

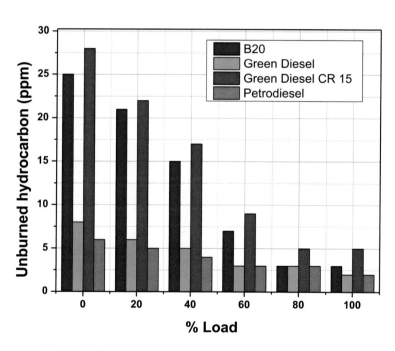

Fig. 7 Unburned hydrocarbon emission at different load conditions on CI engine for test fuels

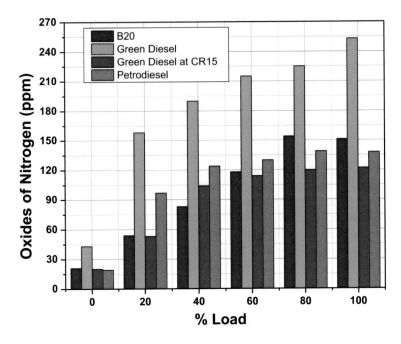

Fig. 8 Oxides of nitrogen emission at different load conditions on CI engine for test fuels

3.4.3 Oxides of Nitrogen Emission (NOₓ)

Figure 8 has depicted the oxides of nitrogen (NO$_x$) emissions of green diesel in comparison with B20, green diesel at CR 15, and petrodiesel. The NO$_x$ emission increases w.r.t increase of load. Green diesel has shown the highest NO$_x$ emission in comparison to other test fuels for the entire load conditions. As mentioned above, green diesel has contained higher cetane number and gross calorific value along with higher oxygen content that accelerated the NO$_x$ production. The green diesel at CR 15 has shown the lowest NO$_x$ emission which was resulted due to incomplete combustion.

4 Conclusions

The experimental investigation of CI engine fueled with MFL based green diesel and its comparison to that of B20, green diesel at CR 15, and petrodiesel has been discussed and the subsequent conclusions have been given below:

- Green diesel has shown shorter ignition delay in comparison with petrodiesel and other test fuels. Therefore, higher BTE, lower BSFC, and lower CO and UHC were observed for green diesel in comparison with other test fuels.

- Green diesel has shown maximum CP and ROPR, near TDC in comparison with petrodiesel and other test fuels. It was resulted due to ignition advance and superior fuel properties of green diesel and this reflects the improved performance of CI engine to other test fuels.
- Green diesel has shown the highest reduction in CO and UHC emission in comparison with petrodiesel and other test fuels which was resulted due to efficient combustion of green diesel in CI engine combustion chamber.
- Green diesel has shown the highest NOx emission in comparison with petrodiesel and other tested fuel, while green diesel at CR 15 has shown the lowest NO_x results of incomplete combustion.

It has been concluded that green diesel shows the best results (improved combustion, performance, and emission characteristic) and it can be recommended for high power rating CI engine. However, except NO_x emission, all the emissions are very low and lie under the emission norms because green diesel shows the characteristic of ultra-low sulfur diesel. It has been concluded that the emission of NO_x can be reduced by adding after-treatment device to the CI engine. This can be treated as future work of this research.

References

1. Aslam M, Konwar LJ, Sarma AK, Kothiyal NC (2015) An investigation of catalytic hydrocracking of high FFA vegetable oils to liquid hydrocarbons using biomass derived heterogeneous catalysts. J Anal Appl Pyrolysis Elsevier B.V. 115:401–409. https://doi.org/10.1016/j.jaap.2015.08.015
2. Aslam M, Kothiyal NC, Sarma AK (2015) True boiling point distillation and product quality assessment of biocrude obtained from Mesua ferrea L. seed oil via hydroprocessing. Clean Technol Environ Policy 17(1):175–185. https://doi.org/10.1007/s10098-014-0774-z
3. Babu V, Murthy M (2017) Butanol and pentanol: the promising biofuels for CI engines – A review. Renew Sustain Energy Rev Elsevier Ltd 78:1068–1088. https://doi.org/10.1016/j.rser.2017.05.038
4. Bezergianni S, Dimitriadis A (2013) Comparison between different types of renewable diesel. Renew Sustain Energy Rev Elsevier 21:110–116. https://doi.org/10.1016/j.rser.2012.12.042
5. Coufalík P, Sikorová J, Vojtisek-lom M, Beránek V, Mikuška P, Krumal K, Topinka J (2017) Blends of butanol and hydrotreated vegetable oils as drop-in replacement for diesel engines: effects on combustion and emissions. Fuel Elsevier 197:407–421. https://doi.org/10.1016/j.fuel.2017.02.039
6. Kham-or P, Suwannasom P, Ruangviriyachai C (2017) Environmental effects effect of agglomerated NiMo HZSM-5 catalyst for the hydrocracking reaction of Jatropha curcas oil. Energy Sources Part A Taylor & Francis 38(24):3694–3701. https://doi.org/10.1080/15567036.2016.1166165
7. Narayanan JHKV (2016) Production of green diesel by hydrotreatment using Jatropha oil: Performance and emission analysis. Waste Biomass Valorization Springer, Netherlands. https://doi.org/10.1007/s12649-016-9729-4
8. Kumar H, Sarma AK, Kumar P (2020) A comprehensive review on preparation, characterization, and combustion characteristics of microemulsion based hybrid biofuels. Renew Sustain Energy Rev 117:109498. https://doi.org/10.1016/j.rser.2019.109498

9. Kumar H, Sarma AK, Kumar P (2019) A novel approach to study the effect of cetane improver on performance, combustion and emission characteristics of a CI Engine fuelled with E20 (diesel – bioethanol) blend. Sustain Chem Pharm 2019(114):100185. https://doi.org/10.1016/j.scp.2019.100185

10. Kumar H, Sarma AK, Kumar P (2021) Experimental investigation of 2-EHN effects upon CI engine attributes fuelled with used cooking oil-based hybrid microemulsion biofuel. Int J Environ Sci Technol. https://doi.org/10.1007/s13762-021-03751-y

11. Ogunkoya D, Roberts WL, Fang T, Thapaliya N (2015) Investigation of the effects of renewable diesel fuels on engine performance, combustion, and emissions. Fuel Elsevier Ltd 140:541–554. https://doi.org/10.1016/j.fuel.2014.09.061

12. Singh D, Sandhu SS, Sarma AK (2019) Environmental effects a comprehensive study for setting up mini- biorefinery pilot plant for biodiesel, hybrid fuel, and hydroprocessed fuels derived from waste cooking oil. Energy Sources Part A Taylor & Francis 1–14. https://doi.org/10.1080/15567036.2019.1587089

13. Singh D, Subramanian KA, Singal SK (2015) Emissions and fuel consumption characteristics of a heavy-duty diesel engine fueled with hydroprocessed renewable diesel and biodiesel. Appl Energy Elsevier Ltd 155:440–446. https://doi.org/10.1016/j.apenergy.2015.06.020

14. Sonthalia A, Kumar N (2017) Hydroprocessed vegetable oil as a fuel for transportation sector: a review. J Energy Inst Elsevier Ltd 1–17. https://doi.org/10.1016/j.joei.2017.10.008

15. Zhao X, Wei L, Cheng S, Kadis E, Cao Y, Boakye E, Gu Z, Julson J (2016) Hydroprocessing of carinata oil for hydrocarbon biofuel over Mo-Zn/Al2O3. Appl Catal B Elsevier B.V. 196:41–49. https://doi.org/10.1016/j.apcatb.2016.05.020

16. Kumar H, Konwar LJ, Aslam M, Sarma AK (2016) Performance, combustion and emission characteristics of a direct injection VCR CI engine using a Jatropha curcas oil microemulsion: a comparative assessment with JCO B100, JCO B20 and petrodiesel. RSC Adv 6(44):37646–37655

17. Singh N, Kumar H, Jha MK, Sarma AK (2015) Complete heat balance, performance and emission evaluation of a CI engine fuelled with Mesua ferrea methyl and ethyl ester's blends with petrodiesel. J Therm Anal Calori 122:907–916. https://doi.org/10.1007/s10973-015-4777-8

Chapter 13
A Techno-Economic and Environmental Perspective on the Role of Green Diesel in a Prospective Fuel Production Mix for Road Transport

Zaira Navas-Anguita, Mario Martín-Gamboa, Pedro L. Cruz, Diego García-Gusano, and Diego Iribarren

Abstract The future road transport fleet is expected to involve the need for a mix of fuels (hydrogen, electricity, liquid biofuels, etc.). Among the biofuel alternatives, green diesel–also known as hydrotreated vegetable oil (HVO)–could play a significant role within prospective fuel production mixes. This chapter presents an energy systems optimisation model for an exploratory scenario on the deployment of green diesel in Spain, involving technology characterisation, carbon footprint behaviour, and long-term envisioning. The results show that HVO based on second-generation biomass would lead the deployment of green diesel as an alternative fuel, requiring the installation and use of new production plants to fulfil the ambitious demand expected for the road transport sector. Despite such a potentially high demand, the associated carbon footprint would remain low over the considered time frame. In fact, the total carbon footprint associated with second-generation biofuels would be below 5 Mt CO_2 eq in 2050. It is concluded that future decisions on the decarbonisation of the transport sector could promote the deployment of cost-effective technologies for the processing of second-generation biomass, including the production of advanced HVO.

Keywords Carbon footprint · Energy systems optimisation model · Hydrotreated vegetable oil · Prospective assessment · Road transport

Z. Navas-Anguita · P. L. Cruz · D. Iribarren
Systems Analysis Unit, IMDEA Energy, Móstoles, 28935 Madrid, Spain

M. Martín-Gamboa (✉)
Chemical and Environmental Engineering Group, Rey Juan Carlos University, Móstoles, 28933 Madrid, Spain
e-mail: mario.mgamboa@urjc.es

D. García-Gusano
TECNALIA, Basque Research and Technology Alliance (BRTA), 48160 Derio, Spain

© The Author(s), under exclusive license to Springer Nature Singapore Pte Ltd. 2022   339
M. Aslam et al. (eds.), *Green Diesel: An Alternative to Biodiesel and Petrodiesel*,
Advances in Sustainability Science and Technology,
https://doi.org/10.1007/978-981-19-2235-0_13

1 Introduction

Despite the COVID-19 pandemic, oil demand rebounded in 2021, yet below 2019 levels [1]. In particular, oil consumption in the road transport sector is recovering faster than in other sectors (e.g., aviation), with both diesel and gasoline recovering pre-pandemic levels in 2022. In spite of the growing deployment of electric vehicles [2–5], the use of liquid fuels is expected to remain relevant in the medium-to-long term, especially for long-distance road transport and aviation [1].

Regarding environmental issues, the transport sector is responsible for 27% of the global greenhouse gas (GHG) emissions [6]. Road transport accounts for the highest contribution, mainly associated with the combustion of gasoline and diesel from all forms of vehicles [7]. In the EU, in order to achieve climate neutrality, a 90% reduction in transport emissions is needed by 2050 [8]. In this sense, the latest directive on the promotion of the use of energy from renewable sources establishes that the share of renewable energy within the final consumption of energy in the transport sector must be at least 14% by 2030, with a contribution of advanced biofuels of at least 3.5% [9].

Currently, liquid biofuels hold a share of more than 3% of the energy needs in the transport sector and have experienced a growth of 13% from 2000 to 2018 [10]. However, 88% of the contribution corresponds to ethanol and biodiesel, often produced from biomass competing for land with food production—first-generation (1G) biomass—and therefore associated with environmental, social and strategic issues [11]. Hence, alternative types of biofuels may be called to play a relevant role. In the case of diesel, mostly used in heavy-duty vehicles [12], green diesel may be one of the feasible alternatives, showing better fuel properties than conventional (1G) biodiesel given its higher cetane number [13]. Green diesel, also referred to as hydrotreated vegetable oil (HVO), is typically produced through hydrodeoxygenation or deoxygenation technologies using vegetable oils and solid catalysts [14, 15]. It has attracted the attention of the research community because of its strategic significance, its potentially suitable environmental performance, the variety of sources to produce it, and its potential cost-effectiveness [15–18].

Several alternatives are likely to compose the future transport fleet, involving a mix of transportation fuels (green liquid fuels, electricity, hydrogen, etc.) and different transition scenarios to be explored. In this context, long-term energy planning considers the quantitative analysis of scenarios, usually aided by energy systems modelling, to contribute to national or regional targets, policies and investment strategies [19]. The long-term energy planning process includes a number of elements, such as institutional arrangements, modelling capabilities and methodologies, and scenario use and communication [20]. Hence, instead of accurately predicting energy trends (for instance, demands, technologies and impacts), energy planning and modelling can help facilitate decision-making processes by orienting them towards the overall sustainability goal under a number of exploratory scenarios.

With the aim of achieving the GHG reduction targets in the transport sector, robust roadmaps supporting decision-making processes are needed. In this regard, and focusing on liquid biofuels, the potential role of green diesel (i.e., HVO) within prospective fuel production mixes can be assessed by energy modelling. In particular, this chapter presents an energy systems optimisation model (ESOM) for an exploratory scenario on the deployment of green diesel in Spain, involving technology characterisation, carbon footprint behaviour, and long-term envisioning. Though performed for the Spanish case study, the approach and key technology trends might be extrapolated to a larger geography facing similar targets and concerns.

After this introduction to HVO as an alternative fuel for future road transport, Sect. 2 presents the methodological framework followed to explore its role in a prospective national fuel production mix for road transport, which is based on energy systems optimisation modelling. Afterwards, Sect. 3 presents the main techno-economic and environmental results in terms of prospective HVO production mix (Sect. 3.1) and prospective carbon footprint (Sect. 3.2), as well as final remarks on potentials, concerns and future research directions (Sect. 3.3). Finally, conclusions are drawn in Sect. 4.

2   Methodological Framework

Spain represents a country with both HVO production and demand. Therefore, it has been decided to take this country as a representative case study due to the multiple interests that such biofuel involves when looking at the European and Spanish long-term plans to decarbonise the entire energy system, particularly the transport sector [19]. According to the National Commission of Markets and Competition [21], HVO upsurge took place in 2011 once biodiesel production was well established. The statistics presented in Table 1 show HVO production besides imports and exports.

This upsurge of HVO came as a result of the PIIBE project (research project for the support of biodiesel in Spain, funded by the CENIT 2006–2009 programme and the Spanish Ministry of Science and Research). Afterwards, the expansion was

Table 1 HVO balance in Spain according to [21]

Year	Production (m^3)	Import (m^3)	Export (m^3)
2011	101,410	911,454	1,779
2012	72,915	808,280	1,779
2013	178,632	51,859	1,771
2014	376,944	8,076	0
2015	261,934	72,874	24,413
2016	417,706	0	57,309
2017	464,975	0	73,864
2018	482,117	0	100,041
2019	544,908	0	175,061
2020	259,933	0	65,552

significantly curbed down mostly due to new European regulations on sustainability standards. It is important to remark that the initial approaches to the biofuel-based transformation of the Spanish oil refinery sector were affected by the life-cycle controversy on the origin of the biomass feedstock used for energy purposes. In the Spanish case, palm oil is still used at 97–99% rates, which opens the door to the biomass transition debate, moving from 1G biomass to second- (2G) and third-generation (3G) ones. This is associated with the land use changes and environmental and social problems of 1G biomass (such as palm), which, in the Spanish case, is collected from Indonesia and Malaysia, with substantial socioenvironmental concerns. Accordingly, the transition to alternative sources is a topic that Spanish biofuel producers have to address without further delay.

In a previous study, Navas-Anguita et al. [22] developed an ESOM to evaluate the long-term behaviour of different fuel production technologies which could satisfy a set of energy demands in the road transport sector in Spain. The analysis was oriented towards 2050 as the time horizon and divided into different groups of production technologies for different types of alternative fuels (electricity, hydrogen, biodiesel, bioethanol, biomethane, HVO, etc.) as well as conventional fuels such as gasoline and diesel. Thus, for a predefined demand for each of those fuels (discussed and assumed based on a set of policy, technical and environmental hypotheses), the model provides a set of technology mixes for the production of each fuel from 2025 to 2050 from a techno-economic (and carbon footprint) point of view. Tables 2 and 3 present, respectively, the investment costs and efficiencies assumed in the model for the biofuel production technologies.

Table 2 Investment cost assumed for biofuel production technologies according to [22, 23]

Technology	Investment cost ($€_{2018}·GJ^{-1}·year$)			
	Base	Year 2030	Year 2040	Year 2050
Sugar fermentation and distillation (1G)	21.8	18.6	15.4	12.2
Starch saccharification, fermentation and distillation (1G)	28.3	24.3	20.3	16.3
Oil extraction and transesterification (1G)	27.2	22.8	18.4	14.0
Oil hydrogenation and isomerisation (1G/2G)	31.3	27.5	23.7	19.9
Biomass fermentation (2G)	29.4	26.3	23.1	20.0
Pyrolysis and upgrading (2G)	20.1	19.6	18.7	18.0
Gasification and Fischer–Tropsch synthesis (2G)	22.5	20.6	18.5	16.8
Microalgae transesterification (3G)	54.7	53.1	51.6	50.0

Table 3 Efficiency assumed for biofuel production technologies according to [22, 23]

Technology	Efficiency (%)	
	Base	Year 2050
Sugar fermentation and distillation (1G)	54	60
Starch saccharification, fermentation and distillation (1G)	59	64
Oil extraction and transesterification (1G)	66	70
Oil hydrogenation and isomerisation (1G/2G)	74	80
Biomass fermentation (2G)	40	43
Pyrolysis and upgrading (2G)	52	55
Gasification and Fischer–Tropsch (2G)	47	58
Microalgae transesterification (3G)	3	5

In terms of capacity, 1G bioethanol and biodiesel production plants can annually produce 300,000–420,000 tonnes (i.e., around 8.1–15.5 TJ per year). HVO annual production varies between 90,000 and 170,000 tonnes (i.e., 4.0–7.5 TJ per year). The size of 2G biomass-to-liquid plants is usually lower, accounting for an annual production around 2.5–6.3 TJ. Efficiencies often differ considerably from case to case. One of the reasons is the specific biomass used. For instance, in sugar fermentation, sugarcane extraction involves higher energy consumption than the extraction of sugar beet. Moreover, in other cases such as oil transesterification, efficiency scores involve great divergence depending on the specific reference and feedstock considered (e.g., raw biomass vs. processed oil). For 2G biofuel production, efficiency scores involve high deviation since the considered technologies are not yet mature or fully deployed.

Within the ESOM developed by Navas-Anguita et al. [19, 22], it should be noted that 1G biofuel production technologies (thus including 1G HVO production options) are set to disappear in the medium term. This modelling choice responds to the need to consider that their feedstock (i.e., 1G biomass) directly competes for land with the food sector and, according to the European Commission, biofuels produced from feedstocks with high land use change emissions (e.g., palm oil) will not be considered as renewable from 2030 [24].

Additionally, this ESOM allows analysts to explore the environmental performance of the fuel production technology mixes in terms of carbon footprint (i.e., life-cycle GHG emissions) [22]. Overall, Fig. 1 represents the modelling scheme implemented in Navas-Anguita et al. (2020) while placing the focus on HVO as an alternative fuel for road transport.

Within the context of the prospective model for different types of fuels carried out by Navas-Anguita et al. [19, 22], Fig. 2 shows the evolution assumed for the fuel demands under the exploratory scenario considered in this chapter. In this work, the focus is placed on discussing the role of HVO in the prospective fuel production mix for road transport in Spain from a techno-economic and environmental perspective.

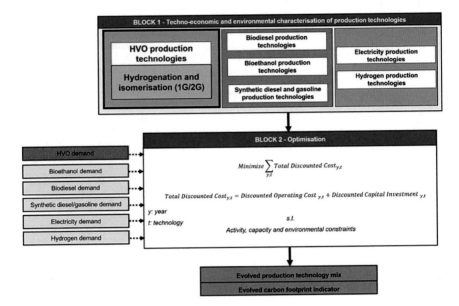

Fig. 1 Methodological framework of the modelling study on prospective technology mixes for the production of alternative transportation fuels, with a focus on HVO

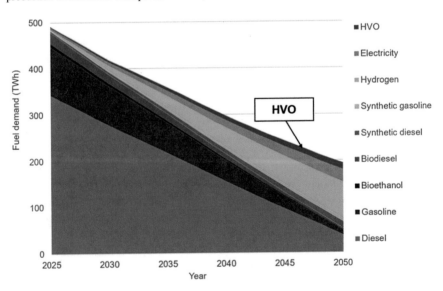

Fig. 2 Prospective demand for transportation fuels assumed in this study

Hence, the role of green diesel (i.e., HVO) is explored from a techno-economic approach based on an ESOM with time horizon 2050 and also including an environmental perspective given by the integration of carbon footprints. While Fig. 2 shows a modest weight of HVO when compared to other fuels within the mix, its role could be crucial to support sectoral decarbonisation. In fact, HVO would play a role not only in the Spanish transport sector but also in other sectors currently using diesel. As regards transport, taking into account the difficulties in electrifying the entire vehicle fleet (in Spain it means more than 34 million vehicles in 2021) and the consideration of 5 million electric vehicles by 2030 within the Spanish Integrated Energy and Climate Plan 2030 (PNIEC 2030), it can be deduced that the decarbonisation of the transport sector will require a great effort to expand all of the available options (Fig. 2). The role of HVO in contributing to this joint effort at a realistic pace in line with a sustainable transition is explored in Sect. 3 from the point of view of a prospective HVO production mix and a prospective carbon footprint.

3   Results and Discussion

3.1   Prospective Green Diesel Production Mix

The optimisation of the HVO production technology mix is shown in Fig. 3. These results show the transition from existing biomass hydrogenation plants to new

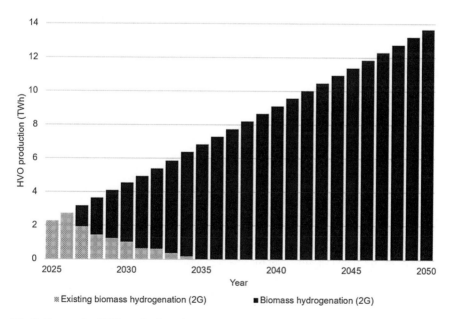

Fig. 3 Prospective HVO production mix

biomass hydrogenation plants, both fed by 2G biomass. In this sense, though initially considered by the model, the options fed by 1G biomass do not appear as an optimisation outcome at any moment, neither during the first years nor beyond 2030 (when 1G biomass is set to be forbidden). In order to understand the conversion to energy units, it is important to take into account that the lower heating value of HVO is 34.4 MJ/l (9.56 kWh/l). It should also be noted that the results in Fig. 1 are displayed from the year 2025 to avoid the COVID-19 effects (not modelled in the original model [19, 22]).

Hence, the first key finding involves that HVO from existing hydrogenation plants (fed by 2G biomass) will not be able to fulfil the demand beyond 2036. From 2027 onwards, HVO with a 2G origin would guide the system, requiring the installation and use of new production plants to fulfil the ambitious demand associated with the road transport sector. Thus, this result is inherently linked to the predefined growing demands for HVO.

Overall, the deployment of technologies to produce advanced (2G) biofuels—rather than 1G biofuels—arises as a key finding, along with a low sensitivity of the biofuel production technology trends to different investment costs. Besides, some aspects such as biomass prices could play a relevant role. For instance, concerning other biofuels, affordable 2G biomass prices under 9 € per tonne are found necessary to feasibly produce bioethanol via 2G fermentation [19, 22]. The significantly high biofuel demands in this study (especially regarding synthetic fuels and HVO) involve the need to guarantee a steady and reliable supply of biomass feedstock for biofuel production, calling for national bioenergy policies with well-defined and realistic targets and financial incentives to collect feedstock for energy purposes.

It is also important to explore the environmental side of HVO penetration. The socioenvironmental impacts associated with the use of 1G biomass to produce HVO and other biofuels could affect long-term planning in case of new European or global constraints. Competition for land with food crops and other aspects such as deforestation and doubtful sustainability standards in developing countries place the focus on using sustainable 2G and 3G biomass. An evaluation of the prospective carbon footprint of the national fuel production mix is carried out in the next section.

3.2 Prospective Carbon Footprint

With the aim of exploring in depth the future role of HVO in the Spanish road transport sector, a prospective environmental assessment should complement the previous techno-economic results, thereby further supporting decision- and policy-making processes. In this regard, the environmental life-cycle performance of HVO production and consumption over the selected time frame is herein presented in terms of carbon footprint, which is aligned with the actual focus of current regulations such as the European Renewable Energy Directive (RED II). To that end, the carbon footprint indicator of each fuel option was endogenously integrated into the transport model of Spain by following the procedure described in García-Gusano et al. [25].

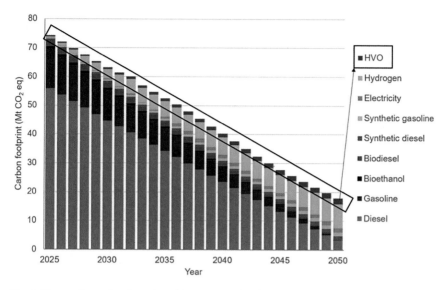

Fig. 4 Prospective carbon footprint of the transportation fuel production mix, with a focus on HVO

In particular, the following carbon footprints were used for the HVO-oriented study (values in kg CO_2 eq per kilogramme of fuel): 1.94 for 1G HVO and 1.78 for 2G HVO [26]. These values involve a "well-to-wheels" approach, thus including the GHG emissions related to cultivation, raw material transport, processing steps, fuel transport, distribution and final use of the HVO fuel according to the scope in Navas-Anguita et al. [19, 22]. Figure 4 shows the evolution of the carbon footprint of the prospective fuel production mix for Spain, placing the focus on HVO.

According to the overall results, the highest carbon footprint values are found in the first years. This is closely linked to the high carbon footprint per functional unit of fossil fuels, which account for significant contributions to the fuel production mix in the short term. In this respect, the initial year (i.e., 2025) involves a value around 75 Mt CO_2 eq, which is in line with the evolution of GHG emissions for the Spanish road transport sector described in the national energy and climate plan (i.e., PNIEC 2030). In the medium-to-long term, fossil fuel divestment and the penetration of alternative fuels would lead to a significant carbon footprint reduction (achieving in 2050 around a quarter of the value in 2025).

When placing the focus on the role that HVO will play in the environmental performance of the transportation fuel mix, the findings are potentially favourable. Despite the significantly high demand for biofuels considered in this study (especially in terms of HVO and synthetic fuels), the total carbon footprint associated with these 2G biofuels remains low over the considered time frame of the model (below 5 Mt CO_2 eq in 2050). These positive findings are directly related to the environmental benefits of using 2G biomass (e.g., GHG emission savings above 80% compared to fossil fuels in the case of HVO from waste oil or animal fats) [9]. Therefore, within the context of biofuels, future decisions aimed at the decarbonisation of the transport sector should promote the deployment of cost-effective technologies for the processing of 2G biomass, including the production of advanced HVO. These measures could be combined, for instance, with the use of green electrolytic hydrogen to further enhance the renewability of HVO production systems.

3.3 Final Remarks

As a key potential, this work allows energy modellers, analysts and planners to consider—in future studies—different exploratory scenarios on fuel production mixes for road transport (e.g., scenarios on different demands for alternative road transportation fuels or hypothetical policy measures on their promotion). Moreover, the key findings highlighted in the previous sections can be extrapolated to a large number of countries addressing similar challenges within a similar geopolitical context. Nevertheless, further advancements on the relevant energy systems model would enhance its usefulness for energy planning and decision-making purposes. For instance, these advancements could address further itemisation of HVO production technologies along with their further techno-economic and environmental specification (including refined carbon footprints).

The decarbonisation of the transport sector through, among other measures, the use of advanced biofuels must be accompanied by the implementation of a sustainable bioeconomy. In order to effectively achieve this objective, prospective analyses should broaden their focus to include not only a more complete set of environmental indicators (beyond carbon footprints) but also indicators belonging to other sustainability dimensions (e.g., the social pillar). In this sense, broadening the range of indicators will allow avoiding burden shifting at an intra- and inter-dimensional level, thereby facilitating decision-making processes actually oriented to the sustainability goal.

Future studies could take advantage of the progress in life cycle sustainability assessment in order to improve and complete prospective analyses of biofuels and, in particular, of HVO. For instance, the consideration of new and/or improved life cycle impact assessment methods (e.g., regionalised methods) and indicators (loss of biodiversity, criticality, etc.) could pave the way towards absolute environmental sustainability assessments. This involves the comparison between the impacts of the fuel mix and the regional or global limits of the environment, depending on the scope of the energy model [27]. With respect to the social dimension, Social Life Cycle Assessment (S-LCA) emerges as an approach that could be used to verify that the change towards a sustainable transport system is carried out according to the principle of "just transition". The recent S-LCA guidelines [28] and the increasing availability of social databases and assessment methods can contribute in this direction.

4 Conclusions

HVO based on 2G biomass is expected to lead the deployment of green diesel as an alternative fuel. This would require the installation and use of new production plants to fulfil the ambitious demand expected for the road transport sector. Despite such a potentially high demand, the associated carbon footprint would remain low over the considered time frame (with 2050 as the time horizon). In fact, the total carbon footprint associated with 2G biofuels would be <5 Mt CO_2 eq in 2050. Hence, future decisions on the decarbonisation of the transport sector could promote the deployment of cost-effective technologies for the processing of 2G biomass, including the production of advanced HVO. According to these findings, and since the decarbonisation of the transport sector will require a great effort to expand all of the available options, it is concluded that there is room for a significant role of HVO in sustainably contributing to this joint undertaking.

Acknowledgements Dr. Martín-Gamboa would like to thank the Regional Government of Madrid for financial support (2019-T2/AMB-15713).

References

1. IEA (2021) World energy outlook 2021. Paris
2. McKinsey & Company (2010) A portfolio of power-trains for Europe: a fact-based analysis. The role of battery electric vehicles, plug-in hybrids and fuel cell electric vehicle
3. European Environment Agency (2016) Electric vehicsles in Europe. https://doi.org/10.2800/100230
4. IEA (2020) Tracking transport 2020. Paris
5. IEA (2021) Global EV outlook 2021. Paris
6. IEA (2021) Greenhouse gas emissions from energy: overview. Paris
7. Our World in Data (2021) Emissions by sector 2021. https://ourworldindata.org/emissions-by-sector. Accessed 25 Oct 2021
8. European Commission, Secretariat-General (2019) The European green deal 2019
9. European Parliament, Council of the European Union (2018) Directive (EU) 2018/2001 of the European Parliament and of the Council of 11 December 2018 on the promotion of the use of energy from renewable sources 2018
10. WBA (2020) Global bioenergy statistics 2020
11. Fischer G, Hizsnyik E, Prieler S, Shah M, van Velthuizen HT (2009) Biofuels and food security. final report to sponsor: the OPEC fund for international development (OFID). International Food Policy Research Institute (IFPRI), Vienna
12. ERTRAC (2019) Long distance freight transport. A roadmap for system integration of road transport. Brussels
13. Chen S, Zhou G, Miao C (2019) Green and renewable bio-diesel produce from oil hydrodeoxygenation: strategies for catalyst development and mechanism. Renew Sustain Energy Rev 101:568–589. https://doi.org/10.1016/j.rser.2018.11.027
14. Arun N, Sharma RV, Dalai AK (2015) Green diesel synthesis by hydrodeoxygenation of bio-based feedstocks: Strategies for catalyst design and development. Renew Sustain Energy Rev 48:240–255. https://doi.org/10.1016/j.rser.2015.03.074

15. Mahdi HI, Bazargan A, McKay G, Azelee NIW, Meili L (2021) Catalytic deoxygenation of palm oil and its residue in green diesel production: a current technological review. Chem Eng Res Des 174:158–187. https://doi.org/10.1016/j.cherd.2021.07.009

16. Datta A, Mandal BK (2016) A comprehensive review of biodiesel as an alternative fuel for compression ignition engine. Renew Sustain Energy Rev 57:799–821. https://doi.org/10.1016/j.rser.2015.12.170

17. Raheem A, Prinsen P, Vuppaladadiyam AK, Zhao M, Luque R (2018) A review on sustainable microalgae based biofuel and bioenergy production: Recent developments. J Clean Prod 181:42–59. https://doi.org/10.1016/j.jclepro.2018.01.125

18. Shamshirband S, Tabatabaei M, Aghbashlo M, Yee PL, Petković D (2016) Support vector machine-based exergetic modelling of a DI diesel engine running on biodiesel–diesel blends containing expanded polystyrene. Appl Therm Eng 94:727–747. https://doi.org/10.1016/j.applthermaleng.2015.10.140

19. Navas-Anguita Z (2021) Prospective assessment of alternative fuel production technologies for decarbonising road transport in Spain. Rey Juan Carlos University

20. IRENA (2021) Energy planning support 2021. https://www.irena.org/energytransition/Energy-Planning-Support. Accessed 25 Oct 2021

21. CNMC (2021) Estadística de biocarburantes|CNMC 2021. https://www.cnmc.es/estadistica/estadistica-de-biocarburantes. Accessed 5 Nov 2021

22. Navas-Anguita Z, García-Gusano D, Iribarren D (2020) Long-term production technology mix of alternative fuels for road transport: a focus on Spain. Energy Convers Manag 226:113498. https://doi.org/10.1016/j.enconman.2020.113498

23. Navas-Anguita Z, García-Gusano D, Iribarren D (2019) A review of techno-economic data for road transportation fuels. Renew Sustain Energy Rev 112:11–26. https://doi.org/10.1016/j.rser.2019.05.041

24. European Commission (2019) Draft document - supplementing Directive (EU) 2018/2001 as regards the determination of high indirect land-use change-risk feedstock for which a significant expansion of the production area into land with high carbon stock is observed and the certificatio 2019

25. García-Gusano D, Martín-Gamboa M, Iribarren D, Dufour J (2016) Prospective analysis of life-cycle indicators through endogenous integration into a national power generation model. Resources 5:39. https://doi.org/10.3390/resources5040039

26. Neste Corporation (2020) Neste renewable diesel handbook. Espoo

27. Bjørn A, Chandrakumar C, Boulay A-M, Doka G, Fang K, Gondran N et al (2020) Review of life-cycle based methods for absolute environmental sustainability assessment and their applications. Environ Res Lett 15:083001. https://doi.org/10.1088/1748-9326/ab89d7

28. UNEP (2020) Guidelines for social life cycle assessment of products and organizations 2020

Chapter 14
Policies, Techno-economic Analysis and Future Perspective of Green Diesel

Khursheed B. Ansari, Saeikh Zaffar Hassan, Saleem Akhtar Farooqui, Raunaq Hasib, Parvez Khan, A. R. Shakeelur Rahman, Mohd Shariq Khan, and Quang Thang Trinh

Abstract Green diesel, also known as paraffinic diesel, is a promising next-generation automotive fuel produced from non-edible or waste vegetable oil through alkali catalyzed transesterification, hydrogenation, and supercritical non-catalytic transesterification. The sustainability of green diesel technology relies on robust policy and acceptance among the stakeholders. Further, the economics of green diesel production technologies are strongly affected by the feedstock cost, hydrogen requirement, and the production capacity of green diesel and its by-product yields. Selecting an appropriate manufacturing location would be promising in decreasing the production cost and improving green diesel economics. Integrating a green diesel plant with a petroleum refinery would enable the utilization of process water, hydrogen, heat (for making steam), and other utilities. Further, the production capacity of the green diesel plant must be at least 0.1 million tons/year to avoid negative net return (or annual profit after the tax) and a more extended payback period. Moreover, the sustainability of vegetable waste-to-green diesel technology

K. B. Ansari (✉) · R. Hasib · P. Khan
Department of Chemical Engineering, Zakir Husain College of Engineering and Technology, Aligarh Muslim University, Aligarh, Uttar Pradesh 202001, India
e-mail: akabadruddin@myamu.ac.in

S. Z. Hassan
Department of Petroleum Studies, Zakir Husain College of Engineering and Technology, Aligarh Muslim University, Aligarh, Uttar Pradesh 202001, India

S. A. Farooqui
Hydroprocessed Renewable Fuel Area, Biofuels Division, Indian Institute of Petroleum, Dehradun 248005, India

A. R. S. Rahman
Department of Applied Science, Shri Vile Parle Kelavani Mandal's Institute of Technology, Dhule 424001, India

M. S. Khan
Department of Chemical Engineering, Dhofar University, Salalah 211, Oman

Q. T. Trinh
Cambridge Centre for Advanced Research and Education in Singapore (CARES), Campus for Research Excellence and Technological Enterprise (CREATE), 1 Create Way, 138602, Singapore

© The Author(s), under exclusive license to Springer Nature Singapore Pte Ltd. 2022   351
M. Aslam et al. (eds.), *Green Diesel: An Alternative to Biodiesel and Petrodiesel*,
Advances in Sustainability Science and Technology,
https://doi.org/10.1007/978-981-19-2235-0_14

also needs the policymakers' fruitful decision on subsidy and tax exemption for green diesel. With the new specification of paraffinic diesel (EN 15940) by European countries, researchers and industries around the globe are looking for the best technological options for a suitable choice for its commercial production.

Keywords Green diesel · Policy · Technological feasibility · Process economics · Case study · Future perspective

1 Introduction

Green diesel is prepared through hydrodeoxygenation, decarbonylation, and decarboxylation (DCO2) reactions of vegetable oil. It finds reasonable proximity for blending with petroleum diesel. One of the motivations of green diesel is the production of HEFA$^+$ (hydro-processed esters and fatty acids+), utilized as a petroleum jet fuel substitute. However, the integration of green diesel into existing petroleum fuel infrastructure needs robust strategies and appropriate policies leading to attractive economics. It is known that fossil fuel consumption has increased tremendously in the recent past, accounting for 80% of global energy consumption and expected to further rise at a 0.9% annual rate until 2030, according to the International Energy Agency (IAE) [1]. The dependence of the worldwide population over liquid fossil fuels (such as diesel, gasoline, fuel oil, and jet fuel) could be a reason for the enhanced consumption, especially in the transportation sector [2]. Diesel meets about 75% of the heavy-duty transportation energy demand among the liquid fuels because of its universal availability, stability, and excellent capability for diesel engine technology carrying heavy loads. However, it results in a significant amount of carbon dioxide (CO_2) mission (10.08 kg CO_2 per gallon diesel combustion) [3].

Further, the widespread and unregulated use of fossil fuels had a detrimental environmental impact, resulting in a significant increase in greenhouse gas emissions [4]. On the other hand, biofuels (i.e., biodiesel/green diesel), being a renewable and environmental-friendly form of energy, may improve energy security and cause a minimal effect on the environment. With negligible sulfur content, the biodiesel, after burning into the internal combustion engine, results in zero emissions of SOx and stimulates vehicle performance. Further, it also cuts brown mist by 40% and lowers the carbon monoxide (CO) levels to a greater extent during combustion [5]. Thus, attempts have been made worldwide to boost biodiesel/green diesel-producing technologies fulfilling the growing demand for transportation fuel.

1.1 Policies for Green Diesel: Global and Indian Context

The expanding international market for biofuels (especially biodiesel/green diesel) is driving national policy and legislative attempts to stimulate bioenergy production.

Over 30 nations worldwide have already implemented or are constantly working to improve the biofuel production programs [6]. The growth of biodiesel production worldwide was first driven only by the agricultural market regulations in the 1990s. Notably, the global market of biodiesel is significantly younger than the ethanol market. The European Union (EU) has been producing biodiesel on an industrial scale since 1992 and remains the field leader. Apart from the EU, Indonesia, Brazil, Argentina, Malaysia, Australia, and the United States are all making significant policies, investments, and efforts in the biodiesel business, producing it sustainably, and becoming major producers in the near future [7]. These nations utilize palm and soybeans, the two most common oilseed crops for biodiesel production, nearing about 51 billion liters per year [8]. Further, several African and Asian nations are also investing in biodiesel production through their Jatropha plant and have developed policies for biofuel.

India's oil consumption remains the third highest globally, with the transport sector accounting for about 70% of diesel and 99.6% of gasoline [9]. In 2015–16, India's crude oil consumption was about 249 million tons (Mt). In contrast, its production was only 36.95 Mt., indicating the indigenous crude oil production can only meet 15% of its demand, with 85% being fulfilled through imports [10]. Several initiatives have been taken by the Government of India (GoI) over the last few decades to increase the production and usage of biofuels. In terms of promoting biofuels in the country, the Ministry of New and Renewable Energy launched a Nationwide Policy on Biofuels in 2009 called *National Biodiesel Mission* (NBM) [9]. It promoted fuel production (or biofuel) from non-edible plants and waste biomasses. This policy aimed to boost rural development and provides jobs in rural areas and remains different from the existing international practices, potentially avoiding food security risks. It is primarily based on non-food feedstocks cultivated on degraded or inadequate agricultural areas, eliminating a potential conflict between fuel and food security. It is inclined to support and promote the effective implementation and usage of indigenous biomass materials for biofuel production (i.e., biodiesel and green diesel) in the framework of international views and national requirements. As a result, NBM policy mainly concentrated on the growth of the Jatropha plant throughout India, which is a non-edible seed rich in oil (40%) and is considered an excellent feedstock for biodiesel production [11]. However, only a few states within India, i.e., Andhra Pradesh, Rajasthan, Chhattisgarh, and Tamil Nadu, have played an active role in the biodiesel mission and encouraged Jatropha cultivation through different incentives and policies. Under the NBM Phase-I, which consists of experimentation and demonstration, the Indian government attempted to secure a 5% blend of petroleum gasoline with biodiesel by the year 2007. However, the intended objective could not be met, and the NBM Phase-I failure needed a thorough examination of all probable causes to ensure the mission's long-term viability. According to government research, the farmers failed to use scientific cultivation and maintenance procedures for Jatropha plantations, resulting in low Jatropha seed yields at multiple geographical locations within India [12]. The policy was further revised and approved by the Union Cabinet in May 2018 to achieve the target of 20% blending of biofuel by 2030. The government has defined several goals, including (i) 24 × 7 electricity to all census villages by 2019; (ii) 175 gega watts (GW) of electricity generation through renewable sources such as solar, biomass, and wind by 2022; (iii) reduction in pollutant emissions 33–35% during renewable energy generation by 2030; and (iv) producing 40% of the electricity from renewable resources of energy by 2030. These goals indicated the government's attempt to reinforce renewable energy infrastructure of the country while promoting the agenda of sustainability.

Apart from the conventional routes, the progress in microalgae-to-biofuel research has shown a new path for sustainable biodiesel production [13]. Algae are of specific importance because of their unique ability to thrive in all sorts of aquatic environments and aptitude for cultivation at any season. Further, it remains the noncompetitive, safest, and fastest growing microorganism employed for biodiesel production through the transesterification process, thus avoiding food scarcity, as seen with other techniques [13]. Algae can produce $18.92 \, m^3$–$75.70 \, m^3$ of biodiesel per acre, relative to $2 \, m^3$–$2.4 \, m^3$ of biodiesel derived from the edible and non-edible oils [14]. India consumes about 200,000 million gallons of biodiesel per year [15], requiring around 384 million hectare area for Jatropha crop cultivation, which is larger than India's geographical area. Microalgae, on the other hand, uses far less space (i.e., 5.4 million hectare area) to produce a similar quantity (200,000 million gallons) of biodiesel, which is less than 2% of India's total land area [15]. Thus, the national-level policy promoting algae-to-biofuel generation would be the need of the hour, and stakeholders and policymakers are taking appropriate steps for it. In this context, the utilization of sewage effluent for algae growth and nutrients (like nitrogen and phosphorus) would be promising. It will minimize the sewage accumulation and offer sustainable bioremediation of such effluents, observed in a large quantum in developing nations like India [16].

1.2 Commercial Technologies of Green Diesel

The worldwide production of green diesel has grown to a greater extent expanding its production volume from 1.5 million m^3 in 2011 to 43.36 million m^3 in 2017 [17]. The technologies involved in green diesel production include Neste NExBTL, Universal Oil Products (UOP)/Eni Ecofining™, UPM BioVerno, governed by different companies globally. An individual plant making green diesel utilizes vegetable oil/animal fats/used cooking oil/fatty acids as a feedstock. The feedstock undergoes multiple processes, including cleaning and pre-treatment followed by deoxygenation (hydrotreatment) in a hydro-isomerization reactor and product (or green diesel) isolation in the separator. Table 1 lists several technologies of green diesel production across the globe. The highest production of green diesel per year is from NExBTL technology adopted by Neste in the Netherlands and Singapore, followed by Ecofining™ and Dynamic Fuels LLC processes used in the United States of America (USA).

The NExBTL technology produces green diesel suitable to be utilized directly for combustion engines instead of blending into conventional petroleum diesel [18,

Table 1 Green diesel technology and producer country-wise [17]

Feedstock	Technology	Production capacity (tons/year)	Location	Company
Vegetable oil/waste animal fat	NExBTL	1,000,000	Netherland/Singapore	Neste
Vegetable oil/waste animal fat	NExBTL	380,000	Finland	Neste
Non-edible vegetable oils/animal fats	Ecofining™	900,000	USA	Diamond green diesel
Vegetable oils, animal fats/used cooking oils	Ecofining™	780,000	Italy	UOP/Eni
Non-edible natural oils and agricultural waste	Ecofining™	780,000	USA	AltAir Fuels
High and low free fatty acid	Dynamic Fuels LLC	250,000	USA	Renewable Energy Group Inc
Crude tall oil	UPM BioVerno	100,000	Finland	UPM Biofuels

19]. NExBTL hydrotreats the vegetable oils or waste fats and makes the green diesel suitable for car and truck engines [19]. The process involved in NExBTL technology is demonstrated in Fig. 1. The NExBTL green diesel is sulfur-free, oxygen-free, and aromatic-free, does not require any additional maintenance, and can last for a longer time with almost negligible deterioration in the quality or water accumulation. Further, the high octane number (i.e., 75–95) enables effective and clean combustion, ensuring the extra combustion power compared to biodiesel (fatty acid methyl ester (FAME)). Because of the greater extent of deoxygenation, green diesel exhibits excellent storage ability and does not show a cleaning effect. Moreover, the comparable cloud point of green diesel with petroleum diesel makes it a potential candidate for blending and transporting in the pipelines [20].

Ecofining™ process governed by Honeywell UOP transforms natural non-edible oils/animal fats through deoxygenation and hydrocracking into Honeywell Green Diesel (Fig. 2). Honeywell UOP strives hard to expand green diesel capacity to 1.1 billion dollars, fulfilling its growing fuel in Europe and North America. One of the largest plants of green diesel is scheduled to be set up in Norco, Louisiana, USA to increase the annual production of green diesel by 150% [22]. Interestingly, Ecofining™ process can be integrated with the petroleum refinery by locating its plant in nearby vicinity [23]. The utilities such as process water and hydrogen can be coupled with the petroleum refinery. Further, the by-product of Ecofining™ process,

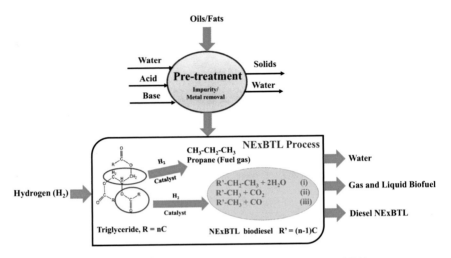

Fig. 1 NExBTL process of catalytic hydrotreating of oils/fats into biodiesel [21]

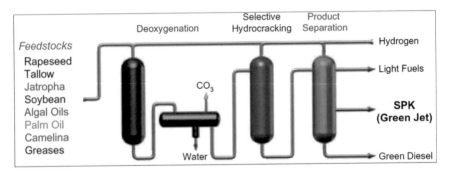

Fig. 2 Schematic of Ecofining™ process generating green diesel (Reproduced from Ref. [23] with permission from Elsevier, Copyright [2011] [23])

such as liquefied petroleum gas and naphtha, can be reused in petroleum refinery. Moreover, owing to the higher cetane value (75–90), green diesel itself may be used to upgrade heavy oils into diesel and enlarge the pool of conventional petroleum diesel [23].

UPM BioVerno technique utilizes wood-derived oil as feedstock and converts it to green diesel through hydrotreatment. The process was first developed at the UPM Research Centre in Lappeenranta, Southeast Finland. The technology is contributing significantly to the forest industry by utilizing wood residue, thus capitalizing on the lifespan of wood-based feedstock substituting fossils.

1.3   Process Simulation and Techno-Economic Approach

The production cost of green diesel can be estimated through process simulation and techno-economic feasibility analysis. In this context, heat integration routines and pinch analysis contribute positively to the minimum energy exploiting process (i.e., heat exchange networks). The maximum level of heat recovery obtained through pinch analysis leads to reduced production costs. Green diesel production technology has been investigated and patented by many research groups and companies for the past two decades. The ease of integrating green diesel technology into existing petroleum refineries makes it an ideal choice for the stakeholders, as discussed earlier in this chapter. In the integrated scheme, vegetable oils are added into the fuel mixtures before feeding them to hydrotreaters, which leads to lower investment and revamping; however, at the same time, the possibility of a 100% renewable fuel is eliminated. Thus, independent processing of hydrotreated vegetable oils into green diesel would be more viable, enabling the plant capacity between 170,000 and 800,000 tons/year. Recently, several feasibility studies have been conducted to produce green diesel from catalytic decarboxylation of rubber seed oil [24], decarboxylation of palm stearin basic soaps [25], hydrodeoxygenation of Karanja oils [26], hydrodeoxygenation of palm oil, hydrogenation of waste vegetable oil [27], etc. Green diesel produced from these routes can be directly utilized in existing internal combustion engines. However, due to the vast category of raw materials and processing routes, a techno-economic analysis for green diesel production becomes essential to determine the best combination of raw materials and processes. Process simulation helps to easily design and determine the process flowsheet for feasibility via enormous routes. Therefore, numerous research exists on process simulation and techno-economic methods for the production of green diesel [1, 24, 26–28].

1.3.1   Process Simulation

The first step in process simulation includes selecting a process and investigating the detailed process flowsheet. A self-coded computer program or available commercial software such as Aspen One, SuperPro Designer, UniSim, etc. can be applicable for this. The software-based process simulation of green diesel production remains attractive in research since it avoids the tedius task of coding and user friendly nature [1, 24, 26, 27]. A typical process simulation procedure consists of (i) selecting and defining components, (ii) choosing an appropriate thermodynamic model, (iii) selecting the required equipment, and (iv) specifying operating conditions, which results in material balance, energy balance, and equipment sizing as an outcome. In the case of the green diesel process flowsheet, the properties of triglycerides (raw material) needed for the simulation are estimated by the fragment approach method [29] or the group contribution method [30]. However, numerous property estimation methods exist, including Redlich Kwong equations of state for high-pressure and high-temperature operation, Peng Robinson equations of state for the low-pressure hydrogenation nearing atmospheric pressure, and the NRTL for aqueous systems [31]. The fragment approach method allows the calculation of thermodynamic and transport properties. The simulation model also includes the kinetic data obtained through the reaction of triglyceride molecules over the catalyst's surface at a given temperature and pressure (if available). The kinetic model depends upon the choice of catalysts, for example, sulfided nickel-molybdenum (S-NiMo) and sulfided copper-molybdenum (S-CoMo), etc.) based processes follow simple power kinetic model [32–35]. The selected kinetic model must be sophisticated, robust, and accurate enough to explain the mechanism of the hydrotreatment of vegetable oils [28, 36–38]. Apart from the kinetic model, the heat integration of the process through the pinch approach (maximizing heat recovery and utilization) enables optimal energy utilization in a plant. There are well-established methods to construct a heat exchange network (HEN) and apply it to optimize energy requirements during the green diesel production process [26]. Simulation software such as Aspen One offers an energy analysis package, making it easy to optimize the energy requirements using a heat exchanger network (HEN). However, in the case of the highest pinch temperature of the process, the HEN fails to offer many optimal designs [28].

1.3.2 Techno-Economic Analysis

The economics of the green diesel-producing plant is determined by fixed (or capital) and variable costs. The economic evaluation of the plant can be carried out using the method described by Peters, Timmerhaus, and West [39] or by Towler and Sinnot's [40]. The preliminary economic design determines optimal operating conditions and a process design through process simulation software such as Aspen One and UniSim. An overall methodology approach for techno-economic analysis is given in Fig. 3. The general assumptions of economic analysis for estimating capital cost and the variable operating cost of green diesel plant are given in Table 2. The net present value (NPV) and payback period (PBP) are calculated in economic study, and their sensitivity analysis is performed, aiming to identify the main contributors to plant cost and their influence on the profitability.

Capital Investment

Once the process simulation is complete, it gives material and energy balance of process plant as well as capacities and sizes of major process equipment as output. This output data is used to calculate the purchased cost of major equipment (Fig. 3). Commonly available textbook equations are used to calculate the approximate purchased cost of each piece of equipment [1], as shown in Table 2. The plant's capital includes the direct (equipment cost along with piping and installation, instrumentation, buildings, land, yard improvements, electrical, service facilities, and miscellaneous expenses) and indirect (engineering and supervision, construction expenses, contractor's fee, contingency, start-up expenses) costs and

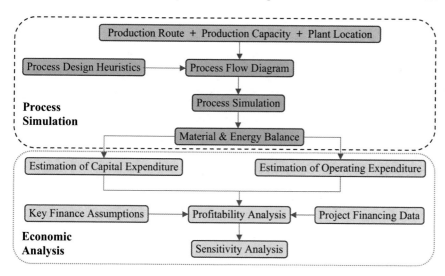

Fig. 3 Methodology for techno-economic analysis of green diesel production [24]

Table 2 Approximate purchased cost of equipment in a green diesel plant

Equipment type	Purchase cost equation ($)	Notes
Reactor	$C^0_{Reactor} = 236471\left(1.25\tau_R \dot{Q}_R\right)^{0.5299}$ where $\dot{Q}_R$ is volumetric flow rate in $m^3 s^{-1}$ and τ_R is residence time in s Reactor volume range is $V_R = 1.25\tau_R \dot{Q} = 0.1–30\ m^3$	Jacketed stainless steel [26]
Distillation column	$C^0_{DistilColumn} = 4555 H^{0.81} D^{1.05}$ where H is column height; D is column diameter in m	Carbon steel [26]
Heat exchanger	$C^0_{HeatX} = 25615 A^{0.575}$ where A is the heat transfer area in m^2 (floating head type shell and tube). For a basic design, $A < 18\ m^2$ and pressure < 4 MPa	Stainless steel [19, 21]
Flash drum	$C^0_{FlashDrum} = 12685\left(\tau_F \dot{Q}\right)^{0.3641}$ where $\dot{Q}$ is volumetric flow rate in $m^3 s^{-1}$ and τ_F is residence time in s. The volume range is $V_F = \tau_F \dot{Q} <10\ m^3$	Stainless steel [19]
Centrifugal pump	$C^0_{Pump} =$ $5.4 e^{\left(9.7171 - 0.6019\ln\left(15850\dot{Q}_P \sqrt{h/0.3048}\right) + 0.0519\left(\ln\left(15850\dot{Q}_P \sqrt{h/0.3048}\right)\right)^2\right)}$ where volumetric flow rate, $\dot{Q}_P$ is in $m^3 s^{-1}$ and pump head, h is in m of flowing fluid. The applicable range is $400 <$ $15850\dot{Q}_P \sqrt{h/0.3048} < 100{,}000$	Cast iron [26]
Fired heater	$C^0_{FiredHeater} = 184967 H_D^{0.7636}$ where heat duty H_D is in MW and pressure range < 5 MPa	Stainless steel [26]

working capital (capital needed for the initial operation of the plant, i.e., raw material and supplies, cash for operating expenses) costs. Direct cost is further classified as inside battery limits (cost of procuring and installing all process equipment) and outside battery limits cost (cost associated with off-site developments that require the plant to run), whereas indirect cost is classified as engineering and contingency costs. The investment cost is adjusted and corrected using the CEPCI ratio (as per chemical engineering's plant cost index) to estimate all investment costs on an equal time basis. Inside battery limits costs are estimated as the sum of equipment costs multiplied by Lang installation factors. In contrast, outside battery limits, engineering, and contingency costs are calculated as percentages of the inside battery limits cost (around 40, 30, and 10%, respectively) (Table 3).

Notably, the major contributors to capital investment are reactor costs [28]. Hydrotreating reactor holds the largest share of the total equipment cost, closely followed by hydrogen compressor and fractional distillation columns [24]. Hydrotreating reactor as the prime capital cost contributor can be explained by its relatively long reaction residence time, requiring a larger vessel volume.

Production Cost

The production cost includes the contribution of direct and indirect costs. The direct operating costs include labor, supervision, maintenance, operating and plant overhead charges, administration and general expenses, utilities (electricity, steam, and cooling water), and raw materials (vegetable oil, hydrogen, water, and catalyst) as well as waste stream treatment. The indirect operating costs include equipment depreciation, interest on the initial investment, royalty fees, and insurance and taxes. The net production cost of green diesel does not include the revenue generated by co-products.

Catalyst prices can be estimated based on their nickel, molybdenum, and platinum content, with a typical hydrodesulfurization catalyst lifecycle of around 1.5 years [28]. Raw materials constitute the most significant fraction of the overall production cost accounting for more than half of the production costs [24, 27, 28, 40, 41]. Given the volatility of vegetable oil (raw material of green diesel production) price and its effect on production costs, the researchers recommend using cheap oils such as waste vegetable oil or animal oil for future investigations. In terms of utility expenses in a green diesel plant, electricity cost makes up the largest portion, closely followed by cooling water and lastly pressure steam. High electricity demands are attributed to the energy-intensive hydrogen compressor unit and fractional distillation columns. Product price and revenue can be calculated by adding margins to production costs and comparing the result with current market values for biodiesel to ensure the competitiveness of the product. In the process industries, plant capacity also signif-icantly impacts overall manufacturing costs. Up to 25% of manufacturing costs are reduced by doubling the plant capacity [42]. The analysis of investment cost and annual production cost calculation by Glisic et al. [1] has indicated that due to the higher investment cost of hydrogenation unit, the price of green diesel produced by a

Table 3 Assumptions for economic analysis of a green diesel plant (reproduced with permission from Mailaram and Maity (2019) [26])

Particulars	Values
Number of years for analysis	10
Operating hours (h/year)	8760
Number of working days (day/year)	365
Depreciation method	Straight line
Installation	52% of total equipment cost
Instrumentation and control	30% of total equipment cost
Piping	75% of total equipment cost
Electrical	12% of total equipment cost
Building	20% of total equipment cost
Yard improvement	11% of total equipment cost
Service facilities	77% of total equipment cost
Land	5% of total equipment cost
Engineering and supervision	9% of direct fixed cost
Construction expenses	11% of direct fixed cost
Legal expenses	1% of direct fixed cost
Contractor fees	6% of direct fixed cost
Contingency	12% of direct fixed cost
Working capital	5% of fixed capital investment
Operating charges	25% of operating labor cost
Plant overhead charges	50% of operating labor and maintenance cost
Maintenance cost	6% of fixed capital investment
General and administration expenses	8% of subtotal operating cost
Insurance and taxes	1.5% of total equipment cost
Salvage value	20% of initial capital investment
Interest rate	5.5% of total capital investment

standalone process unit is very high. Therefore, it does not fulfill economic feasibility at this point.

Profitability Analysis

To evaluate the economic performance of the process, a discounted cash flow analysis, which considers the time value of money, is conducted. The primary parameters of evaluating economic performance are net present value (NPV) and internal rate of return (IRR). Further, to calculate the profitability of the plant, annual cash flow is determined considering the earnings before interest and taxes (EBIT). EBIT is obtained after subtracting the production costs to revenues and taxation of 15% (may change with country) over profits. The profitability analysis of green diesel production from different vegetable oils has shown a money-making process after the fourth [24], fifth [26], or seventh [28] years. However, diesel and vegetable oil prices unpredictably make it a risky investment. Therefore, governmental incentives and funding is required to improve this situation.

Sensitivity Analysis

The parameters such as prices of raw materials and utilities could fluctuate in the future and affect the minimum fuel selling price. Therefore, a sensitivity analysis becomes necessary to analyze fluctuations over an extensive range. It is conducted to understand better the effects of uncertain variables on the production of green diesel that could affect production cost and revenues. Using impact analysis, the sensitivity study is performed to estimate the most significant parameter that influences the net present value (NPV) [22]. For example, the prices of raw materials such as vegetable oil, hydrogen, and utilities are varied by ±50%. Low prices of the raw materials are desirable, and the sensitivity analysis helps to determine the effect of low or high raw material prices on profitability. A similar approach is used for other entities concerning green diesel production. Further, equipment cost is also varied in the analysis by considering the accuracy of the cost estimation method used. Income tax is also included in sensitivity analysis depending upon the possible changes in the investment policy of a country. The financial interest varies between 50 and 150% of the fixed capital investment and depends on a country's economic view. Moreover, the plant overhead and administration expenses (general expenses), start-up period, and maintenance work (variable cost parameters) are also captured in the sensitivity analysis and investigated thoroughly for the profitability of the technology. In producing green diesel from vegetable oils, raw material price is the most sensitive parameter in affecting the production cost. It, therefore, plays an essential role in determining the commercialization of green diesel production. The sensitivity analysis performed on hydroprocessing located in a refinery suggested that the NPV remains highly sensitive to variations in feedstock prices and process unit capacity; when plant capacity becomes doubled, NPV increases sevenfold for hydroprocessing [1]. Thus, the sensitivity analysis highlights significant capital investment and operating cost reduction opportunities. Several factors affecting the economics of green diesel technology are discussed in the next section.

2 Factors Affecting the Economics of Green Diesel

2.1 Feedstock Cost

The production and use of green diesel ramp up to meet the clean and renewable energy requirements and comply with the environmental regulations. EU alone witnessed 90% of renewable energy use in the transportation sector, and much of its share coming from green diesel [43]. Until 2016, the world's major biofuel was produced using edible vegetable oil (such as rapeseed/soybean/sunflower) with severe food security implications. This triggers various decisions to limit edible raw biomass for green diesel production and promote second-generation feedstock, viz., lignocellulosic biomass [44]. Table 4 depicts major green diesel producers and the feedstock they employ.

The green diesel production cost is the accumulated cost of feedstock, conversion cost (for enzymes/catalysts), and operation and maintenance cost minus the value of co-products over time [45]. Several countries are subsidizing ethanol production to boost the production of green diesel, making green diesel more attractive. These reductions in feedstock cost are generally offset with longer average transport distance associated with higher output feed size [46]. Even with subsidiaries, the lack of feedstock land and moderate environmental efficiencies remains the critical problem of green diesel generation [47]. Regardless of these challenges, the main incentive of producing biodiesel is achieved through reduced CO_2 emissions. Kumar et al. [48] compared the price of petroleum-derived liquid fuel and biodiesel made from slurry and reported 85% less CO_2 emissions. Despite proven benefits and substantial subsidiaries, the cost of green diesel has not dramatically decreased to be in comparison with fossil fuel, and now the efforts are focused on second-generation feedstocks than on conventional feed [49]. Table 5 gives the cost of different feedstocks needed for ethanol production.

Table 4 Main green diesel produced and their feedstock type

Feedstock type	Producers
Vegetable oil	Neste, UOP, Diamon Green
Waste animal fat	UOP, Neste
Non-edible vegetable oils	AltAir Fuels
Used cooking oil	UOP/Eni
Crude tall oil	UPM Biofuels
Fatty acid	Energy Group (REG) Inc
Non-edible jatropha oil, used cooking oil, waste palm oil	CSIR-Indian Institute of Petroleum

Table 5 Cost of biofuels produced from different feedstocks

Feedstock type	Cost (Euro)	Remark
Sugar cane—Ethanol	€ 0.5/l	IEA 2013
Corn/sugar beet/wheat—Ethanol	€ 0.9–€ 1.6/l	IEA 2013, No subsidiaries
Lignocellulosic—Ethanol	€ 4/l	Salvi and Panwar [50]
Conventional green diesel—Ethanol	€ 0.6–€ 1.1/l	Silalertruksa and Gheewala [51]
Algae (photobioreactors)	€ 0.63- € 6.6/l	Borowitzka and Moheimani [49]

2.2 Transportation Cost

Enhanced rise in transportation fuel use has elevated interest in using diverse sources for biodiesel production. Diesel production is divided into several stages: raw material extraction/transportation/processing to the final product. Transportation cost of feedstock affects the efficiency and profitability of biomass harvest and collection operations. To keep the low transportation cost, supply chain logistics has to manage biomass shortages/higher drought prices well. Cost-effective transportation makes shipping feedstock from regions spared by droughts and high availability to the area where feedstock is limited. Optimizing the feedstock material density, particle size, moisture, and ash content could potentially reduce the transportation cost as much by 21% [52].

2.3 Catalyst Cost

Green diesel production from waste vegetable oil can take place under homogeneous or heterogeneous catalysts. Among homogenous categories, alkaline (i.e., sodium hydroxide (NaOH), potassium hydroxide (KOH), sodium butoxide, sodium methoxide (CH_3ONa), and potassium methoxide (CH_3OK) and acidic (sulfuric acid, ferric sulfate, hydrochloric acid, sulfonic acid, and organic sulfonic acid) catalysts exist [53–59]. Notably, the addition of catalysts enhances the overall production cost. Further, due to liquid form, these catalysts must be removed from the final product mixture, including separation cost. High ash level is achieved in alkaline catalysts, whereas acidic catalysts cause corrosion [60]. The costly separation process of homogenous catalysts enables the choice of heterogeneous catalysts for green diesel production more attractive [61]. Heterogeneous catalysts are available in powder and pellet forms, and as a result, the catalyst and reaction system includes multiple phases, and therefore the catalyst may be easily separated once the reaction is finished [62]. Heterogeneous catalysts have been shown to enhance the transesterification process by removing the additional catalyst processing/separation costs and lowering pollution emissions. Metal oxides (Ni-Co supported on multi-walled carbon nanotube) [63]; platinum catalyst, viz., Pt/HZSM-22/Al_2O_3 [64]; Ni/γAl2O3 [65]; and zeolites are commonly used as heterogeneous catalysts for green diesel production. Heterogeneous catalysts offer a non-corrosive, environmental-friendly nature, fewer disposal issues, and high thermal stability. However, it requires high-temperature and high-pressure operation and a relatively high methanol-to-oil ratio for the transesterification process. In addition, the solid catalyst-to-feedstock ratio remains substantially lower than the homogeneous catalyst-to-feedstock ratio. For instance, 5.7 tons of solid-supported magnesium oxide (MgO) catalyst is sufficient to produce 100,000 tons of biodiesel [65].

3 Technological Feasibility in Green Diesel Production

3.1 Catalytic Hydrodeoxygenation (HDO) Route

The catalytic hydrodeoxygenation (HDO) technique is one of the most promising routes to produce green diesel [37, 66, 67]. The abundant hydrogen and heterogeneous metal catalyst enabled non-edible or waste vegetable oils to convert into liquid hydrocarbon compounds (*called* green diesel) at elevated temperatures (300–425 °C). Green diesel produced from this technique consists of n-alkanes, similar to the chemical composition of fossil-derived diesel [68]. Neste Oil has patented the HDO technique for green diesel production at the commercial level [69]. The large-scale catalytic HDO process is controlled by temperature, hydrogen pressure, reaction environment, catalyst, and substrate. These operating parameters significantly affect the selectivity, conversion, final product distribution, and catalysts' deactivation [29, 70–74]. Temperature affects the reaction rate, fatty acids conversion, and hydrocarbon distribution. An increase in reaction temperature from 300 °C to 360 °C increases the conversion by fourfolds while simultaneously increasing the aromatic content in the product mixture. Higher hydrogen pressure causes higher conversion but enables a lower concentration of aromatic compounds. Excess hydrogen in the reaction environment accelerates the hydrogenation of unsaturated triglycerides and therefore increases the concentration of saturated hydrocarbon in the produced green diesel. Further, a good selection of catalysts decreases the reaction's thermal energy barrier, making the process energy-effective and cost-effective. The most commonly used catalysts are sulfided transition metals such as molybdenum (Mo) and tungsten (W) doped with nickel (Ni) or cobalt (Co).

One of the problems with green diesel obtained from the HDO process is poor cold flow properties. In this context, the hydro-isomerization process improves the cold flow properties of green diesel by in situ converting HDO-derived n-alkanes into iso-alkanes. Thus, hydro-isomerization would be essential in green diesel-producing plants located in cold regions. Further, hydrodeoxygenation remains the most favorable route to produce green diesel from the point of view of carbon atom economy as it produces hydrocarbon of the same number of fatty acids and water. However, water can cause the deactivation of catalysts which adds extra cost and effort to the technological development of green diesel.

For the HDO process, the trickle bed reactor in semi-batch operation gives the best performance. Process capacity needs to be designed carefully to make the green diesel plant profitable. For a profitable venture, it is proposed that a standalone unit should produce nearly 500,000 tons per year, which will consume around 600,000 tons of vegetable oil per year. The hydrogen availability and production cost are other factors in the HDO process's success. Another option to make the HDO process economically feasible is the co-processing of green diesel at a conventional refinery (as discussed earlier), where already available hydrotreatment units can be utilized with a slight modification in catalysts bed.

3.2 Fischer–Tropsch Process

Fischer–Tropsch process (FTP) is an established technology for liquid hydrocarbon fuel (i.e., green diesel, naphtha, jet fuel, etc.) generation from syngas (CO and H_2). Various reactions involved in the FTP process include partial oxidation, reforming, or gasification of carbonaceous feedstock, as listed below:

$C + \frac{1}{2} O_2 \rightarrow CO$ $\Delta H = -111$ kJ/mol (partial oxidation).

$C + H_2O \rightarrow CO + H_2$ $\Delta H = +131$ kJ/mol (water–gas reaction).

$C + CO_2 \leftrightarrow 2CO$ $\Delta H = +172$ kJ/mol (Boudouard reaction).

$CH_4 + \frac{1}{2} O_2 \rightarrow CO + 2H_2$ $\Delta H = -36$ kJ/mol (partial oxidation).

$CH_4 + H_2O \leftrightarrow CO + 3H_2$ $\Delta H = +206$ kJ/mol (steam reforming).

$CH_4 + CO_2 \leftrightarrow 2CO + 2H_2$ $\Delta H = +247$ kJ/mol (dry reforming).

$2CH_4 + CO_2 + H_2O \leftrightarrow 3CO + 5H_2$ $\Delta H > 0$ (combined steam and dry reforming).

$CO + H_2O \leftrightarrow CO_2 + H_2$ $\Delta H = -41$ kJ/mol (water–gas shift reaction).

In FTP, surface polymerization reaction converts syngas into saturated or unsaturated hydrocarbons (carbon number $\approx C_1 - C_{40}$) over iron (Fe)- and cobalt (Co)-based catalysts. Ruthenium- and nickel-based catalysts are also promising, but ruthenium is less abundant and expensive, while nickel which is prone to coke poisoning and promotes methanation is not the preferred choice for FTP in the industry [75, 76]. The surface polymerization reaction conducted at 250–350 °C and 10–60 bar remains highly exothermic [43]. The following reactions take place during FTP:

FTP reactions:	Paraffins (alkanes)	$(2n + 1)H_2 + nCO \rightarrow C_nH_{2n+2} + nH_2O$
	Olefins (alkenes)	$2nH_2 + nCO \rightarrow C_nH_{2n} + nH_2O$
	Alcohols	$2nH_2 + nCO \rightarrow H(CH_2)_nOH + (n - 1)H_2O$
Water–gas shift reaction:		$nCO + H_2O \leftrightarrow CO_2 + H_2$

The syncrude produced from FTP mainly consists of paraffinic and olefinic hydrocarbons. Industrial reactors usually operate at high-temperature FT (HTFT)

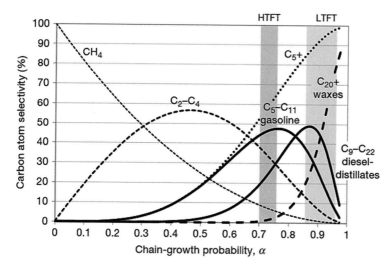

Fig. 4 Distribution of hydrocarbon products during FTP with varying α-value. HTFT corresponds approximately to 0.70 < α < 0.75, and LTFT to about 0.85 < α < 0.95 (Reproduced from Ref. [78] with permission from Elsevier, Copyright [2013]) [78]

at 300–350 °C or low-temperature FT (LTFT) at 200–250 °C. The distribution of hydrocarbon products as a function of chain growth probability factor (α) is shown in Fig. 4. HTFT reactors usually run on Fe catalyst, which promotes olefins and naphtha production. In contrast, LTFT reactors can run on Fe- and Co-based catalysts and produce diesel and wax fractions [77]. The wax thus generated is subsequently hydrocracked to convert into diesel and some naphtha.

Several reactors used in FTP are depicted in Fig. 5 [78]. Diesel production in LTFT reactors operates under a three-phase reaction system (i.e., gaseous reactants–gas and liquid products–solid catalyst). Some drawbacks associated with trickle-mode fixed-bed reactors are pressure drop and intra-particle diffusion limitations, adversely affecting the reactor performance. The slurry bubble-column reactor is developed to resolve some issues of fixed-bed reactors. Still, some pre-requisites like efficient solid–liquid separation to remove wax from the reactor to ensure slurry process and mechanical stability of catalyst in the moving bed environment are to be maintained. More recently developed micro-channel reactor has high mass and heat transfer and overcomes the problems associated with fixed-bed and slurry bubble-column reactors. Some companies commercially pursue this technology [79].

3.2.1 Commercial Status of FTP

Coal-based Sasol plant in South Africa with 170,000 barrel per day (bpd) capacity [80] and natural gas-based Pearl GTL (gas-to-liquid) plant in Qatar with 14,000 bpd capacity [81] are the examples of FTP technology proven at the commercial level.

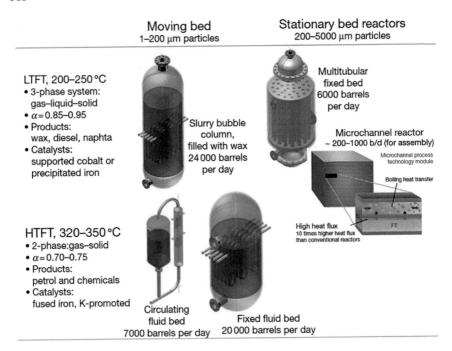

Moving bed
1–200 μm particles

Stationary bed reactors
200–5000 μm particles

LTFT, 200–250 °C
- 3-phase system:
  gas–liquid–solid
- α=0.85–0.95
- Products:
  wax, diesel, naphta
- Catalysts:
  supported cobalt or
  precipitated iron

Slurry bubble
column,
filled with wax
24 000 barrels
per day

Multitubular
fixed bed
6000 barrels
per day

Microchannel reactor
~ 200–1000 b/d (for assembly)

Microchannel process
technology module

Boiling heat transfer

HTFT, 320–350 °C
- 2-phase:gas–solid
- α=0.70–0.75
- Products:
  petrol and chemicals
- Catalysts:
  fused iron, K-promoted

High heat flux
10 times higher heat flux
than conventional reactors

FT

Circulating
fluid bed
7000 barrels per day

Fixed fluid bed
20 000 barrels per day

Fig. 5 Different types of reactors used in FTP (Reproduced from Ref. [78] with permission from Elsevier, Copyright [2013]) [78]

In 2013, Sasol's CTL (coal-to-liquid) facility at Secunda reported an annual profit of $2 billion. Sasol-Qatar Petroleum's Oryx-GTL facility reported an annual profit of $500 million [78]. However, in terms of an environmental impact compared to crude oil, CTL technology has no clear advantages over crude oil emission, and therefore the CTL facility needed to be integrated with the carbon capture and storage (CCS) facility. Biomass-to-liquid (BTL) technology is the upcoming FTP technology based on biomass feedstock. An EU-funded international COMSYN project (partners from Finland, Germany, Czech Republic, and Italy) suggests an objective to demonstrate the production of liquid fuels from biomass at low cost and later use the test data to scale the industrial production unit. Further, the feasibility study for the optimized production with maximum integration benefits is also included [82]. The Lakeview project by Red Rock biofuels produces 16.1 million gallons per annum of low-carbon FT liquid (renewable jet and diesel fuels) from approximately 166,000 dry tons of waste woody biomass [83]. AgBioEn, Australia's ground-breaking bioenergy facility, combines pyrolysis and FTP to produce diesel and jet fuel from biomass [84]. Sierra biofuel plant (by Fulcrum bioenergy) is the first commercial-scale plant constructed in July 2021 to convert 350 kTPY of municipal solid waste (MSW) into 175 kTPY of syncrude as the feedstock for the biorefinery processing to produce about 42 million liters of liquid fuels [85].

Several techno-economic feasibility reports are published on the viability of BTL technology. Techno-economic assessment (TEA) of FT diesel production from forest residue using high-temperature entrained flow gasification estimated the cost around 125–130 €/MWh with/without carbon capture and storage (CCS) [86]. Integrated and standalone configurations for Syncrude production via FTP were studied to estimate the cost of production and environmental impact. FT liquid costs around 42–140 €/MWh depending upon feed type and price, process configuration, crude/upgraded product, plant capacity, economic incentive, etc. Therefore, it is not that simple which reactor and catalyst type results in the best process performance. Instead, site-specific assessment considering process configuration, reactor and catalyst type, feedstock type and availability, economy-of-scale, etc. provides a better estimation of production viability and cost feasibility.

4 Economics of Green Diesel Production

This section presents the understanding of green diesel economics through a case study. The most appealing biomass source for producing transportation fuels is triglycerides (i.e., vegetable oils, waste cooking oils, animal fats, and microalgal oils). Compared to cellulosic biomass, triglycerides are simple and easily convertible to hydrocarbon fuel [26]. There exists a significant price differential between the biofuels made from waste oil/animal fats and the biofuels prepared from plant-derived vegetable oil. Renewable diesel made from waste oil/animal fats in Europe is worth \$400–\$600/MT more than diesel made from crop oil feeds. For example, the California market analysis suggested an additional \$0.75–\$0.90 per gallon (\$250–\$300/MT) of incentives for renewable diesel production through cooking oil compared to using virgin soybean oil [87]. Refiners must consider whether their capital expenditure is correctly allocated to capture this additional value with such a large pricing disparity. Answers must be validated to justify the increased capital expenditure required to handle sustainable feeds and avoid being locked out of future regulated marketplaces. Waste oils include nitrogen, metals, and polyethylene as contaminants, which must be removed from the raw feedstocks (i.e., vegetable oils) to increase the catalyst life and to meet the standards of final product with diesel.

In literature, there are two different schemes proposed for hydrodeoxygenation of vegetable oil, in one of the schemes, hydrogen reduces the fatty acids of triglycerides into the fatty aldehydes over the catalyst's metallic core [26]. A decarbonylation (removal of oxygen as CO) procedure is then used to convert the fatty aldehydes to hydrocarbons. The water–gas shift process converts the carbon monoxide produced during the decarbonylation reaction to carbon dioxide. This reaction produces hydrocarbons with fewer carbon atoms than the fatty acids they replace. In the second pathway, fatty acids are reduced to fatty alcohols via intermediate fatty aldehydes under the hydrogen environment on the catalyst's metallic sites. Further, the dehydration is followed by a hydrogenation reaction to convert the fatty alcohols to hydrocarbons. This reaction produces hydrocarbons with the same carbon atoms as the corresponding fatty acids. The conversion rate of fatty acids via this pathway remains modest. Vegetable oil conversion via this technique is estimated to be just 5%. Most of the hydrodeoxygenation (HDO) of vegetable oil is performed in a fixed-bed reactor in the temperature range of 523–773 K and under high hydrogen pressures (>100 bar). The HDO of vegetable oil is highly exothermic reaction and is almost 6–10 times of the typical refinery hydrocracking reactions [88]. For a plant capacity of 0.12 MMT per year, the HDO of vegetable oil is an exothermic process creating 3.89 MW of heat. This available thermal energy can be used to create process steam useful in distillation column reboiler, and the remaining may be utilized to generate electricity.

Green diesel production costs at a capacity of 0.05 MMT per year are USD 0.970 per kg for direct HDO and USD 0.931 per kg for two-step HDO, respectively. Increasing plant capacity to 0.1 MMT per year lowers the production costs by 16.5% and 16.6%, respectively, for direct and two-step HDO. Vegetable oil alone contributes 70% and 75% of the production costs of green diesel, respectively, for the direct and two-step HDO for a plant capacity of 0.12 MMT per year. Thus, the feedstock may claim 70–80% of the production cost. The availability of low-cost feedstock is thus critical for enhancing the economics of green diesel through HDO of vegetable oil. The utility contributes marginally (1.8% for direct process and 4.2% for two-step HDO) of the production cost for a plant capacity of 0.12 MMT per year. Further, HDO of vegetable oil necessitates hydrogen, accounting for 1.7–3.3% of the total manufacturing cost. Moreover, indirect expenses account for 9.8% and 12.1% of total costs for direct and two-step HDO, respectively.

Vegetable oil HDO is relatively simple technology with a lower initial expenditure. This makes HDO from vegetable oil an appealing candidate for green diesel generation in hydrocarbon biorefineries. The co-product credit accounted for a large portion of the entire production cost. Glycerol and propane yielded roughly 6% and 15% credits in direct and two-step HDO, respectively, with a plant capacity of 0.12 MMT per year. On the other hand, the catalyst only accounted for around 2.2% and 2.4% of the direct and two-step HDO manufacturing costs, respectively. For direct and two-step HDO, the minimum selling prices of green diesel were found to be USD 1.2 per kg (USD 3.7 per gal) and USD 1.2 per kg (USD 3.8 per gal), respectively, for an acceptable rate of return of 8.5% [26].

5 Future Perspective and Recommendations

The commercial technologies producing green diesel are mature and have a scope for further expansion. Integrating a green diesel production plant with an existing petroleum refinery would be a crucial step and need of the hour to save energy and hence money with proper heat integration and resource utilization. The worldwide green diesel production is expected to rise to 13 billion liters per year by 2024 from the existing 6 to 7 billion liters per year capacity [89]. Notably, worldwide regulations and policies are crucial for green diesel production at a large scale.

Remarkably, the regulation-driven call in the United States of America (USA) and the European union is anticipated to boost a 5 billion US Dollars of investment (in the next 5 years) in setting up and implementing green diesel plants. The NEXT Renewable Fuels' Oregon project, Diesel's expansion project at Louisiana, World Energy in California, USA, Neste expansion project of green diesel in Singapore, and ENI's Venice project in Italy are a few initiatives toward enhancing the global green diesel production [89]. This intended production of green diesel is likely to grow the global market and fulfill the gap of fossil fuel demand to a greater extent. The lucrative and sustainable market of green diesel would be dependent on (i) ability to determine the production target against the demand and strategies to achieve it, (ii) adopting a low-cost production route through heat integration and process condition optimization, (iii) building a robust feedstock value chain, and (iv) working closely with end users such as shipping and aviation industries where fewer substitutes to decarbonization exist [90].

6 Conclusions

This chapter presents an overview of green diesel production, emphasizing policies, technological advancement, process simulation approach, economic analysis, and future perspectives. Many countries are slowly replacing their conventional fuel markets with renewable and sustainable fuels (i.e., green diesel or renewable diesel). Government regulations should be considered while implementing renewable fuel techniques. The production of green diesel is a need to hour to compensate the fuel requirement in the transportation sector. The green diesel obtained from waste vegetable oils would be a better choice for this purpose. Process design and simulation approach may result in effective technology and better economics; however, a sustainable supply chain of the feedstocks, transport infrastructure, and processing cost are crucial factors determining the feasibility of green diesel production at a large scale. The process profitability is mainly governed by raw material price, and hence the alternative and cheaper raw materials derived from used and waste products (i.e., waste vegetable oil) are essential. Notably, hydrotreatment (or HDO) and isomerization produce green diesel with improved cold flow properties, making it attractive in colder countries. FTP is also an attractive approach for green diesel production; however, it depends strongly on the feed type and price, process configuration, upgrading methods, plant capacity, and economic incentive. The economic analysis of green diesel production suggested 4–7 years to gain profit, although feedstock and product price fluctuations make it a risky investment. Governmental incentives and funding would be essential to improve this situation.

References

1. Glisic SB, Pajnik JM, Orlović AM (2016) Process and techno-economic analysis of green diesel production from waste vegetable oil and the comparison with ester type biodiesel production. Appl Energy 170:176–185
2. Wibowo AA, Mustain A, Lusiani CE, Hartanto D, Ginting RR Green diesel production from waste vegetable oil: a simulation study. In: AIP conference proceedings, vol 2223, p 020008
3. Zhou L, Lawal A (2016) Hydrodeoxygenation of microalgae oil to green diesel over Pt, Rh and presulfided NiMo catalysts. Catal Sci & Technol 6:1442–1454
4. Hughes L, Rudolph J (2011) Future world oil production: growth, plateau, or peak? Curr Opin Environ Sustain 3:225–234
5. Takkellapati S, Li T, Gonzalez MA (2018) An overview of biorefinery-derived platform chemicals from a cellulose and hemicellulose biorefinery. Clean Technol Environ Policy 20:1615–1630
6. Patnaik S, Martha S, Parida KM (2016) An overview of the structural, textural and morphological modulations of g-C3N4 towards photocatalytic hydrogen production. RSC Adv 6:46929–46951
7. Williams JB (2002) Production of biodiesel in Europe — the markets. Eur J Lipid Sci Technol 104:361–362
8. Johnston M, Holloway T (2007) A global comparison of national biodiesel production potentials. Environ Sci Technol 41:7967–7973
9. M.o.P.N. Gas (2018) National Policy on Biofuels 2018, Government of India, Gaz India, pp 123
10. S.I.p.n. gas (2015) Ministry of Petroleum and Natural Gas Economic and Statistics
11. Gmünder S, Singh R, Pfister S, Adheloya A, Zah R (2012) Environmental impacts of Jatropha curcas biodiesel in India. J Biomed Biotechnol 2012:623070
12. Pradhan S, Ruysenaar S (2014) Burning desires: untangling and interpreting 'pro-poor' biofuel policy processes in India and South Africa. Environ Plann A: Econ Space 46:299–317
13. Roberts GW, Fortier M-OP, Sturm BSM, Stagg-Williams SM (2013) Promising pathway for algal biofuels through wastewater cultivation and hydrothermal conversion. Energy Fuels 27:857–867
14. W. L, Biodiesel from Algae oil, (2007).
15. Khan SA, Rashmi, Hussain MZ, Prasad S, Banerjee UC (2009)Prospects of biodiesel production from microalgae in India, Renew Sustain Energy Rev 13:23612372
16. Park JBK, Craggs RJ, Shilton AN (2011) Wastewater treatment high rate algal ponds for biofuel production. Biores Technol 102:35–42
17. D.i.F. 2020 (2018) A global market survey, emerging markets online
18. NEXBTL (2015) Move forward with NEXBTL renewable diesel, pp 18
19. G.C. Congress (2008) Neste oil to build $1B NExBTL renewable diesel plant in Rotterdam
20. Dahiya A (2020) Cutting-edge biofuel conversion technologies to integrate into petroleum-based infrastructure and integrated biorefineries (Chap. 31). In: Dahiya A (ed) Bioenergy, Second. Academic Press, pp 649–670
21. BezergianniS (2013) Catalytic hydroprocessing of liquid biomass for biofuels production, liquid, gaseous and solid biofuels - conversion techniques
22. UOP H (2019) Honeywell ecofining technology helps diamond green diesel become one of the World's largest renewable diesel plants, honeywell ecofining technology
23. Ray A, Anumakonda A (2011) Production of green liquid hydrocarbon fuels (Chap. 26). In: Pandey A, Larroche C, Ricke SC, Dussap C-G, Gnansounou E (eds) Biofuels. Academic Press, Amsterdam, pp 587–608
24. Cheah KW, Yusup S, Gurdeep Singh HK, Uemura Y, Lam HL (2017) Process simulation and techno economic analysis of renewable diesel production via catalytic decarboxylation of rubber seed oil – a case study in Malaysia. J Environ Manag 203:950–961

25. Pratiwi M, Muraza O, Neonufa GF, Purwadi R, Prakoso T, Soerawidjaja TH (2019) Production of sustainable diesel via decarboxylation of palm stearin basic soaps. Energy Fuels 33:11648–11654

26. Mailaram S, Maity SK (2019) Techno-economic evaluation of two alternative processes for production of green diesel from karanja oil: a pinch analysis approach. J Renew Sustain Energy 11:025906

27. Martinez-Hernandez E, Ramírez-Verduzco LF, Amezcua-Allieri MA, Aburto J (2019) Process simulation and techno-economic analysis of bio-jet fuel and green diesel production — minimum selling prices. Chem Eng Res Des 146:60–70

28. Fernández-Villami JM, de Mendoza Paniagua AH (2018) Preliminary design of the green diesel production process by hydrotreatment of vegetable oils., Student Contest Problem, EURECHA, pp 115

29. Snåre M, Kubičková I, Mäki-Arvela P, Chichova D, Eränen K, Murzin DY (2008) Catalytic deoxygenation of unsaturated renewable feedstocks for production of diesel fuel hydrocarbons. Fuel 87:933–945

30. Smejkal Q, Smejkalová L, Kubička D (2009) Thermodynamic balance in reaction system of total vegetable oil hydrogenation. Chem Eng J 146:155–160

31. Torres-Ortega CE, Gong J, You F, Rong B-G (2017) Optimal synthesis of integrated process for co-production of biodiesel and hydrotreated vegetable oil (HVO) diesel from hybrid oil feedstocks. In: Espuña A, Graells M, Puigjaner L (eds) Computer aided chemical engineering. Elsevier, pp 673–678

32. Manco V, Felipe J (2019) Conceptual design of a palm oil hydrotreatment reactor for commercial diesel production

33. Kubička D, Tukač V (2013) Chapter three - hydrotreating of triglyceride-based feedstocks in refineries. In: Murzin DY (ed) Advances in chemical engineering. Academic Press, pp 141–194

34. Landberg K (2016) Experimental and kinetic modelling study of hydrodeoxygenation of tall oil to renewable fuel. Master's thesis. Chalmers University of Technology. SE-412 96 Gothenburg

35. Zhang H, Lin H, Wang W, Zheng Y, Hu P (2014) Hydroprocessing of waste cooking oil over a dispersed nano catalyst: kinetics study and temperature effect. Appl Catal B 150–151:238–248

36. Snåre M, Kubičková I, Mäki-Arvela P, Eränen K, Wärnå J, Murzin DY (2007) Production of diesel fuel from renewable feeds: kinetics of ethyl stearate decarboxylation. Chem Eng J 134:29–34

37. Tirado A, Ancheyta J, Trejo F (2018) Kinetic and reactor modeling of catalytic hydrotreatment of vegetable oils. Energy Fuels 32:7245–7261

38. Li X, Luo X, Jin Y, Li J, Zhang H, Zhang A, Xie J (2018) Heterogeneous sulfur-free hydrodeoxygenation catalysts for selectively upgrading the renewable bio-oils to second generation biofuels. Renew Sustain Energy Rev 82:3762–3797

39. Timmerhaus KD, Peters MS, West RE (2003) Plant design and economics for chemical engineers, vol 4. McGraw-Hill New York

40. Miller P, Kumar A (2014) Techno-economic assessment of hydrogenation-derived renewable diesel production from canola and camelina. Sustain Energy Technol Assess 6:105–115

41. Kantama A, Narataruksa P, Hunpinyo P, Prapainainar C (2015) Techno-economic assessment of a heat-integrated process for hydrogenated renewable diesel production from palm fatty acid distillate. Biomass Bioenergy 83:448–459

42. Lange J-P (2007) Lignocellulose conversion: an introduction to chemistry, process and economics. Biofuels Bioprod Biorefin 1:39–48

43. Douvartzides SL, Charisiou ND, Papageridis KN, Goula MA (2019) Green diesel: biomass feedstocks production technologies, catalytic research, fuel properties and performance in compression ignition internal combustion engines. Energies 12:809

44. Boutesteijn C, Drabik D, Venus TJ (2017) The interaction between EU biofuel policy and first- and second-generation biodiesel production. Ind Crops Prod 106:124–129

45. Kandaramath HariT, Yaakob Z, Binitha NN (2015) Aviation biofuel from renewable resources: Routes, opportunities and challenges. Renew Sustain Energy Rev 42:1234–1244

46. Alemayehu Gashaw AT (2014) Production of biodiesel from waste cooking oil and factors affecting its formation: a review. Int J Sustain Green Energy 3:92–98

47. Vignesh P, Pradeep Kumar AR, Ganesh NS, Jayaseelan V, Sudhakar K (2021) Biodiesel and green diesel generation: an overview. Oil & Gas Sci Technol - Revue d'IFP Energies nouvelles 76:6

48. Kumar S, Shrestha P, Abdul Salam P (2013) A review of biofuel policies in the major biofuel producing countries of ASEAN: production, targets, policy drivers and impacts. Renew Sustain Energy Rev 26:822–836

49. Borowitzka MA, Moheimani NR (2013) Sustainable biofuels from algae. Mitig Adapt Strat Glob Change 18:13–25

50. Salvi BL, Panwar NL (2012) Biodiesel resources and production technologies – a review. Renew Sustain Energy Rev 16:3680–3689

51. Silalertruksa T, Gheewala SH (2012) Environmental sustainability assessment of palm biodiesel production in Thailand. Energy 43:306–314

52. Friedemann AJ (2021) Biodiesel from Algae, in: life after fossil fuels: a reality check on alternative energy. Springer International Publishing, Cham, pp 145–151

53. Wang Y, Pengzhan Liu SO, Zhang Z (2007) Preparation of biodiesel from waste cooking oil via two-step catalyzed process. Energy Convers Manag 48:184–188

54. Sharma YC, Singh B (2009) Development of biodiesel: current scenario. Renew Sustain Energy Rev 13:1646–1651

55. Macario A, Giordano G, Onida B, Cocina D, Tagarelli A, Giuffrè AM (2010) Biodiesel production process by homogeneous/heterogeneous catalytic system using an acid–base catalyst. Appl Catal A 378:160–168

56. Sharma YC, Singh B, Upadhyay SN (2008) Advancements in development and characterization of biodiesel: a review. Fuel 87:2355–2373

57. Patil P, Deng S, Isaac Rhodes J, Lammers PJ (2010) Conversion of waste cooking oil to biodiesel using ferric sulfate and supercritical methanol processes. Fuel 89:360–364

58. Nichmon Puagsang IC, Chantrapromma S, Palamanit A (2020) Production of biodiesel from low-grade crude palm oil using hydrochloric acid. Environ Asia 14:2332

59. Liu F, Ma X, Li H, Wang Y, Cui P, Guo M, Yaxin H, Lu W, Zhou S, Yu M (2020) Dilute sulfonic acid post functionalized metal organic framework as a heterogeneous acid catalyst for esterification to produce biodiesel. Fuel 266:117149

60. Canakci M, Van Gerpen J (1999) Biodiesel production via acid catalysis. Trans ASAE 42:1203–1210

61. Melero JA, Iglesias J, Morales G (2009) Heterogeneous acid catalysts for biodiesel production: current status and future challenges. Green Chem 11:1285–1308

62. Thangaraj B, Solomon PR, Muniyandi B, Ranganathan S, Lin L (2018) Catalysis in biodiesel production—a review. Clean Energy 3:2–23

63. Asikin-Mijan N, Lee HV, Abdulkareem-Alsultan G, Afandi A, Taufiq-Yap YH (2017) Production of green diesel via cleaner catalytic deoxygenation of Jatropha curcas oil. J Clean Prod 167:1048–1059

64. Hancsók J, Krár M, Magyar S, Boda L, Holló A, Kalló D (2007) Investigation of the production of high cetane number bio gas oil from pre-hydrogenated vegetable oils over Pt/HZSM-22/Al2O3. Microporous Mesoporous Mater 101:148–152

65. Dossin TF, Reyniers M-F, Berger RJ, Marin GB (2006) Simulation of heterogeneously MgO-catalyzed transesterification for fine-chemical and biodiesel industrial production. Appl Catal B 67:136–148

66. Kubička D, Bejblová M, Vlk J (2010) Conversion of vegetable oils into hydrocarbons over CoMo/MCM-41 catalysts. Top Catal 53:168–178

67. Rogers KA, Zheng Y (2016) Selective deoxygenation of biomass-derived bio-oils within hydrogen-modest environments: a review and new insights. Chemsuschem 9:1750–1772

68. Vonortas A, Papayannakos N (2014) Comparative analysis of biodiesel versus green diesel, WIREs. Energy Environ 3:3–23

69. Jakkula J, Niemi V, Nikkonen J, Purola V-M, Myllyoja J, Aalto P, Lehtonen J, Alopaeus V (2003) Process for producing a hydrocarbon component of biological origin. EUROPEAN PATENT: EP1396531A2, pp 116
70. Santillan-Jimenez E, Morgan T, Lacny J, Mohapatra S, Crocker M (2013) Catalytic deoxygenation of triglycerides and fatty acids to hydrocarbons over carbon-supported nickel. Fuel 103:1010–1017
71. Mäki-Arvela P, Rozmysłowicz B, Lestari S, Simakova O, Eränen K, Salmi T, Murzin DY (2011) Catalytic deoxygenation of tall oil fatty acid over palladium supported on mesoporous carbon. Energy Fuels 25:2815–2825
72. Madsen AT, Ahmed EH, Christensen CH, Fehrmann R, Riisager A (2011) Hydrodeoxygenation of waste fat for diesel production: Study on model feed with Pt/alumina catalyst. Fuel 90:3433–3438
73. Horáček J, Tišler Z, Rubáš V, Kubička D (2014) HDO catalysts for triglycerides conversion into pyrolysis and isomerization feedstock. Fuel 121:57–64
74. Kubička D, Horáček J (2011) Deactivation of HDS catalysts in deoxygenation of vegetable oils. Appl Catal A 394:9–17
75. Li W-Z, Liu J-X, Gu J, Zhou W, Yao S-Y, Si R, Guo Y, Su H-Y, Yan C-H, Li W-X, Zhang Y-W, Ma D (2017) Chemical Insights into the design and development of face-centered cubic ruthenium catalysts for Fischer-Tropsch synthesis. J Am Chem Soc 139:2267–2276
76. Enger BC, Holmen A (2012) Nickel and Fischer-Tropsch synthesis. Catal Rev 54:437–488
77. Hu J, Yu F, Lu Y (2012) Application of Fischer-Tropsch synthesis in biomass to liquid conversion. Catalysts 2:303–326
78. van de Loosdrecht J, Botes FG, Ciobica IM, Ferreira A, Gibson P, Moodley DJ, Saib AM, Visagie JL, Weststrate CJ, Niemantsverdriet JW (2013) 7.20 - Fischer-Tropsch synthesis: catalysts and chemistry. In: Reedijk J, Poeppelmeier K (eds) Comprehensive inorganic chemistry II, Second. Elsevier, Amsterdam, pp 525–557
79. Deshmukh SR, Tonkovich ALY, McDaniel JS, Schrader LD, Burton CD, Jarosch KT, Simpson AM, Kilanowski DR, LeViness S (2011) Enabling cellulosic diesel with microchannel technology. Biofuels 2:315–324
80. Dry ME (2002) High quality diesel via the Fischer-Tropsch process – a review. J Chem Technol Biotechnol 77:43–50
81. Guettel R, Kunz U, Turek T (2008) Reactors for Fischer-Tropsch synthesis. Chem Eng Technol 31:746–754
82. COMSYN (2022) Next generation bio-fuel technology. Accessed 31 Jan 2022
83. R.R. biofuels (2022) Lakeview project summary
84. AgBioEn (2022) A cleaner energy future. Accessed 31 Jan 2022
85. F. BioEnergy (2022) Sierra biofuels plant bright future. Accessed 31 Jan 2022
86. Tagomori IS, Rochedo PRR, Szklo A (2019) Techno-economic and georeferenced analysis of forestry residues-based Fischer-Tropsch diesel with carbon capture in Brazil. Biomass Bioenergy 123:134–148
87. H. UOP (2022) Feedstock price vs biofuel price – where to invest your project capital, industry solutions - renewable fuels, p 1
88. Anand M, Farooqui SA, Kumar R, Joshi R, Kumar R, Sibi MG, Singh H, Sinha AK (2016) Kinetics, thermodynamics and mechanisms for hydroprocessing of renewable oils. Appl Catal A 516:144–152
89. FutureBridge (2020) Renewable diesel: the fuel of the future
90. BeswetherickT (2022) Renewable diesel: frontier fuel with a future? Accent Energy 1

Printed in the United States
by Baker & Taylor Publisher Services